MONOGRAPHS OF THE CENTER FOR SOUTHEAST AS
KYOTO UNIVERSITY,
ENGLISH-LANGUAGE SERIES, NO. 20

# Tropical Rain Forests
# of Southeast Asia

MONOGRAPHS OF THE CENTER FOR SOUTHEAST ASIAN STUDIES,
KYOTO UNIVERSITY,
ENGLISH-LANGUAGE SERIES, NO. 20

# Tropical Rain Forests of Southeast Asia

## A Forest Ecologist's View

BY

*Isamu Yamada*

TRANSLATED BY

*Peter Hawkes*

UNIVERSITY OF HAWAI'I PRESS
Honolulu

97  98  99  00  01  02      5  4  3  2  1

**Library of Congress Cataloging-in-Publication Data**
Yamada, Isamu, 1943–
[Tōnan Ajia no nettai taurin sekai. English]
Tropical rain forests of Southeast Asia : a forest ecologist's
view / by Isamu Yamada ; translated by Peter Hawkes.
p.   cm. — (Monographs of the Center for Southeast Asian
Studies, Kyoto University ; no. 20)
Includes bibliographical references (p.    ) and index.
ISBN 0–8248–1936–5 (cloth : alk. paper). — ISBN 0–8248–1937–3
(pbk. : alk. paper)
1. Rain forest ecology—Asia, Southeastern.   2. Rain forest
conservation—Asia, Southeastern.   I. Title.   II. Series:
Monographs of the Center for Southeast Asian Studies, Kyoto
University ; 20.
QK360.Y3513   1998
577.34'0959—dc21                    97–22345
CIP

The publication of this book was financed in part by a Grant-in-Aid for Publication of Scientific
Research Results from the Ministry of Education, Science, and Culture of Japan.

University of Hawai'i Press books are printed on acid-free paper and meet
the guidelines for permanence and durability of the Council on Library Resources

The map on the cover is part of the map of Asia by Cornelius de Jode, 1593
(reproduced courtesy of the Kobe City Museum).

Designed by Kenneth Miyamoto

# Contents

*Plates follow page 176*

# Figures

# Tables

ix

# *Plates*

1. Mangrove forest at Temburong, Brunei
2. Peat swamp forest at Badas, Brunei
3. *Alan* forest at Badas, Brunei
4. *Alan bunga* forest at Badas, Brunei
5. Canopy of *alan bunga* forest at Badas, Brunei, seen from below
6. Area defoliated by *ulat bulu* at Badas, Brunei
7. Logging *alan* forest at Badas, Brunei
8. Land reclamation in mixed peat swamp forest along the Belait River, Brunei
9. Extensive swamp on the middle reach of the Sepik River, Papua New Guinea
10. Freshwater swamp forest on the upper Belait River, Brunei
11. Planted sago at the edge of freshwater swamp forest, Belait River, Brunei
12. Natural sago forest on the upper Sepik River, Papua New Guinea
13. Mixed dipterocarp forest at Andulau, Brunei
14. Mixed dipterocarp forest near Balikpapan, Kalimantan, Indonesia
15. Mixed dipterocarp forest at Labi, Brunei
16. Logging area in mixed dipterocarp forest at Andulau, Brunei
17. Logged area in Balikpapan, Kalimantan, Indonesia
18. Canopy of mixed dipterocarp forest in Aceh, North Sumatra, Indonesia
19. Area of repeated swiddening after clearing of the mixed dipterocarp forest, Aceh, North Sumatra, Indonesia
20. Oil palm cultivation in Sarawak, Malaysia
21. *Kerangas* forest, Badas, Brunei
22. Montane forest at an elevation of around 1,500 m, Mount Pangrango, West Java, Indonesia
23. Moss forest at an elevation of around 2,800 m, Mount Pangrango, West Java, Indonesia
24. Monsoon forest at Sarakhet, northeast Thailand
25. Savanna landscape resulting from human intervention in the monsoon forest, Baluran, East Java, Indonesia
26. Botanical garden at Bogor, West Java, Indonesia

# *Preface*

This book is an English translation of my work in Japanese entitled *Tōnan Ajia no nettaitaurin sekai* [The tropical rain forest world of Southeast Asia], which was published in 1991 by Sobunsha. It was written originally for a Japanese readership and has been revised and reorganized with an international audience in mind.

Tropical rain forests have attracted the world's attention in recent years. These most diverse and complex of the world's ecosystems have been disappearing rapidly, at a rate of more than 15 million hectares annually, because of logging, shifting cultivation, conversion to farmland and pasture, and other causes. This in turn has led to the destruction of ecosystems, loss of genetic resources, increased atmospheric concentrations of carbon dioxide and methane, destruction of forest dwellers' life zones, and other problems that have erupted in various parts of the world.

For more than a quarter of a century, I have continued to survey and study these forests from the standpoint of forest ecology. While working at the Forest Tree Breeding Institute of the Ministry of Agriculture, Forestry and Fisheries, I also had the opportunity to consider the problems of gene resources. In my field surveys, I was able to cover the whole range of forest types, from mangrove, freshwater, and peat swamp forests through lowland dipterocarp forest to montane forest, and I gained a broad picture of the form of tropical rain forests. Here, on the basis of my field surveys, I describe the forms of the various forests growing in a tropical rain forest climate.

In a sense, this book charts the tracks of twenty-five years of continuous work in the tropical rain forests of Southeast Asia by one researcher, who entered this world in his youth; this sets it apart from the conventional textbook or monograph. Although several years have passed since I wrote the Japanese version, the basic picture presented here remains essentially unchanged. When I wrote the original version, my focus had already begun to shift toward the relationship between humans and the forest, rather than resting on pure forest ecology. Thus, this book marks the end of one phase of my work.

My first experience of the tropical forests of Southeast Asia was in 1965 when, as a third-year undergraduate student, I visited Thailand, Cambodia, Malaysia, and Singapore with professors Motoji Tagawa and Ryozo Yoshii, whose energy and dedication impressed me greatly. From early morning to sunset they climbed cliffs and descended valleys, laboring to collect specimens; then they worked until late at night organizing the specimens. I still remember clearly Professor Tagawa preparing beautiful specimens of giant tropical plants and telling me, "You must prepare a specimen with a view to taking a good photograph." At that time, I also had occasion to meet Kunio Iwatsuki, who subsequently helped me in various ways.

The Forest Ecology Laboratory to which I belonged at Kyoto University's Faculty of Agriculture was actively pursuing research in forest ecology centered on productivity studies under professors Tsunahide Shidei and Toshio Tsutsumi. Among my seniors in the section was Kazuhiko Ogino, a pioneer who went abroad to study and began tropical research in Thailand, and who remains among my most trusted colleagues.

As a postgraduate student, I went to Indonesia to study and began my work in earnest. That was when I began my association with the Center for Southeast Asian Studies, Kyoto University, where Yoshikazu Takaya has been particularly influential on me.

During my eight-year interlude at the Forest Tree Breeding Institute, I learned what it is to work for the state, being in a totally different research environment from my university days, and also being close to Tokyo. I learned much about the administrative organization that reaches from the national forestry agency to every forest district of the nation, and about the attitudes toward that organization of the people who live in the field, and I am particularly grateful to Kaichiro Kawamura. At this time I had the opportunity to participate in a cooperative venture with the Brunei Forest Research Center, and I am indebted to people from the Japan International Cooperation Agency (JICA) and the Ministry of Foreign Affairs for their kindness. These include Tomoya Kawamura (the former ambassador to Brunei), Katsura Watanabe, Yutaka Shimomoto, and Shigeru Miyawaki.

Most of my working time was spent out-of-doors. In Cibodas, where I made my first prolonged stay, I was helped greatly by Bapak Nurta and his family. Looking back to my youth, I feel I made many unreasonable requests of them, yet they always worked to their utmost for my best interests. Bapak Nurta, who passed away in 1988 at the age of eighty, now rests near Cibodas, keeping watch over Mount Pangrango. Others in Indonesia to whom I am particularly indebted include Sukristijono Sukardjo, who accompanied me in Sumatra, and in Bogor, Otto Soemarwoto, Kuswata Kartawinata, Bapak Nedi, Rusdy Nasution, and Yoshitaka Mangyoku.

In Brunei, I shall always remember the cooperation of Hj. Mahari, Hj. Yassin, J.A.R. Anderson, and others from the start of the project, and Mr. Niga, who always accompanied me in the field.

Others who have had an immeasurable influence on me, either directly or indirectly, include Tatsuo Kira, Tadao Umesao, Fusato Ogawa, Kazutake Kyuma, Kihachiro Kikuzawa, Hiroshi Omura, Takanobu Furukoshi, Kihachiro Ohba, Sumihiko Hatsushima, Takahide Hosokawa, and Gentaro Imadate.

In addition to these people, I encountered many others each time I went forth into the field. In venturing into the tropics, I have been drawn not only by the beauty, vastness, and unknown of the tropical rain forest, but also by the unsurpassed warmth of heart of the people who live there.

I have returned many times to the tropics, where I repeatedly have become seriously ill. The fact that I can still venture forth owes much to the skill and kindness of Dr. Yasumasa Kondo. I have also caused my parents grave concern. My wife, Yoko, who lived with me in Brunei and is one of few women to have entered the tropical peat swamp forest, regards my tropical voyages with something approaching resignation.

Whenever I return to the tropical forest, I am elated; my weariness evaporates, and I feel purified, having somehow entered a fairy-tale world of long ago. In the mountains of Sumatra, I met the Kohar family. Every day the husband gathered rattan from the forest, his wife tended the swiddens, and his daughter fished the river. They allowed our party of six to occupy one of the two rooms in their stilted house, without showing the least displeasure, and we stayed with them for two weeks. The family lived in extreme poverty—the wife and daughter wore rags—but they were always cheerful. It is a joy to know that such truly ingenuous people can still be encountered, in the heart of the mountains, far from the cities.

For publication of the Japanese version of this book, I received in 1990 a Grant-in-Aid for Publication of Scientific Research Results from the Ministry of Education, Science, Sports and Culture in Japan. I am indebted to Masaki Kuboi of Sobunsha for his efforts during the publication process.

For assistance in preparing the English version of this book, I am grateful to a referee at the University of Hawai'i Press for valuable comments, which I believe have made the English version all the more readable. I am also grateful to the translator, Peter Hawkes, who pointed out shortcomings in the Japanese version that, when rectified, resulted in substantial improvement. The publication of the English version was also supported by a Grant-in-Aid for Publication of Scientific Research Results from the Ministry of Education, Science, Sports and Culture.

# Introduction

Global environmental problems have so focused the world's attention on tropical rain forests that the term at least, if not its substance, is generally familiar. Few people, however, know how tropical rain forests are structured or what life-forms can be seen there. These forests have a unique atmosphere that can be encountered nowhere else on earth. They hold the wonder of a world of structural and functional climax that has been established under temperature and humidity conditions that are optimally suited to plants; this is a world comprising a multi-layered society of plants reaching from the ground to the treetops 70 meters above, a world that hosts a stable society in the midst of the most complex diversity on earth. Encircling the earth around the equator, the tropical rain forests constitute the richest and, despite steadily continuing research, least-understood of the world's ecosystems.

The tropical rain forests have a long history of disturbance by humans, who since antiquity have collected forest products. After a long, continuous period of minuscule disturbance, in which a few people extracted small quantities of products from the vast forest, large-scale exploitation of the rain forests began around the mid-twentieth century and is now jeopardizing their existence. From coastal mangrove, freshwater, and peat swamp forests through the heartland of the dry lowland rain forest to montane rain forest, these characteristic vegetation types of the tropical rain forest climate survive in their primitive, untouched state only in a few locations. A major objective of this book is to introduce these flora.

According to Richards (1952), the earliest written description of a tropical rain forest was by Christopher Columbus. Richards also notes that the tropical rain forests of Malaysia have existed in their present form since the Tertiary period. And such hunter-gatherer peoples as the Orang Asli, the Kubu, and the Penan probably have continued to live a lifestyle there that has not changed greatly over time.

The term *tropical rain forest* derives from the German *tropische Regenwald,* coined by Schimper (1891). He is regarded as the first scientist to have described this forest scientifically and given it a name.

Western science, whose development centered on the temperate zones, was severely shaken by its encounter with the enormous size and diversity of tropical vegetation. This engendered a great many mistaken or exaggerated accounts, as well as accurate observations and surveys. From this confusion, based on his research into rain forests on three continents, Richards (1952) compiled accurate scientific knowledge in his *Tropical Rain Forest*. It was the first book to gather together existing knowledge, organizing the results of research on tropical rain forests into a scientific system of community ecology based on Clements' climax theory, which then constituted the mainstream of ecological thinking.

Today, several decades after it was written, Richards' book remains a paradigm. It reveals the author's outstanding skill in extracting only what was necessary from a vast body of literature and incorporating it into his own system, casually correcting sections he thought were wrong and building a coherent theory. What is more surprising is that important research themes anticipated by Richards have until recently gone unresolved. For example, in the chapter "Regeneration," he raises the following questions:

> What is the average age at death of trees in the different strata? What is the normal age-class representation of the chief dominants in a mature undisturbed Rain forest and how does the relative abundance of species differ in different age-classes? How does the growth rate of a large rain-forest tree vary during the successive stages of its development? At what stage does the heaviest mortality (and hence the most intense natural selection) occur? Finally, it may be asked, does the floristic composition of a *small* area of forest remain the same or is a Mixed Rain forest to be regarded rather as a large-scale mosaic, the species on a given area being succeeded not by the same but by a different combination of species?

Richards regarded these problems as among the most urgent facing rain forest ecologists of the time. Regrettably, several decades on, we have not obtained complete data giving definite answers. This is not, of course, due solely to the depth and penetration of Richards' insights: These undoubtedly are basically difficult questions, and it is also true that tropical forest researchers are few. Formerly, research in the temperate zones formed the mainstream, and the tropics were considered a special case. As Van Steenis pointed out, however, in a passage that Richards cites at the start of his book, it is the richly varied tropical vegetation that should be taken as a basis for investigations, not the less complicated, impoverished European vegetation that has risen from it by selection. Although this idea still does not constitute the axis for the mainstream of research, it is now generally recognized. And, at least in the field of forest ecology, the time has come when no researcher can receive a reasonable evaluation without showing data from the tropics. I cannot describe here all the kinds of surveys performed to date, but I shall summarize the main developments.

Considerably after Richards, Whitmore (1975) published an overview of the rain forests of Southeast Asia. Drawing freely on later literature, Whitmore presented faithful descriptions of the rain forests. While Richards' book was more theoretical, Whitmore's was closer to a monograph, but it presented several notable theoretical viewpoints concerning productivity and gaps. In the 1975 edition, the bibliography contained 680 entries, while in the second edition (1982) the number had risen to 1,140, of which 570 had been cited in the first edition and 570 were newly added. Subsequently, he told me that the further burgeoning of the literature, and the appearance of important works on the American and African forests, had made it impossible for him to refer to it all and deterred him from producing a third edition.

The major developments since Richards probably involve work on the productivity and gaps pointed out by Whitmore. Productivity was measured systematically in different world ecosystems under the International Biological Programme (IBP), among which Japan's rain forest study in Malaysia was probably the greatest achievement. Before this, the Japanese productivity group had measured the productivity of forests in various parts of Japan; eventually, through the efforts of a study group centered on Osaka Municipal University and Kyoto University, this work was extended to Southeast Asia, where a survey of the Khao Chong forest of South Thailand was completed.

For the work on tropical rain forests that had been left until last, Pasoh in Malaysia was chosen, and a settled survey was conducted over four years beginning in 1970 in the form of an international cooperative study involving Japan, Malaysia, and the United Kingdom. This work, which involved Japan's forte, was, as Kira (1983) has noted, a new achievement in which Richards was finally overtaken. The work of productivity measurement, however, involving the felling of trees, cutting them into sections, weighing the various parts constituting each layer of the forest, measuring respiration, and so on, requires considerable labor and persistence, and unfortunately, few but the Japanese seem willing to undertake this. In America, a research group under Kyoto University's Shidei provided guidance in the form of joint research, but few results have subsequently appeared. The methodology of productivity study has been epochal in the long history of ecological research, but it has failed to spread worldwide. Japan has undoubtedly pulled ahead of the pack in this field, the substance of which is wide-ranging and encompasses the work on litter fall presented in chapter 3 in this volume.

While productivity survey is at the forefront of developments in ecosystem ecology, gap research lies at the heart of community ecology. Only in the past decade have many young ecologists immersed themselves in gap research. However, tropical gap research centers on work carried out in Central and South America by members of the Organization for Tropical Studies (OTS) and the Smithsonian Tropical Research Institute in Barro Colorado. In tropical Asia there have been fewer studies, although Whitmore's proposal of three phases, namely, a mature phase, a building phase, and a gap phase, has contributed greatly to the analysis of the community structure of rain forest. Whitmore discusses these in relation to the process of regeneration, developing a viewpoint that, according to Richards, originates with Aubreville. Gap research has given rise to many reports but has taught us disproportionately little of substance. We still have no firm answers to the sort of questions that Richards raised in his chapter on regeneration.

Because of the low seasonality in rain forests, the question of when flowers bloom has long attracted interest. During World War II, Koriba produced substantial work with British cooperation in the Singapore Botanic Garden. Phenological research, as will be described later, has centered on Malaysia, but it has not been linked to stratification or the habits of tree species in each layer. The phenological study described in chapter 3 in this volume is characterized by a discussion of growth periods and other factors.

Methods of classifying the vegetation of Southeast Asia vary from one author to another and from one country to another. One scheme that covers the region known botanically as Malesia and takes into account water and soil conditions is that of van Steenis (1957). A slightly modified version appears in Whitmore (1975), but since the original is simpler and clearer, it is reproduced here as figure I-1. Fourteen climaxal vegetation types are divided according to whether the climate is perhumid or seasonal. Monsoon forest is found in a seasonal climate, while the other thirteen types occur in a rain forest climate. These thirteen are divided further into those occurring in locations that are temporarily or permanently immersed in water and those occurring on dry land. The first three chapters of this book treat, from the former, mangrove, freshwater swamp forest, and peat swamp forest, and, from the latter, tropical rain forest and montane rain forest.

Mangrove, although also found in subtropical climates, is distributed mainly in areas of rain forest climate. It is the most seaward of the rain forest types, occurring on lowland inundated to a greater or lesser degree by seawater. Behind the salt to brackish water areas of mangrove lie freshwater swamp forest and peat swamp forest. In these freshwater areas, water is supplied mainly by rivers in the former type and by rainfall in the latter. Mangrove and freshwater swamp forest are influenced to some extent by the tides, while peat swamp forest has stagnant water. Soils consist of mud and sand in mangrove, clay in freshwater swamp forest, and peat in peat swamp forest. In these forest types, trees have adapted to inundation by means of various root formations, including pneumatophores, prop roots, buttress roots, and aerial roots, which are characteristic of individual species. These morphological features are particularly marked in mangrove and have long attracted researchers' interest. Published works on mangrove number

## Climaxal Vegetation Types of Malaysia

| | | | | | | |
|---|---|---|---|---|---|---|
| Climate everwet | Temporarily or permanently submerged | Salt | Submerged | Sand | 1. | Littoral |
| | | | Tides | Mud (or) sand | 2. | Mangrove |
| | | Freshwater | Stagnant | Eutrophic, shallow | 3. | Swamp forest |
| | | | | Oligotrophic, shallow | 4. | Peat forest |
| | | | | Eutrophic, deep | 5. | Hydrophytes |
| | | | Swift running | | 6. | Rheophytes |
| | Dry land | Coastal | | | 7. | Beach vegetation |
| | | Inland to timberline | Siliceous podsol | Mostly lowland | 8. | Siliceous vegetation |
| | | | Limestone | Mostly lowland | 9. | Limestone vegetation |
| | | | Other soils | 2-1,000 m. | 10. | Tropical rain forest |
| | | | | 1,000-2,400 m. | 11. | Montane rain forest |
| | | | | 2,400-4,150 m. | 12. | Subalpine rain forest |
| | | Above timberline | | | 13. | Alpine zone |
| Climate seasonal | | | | | 14. | Monsoon forest |

Figure I-1. Climaxal vegetation types in Malaysia. SOURCE: van Steenis, 1957.

more than seven thousand, while those on the other two swamp forests stop at fewer than a hundred.

Mangrove forest has a simple structure, usually consisting of a single stratum, with individual species forming pure stands and the tallest trees reaching 45 m. Depending on salinity, the period and depth of inundation, tidal strength, and other factors, different species are distributed in successive zones parallel to the coast. Most seaward are *Avicennia* and *Sonneratia*, while farther inland *Rhizophora* and *Bruguiera* are dominant, together with *Ceriops*, *Xylocarpus*, *Lumnitzera*, and other genera. On the ground, *Acrostichum aureum*, *Acanthus ilicifolius*, and other species proliferate; in brackish water, *nipa* palms form large communities. Epiphytes and ferns are uncommon.

Freshwater swamp forest grows on wetland along rivers and is temporarily or permanently inundated by nutrient-rich, fresh water. It is a mixed forest with well-developed stratification, and the tallest trees reach 50 m. Epiphytes and climbers are common. Various species occur, depending on the depth and duration of inundation and other factors; van Steenis lists the following genera and species as typical:

> *Adina, Alstonia, Barringtonia spicata, Campnosperma, Coccoceras, Couthovia, Crateva, Dillenia, Dyera, Elaeocarpus littoralis, Erythrina fusca, Eugenia, Ficus retusa, Glochidion, Gluta renghas, Lophopetalum, Mangifera gedebe, Memecylon, Metroxylon, Nauclea, Pandanus, Parkia, Pentaspadon motleyi, Sapotaceae divers, Serianthes, Shorea belangeran,* other *Shorea, Timonius,* and *Vatica rassak.* (van Steenis, 1957)

Peat swamp forest is a highly distinctive type of forest growing on peat 1–20 m deep. The only water to enter the swamp is nutrient-poor rainwater, while the water draining from it is dark red-brown and acidic. The peat is generally dome shaped, and on it six forest types are distinguishable, distributed in concentric circles. At the periphery is large forest like lowland dipterocarp forest, and toward the center, trees become progressively stunted. In the peat swamp forests of Sarawak and Brunei, *Shorea albida* forms pure stands over extensive areas. Among the main species are the following:

> *Alstonia, Amoora, Anisoptera, Calophyllum, Combretocarpus rotundatus, Cratoxylum, Dryobalanops, Durio carinatus, Eugenia, Gonystylus, Jackia, Koompassia, Litsea, Lophopetalum, Melanorrhoea, Pandanus, Parastemon, Payena, Pholidacarpus, Ploiarium, Salacca, Sapotaceae divers, Shorea, Tetramerista glabra, Tristania,* and *Xylopia.* (van Steenis, 1957)

These three swamp forest types occupy smaller areas than lowland dipterocarp forest, amounting at most to 10–15 percent of the national area of each country in Southeast Asia. Also, except in parts of New Guinea, many of these forests have been subject to development, and they are shrinking fast.

Dry lowland forest is represented by mixed dipterocarp forest, the largest and most extensive forest type of Malesia, which stretches from the upper margin of the wetland to elevations of about 1,500 m. Probably its most striking feature is the tall evergreens that emerge above the closed canopy and rise to heights of up to 80 m. Stratification is well developed, with five distinguishable layers, including the shrub and ground layers. Most of the trees have straight trunks, hemispherical crowns, and large buttress roots. Large woody climbers wind around the trunks, and epiphytes grow on tree branches. Many plant species are present, none of which can be regarded as particularly dominant. This forest type has the richest variety of species, with members of the Dipterocarpaceae prominent. In Sumatra, the Malay Peninsula, Borneo, and Mindanao, these dominate the overstory. Such dominance by a single family (although one that includes more than 500 species) can be said to be characteristic of Southeast Asia. The following other families are comparatively well represented in this forest:

*Anacardiaceae, Annonaceae, Burseraceae, Caesalpineaceae, Celastraceae, Combretaceae, Connaraceae, Dilleniaceae, Ebenaceae, Euphorbiaceae, Guttiferae, Icacinaceae, Lauraceae, Meliaceae, Mimosaceae, Moraceae, Myristicaceae, Myrtaceae, Ochnaceae, Olacaceae, Palmae, Papilionaceae, Rutaceae, Sapindaceae, Sapotaceae, Sterculiaceae, Thymelaeaceae,* and *Tiliaceae.* (van Steenis, 1957)

The most common genera are *Eugenia* and *Ficus,* with more than 500 species; the following genera also have many species represented: *Aglaia, Ardisia, Calamus, Calophyllum, Daemonorops, Diospyros, Dipterocarpus, Dysoxylum, Elaeocarpus, Garcinia, Litsea, Quercus, Schefflera,* and *Shorea.*

Genera represented by large trees are *Agathis, Altingia, Bischofia, Dialium, Duabanga, Dyera, Ficus, Gossampinus, Kickxia, Koompassia, Morrus, Octomeles,* and *Pinus.*

Dipterocarpaceae disappear at altitudes between 1,000 and 1,500 m; above 1,500 m, montane forest appears, in which the principal flora consists of Fagaceae.

Montane forest is characterized primarily by an abundance of epiphytes and climbers. Tree trunks are clothed in mosses and lichens, and wrapped around with epiphytic climbers. In the tree crowns, above the branches, epiphytic orchids and ferns overlap each other densely. With increasing altitude and the transition to moss forest, the volume of mosses increases to twice the thickness of the tree trunks.

Except for the lack of emergents, montane forest at lower altitudes has the same stratification as mixed dipterocarp forest. Tree heights decrease with altitude, and at around 3,000 m, the forest consists of a single layer of trees about 10 m tall. The main groups seen here are Lauraceae, Fagaceae, Cunoniaceae, Monimiaceae, *Acer, Ulmus,* Magnoliaceae, Hamamelidaceae, *Rapanea, Leptospermum,* and Ericaceae.

Herbaceous vegetation appears in the mountains, unlike in the lowland, including species in the following families that can also be seen in subtropical and temperate zones: Boraginaceae, Campanulaceae, Caryophyllaceae, Centrolepidaceae, Cyperaceae, Gentianaceae, Labiatae, Liliaceae, Primulaceae, Ranunculaceae, Umbelliferae, and Violaceae.

When van Steenis made his surveys, primary vegetation remained in abundance, and even around Jakarta there was swamp forest. However, with population growth, the demand for tropical timber, the clearing of forest for cultivation of cash crops, and other factors, the tropical rain forest has changed rapidly, and its very survival is now threatened. Substantial areas of primeval forest today can be found only in the swamps and high mountains.

In this situation, problems have arisen on a global scale involving environmental destruction and the conservation of genetic resources. The environmental problem first attracted attention through movements to protect the ecosystems of the tropics, but as the involvement of other problems became apparent, such as the increase of atmospheric carbon dioxide, its scale became global.

With the disappearance of the rain forest, genetic resource conservation stressed the value of species facing extinction; with the development of biological engineering, it came to encompass the entire process, from the fundamentals of exploration, collection, and identification of genetic resources through to utilization. This development began with the 1972 environmental congress in Stockholm, and since then it has attracted considerable attention.

Against this background, the first three of this volume's four chapters depict the forest types of Southeast Asia's tropical rain forests as seen by someone trekking from the coast up to the mountains. Chapter 4 discusses problems of genetic conservation of forest resources. Although each chapter stands alone in representing one phase of my earlier work, each rests on my firm conviction about the importance of research into the tropical rain forests of Southeast Asia.

Chapter 1 presents a monographic description of the swamp forests occupying the coastal zones of Southeast Asia, based mainly on the literature, and discusses the present situation and future directions of research in light of my own experiences. Although they account for at most 10–15 percent of the total land area in Southeast Asia, these forests are extremely important, because it is here that the most drastic transformations are taking place.

In chapter 2, I analyze the typical forest types of Brunei and the biological developments of trees there, based on data from my research and the material presented in chapter 1. These first two chapters focus mainly on the swamps, making little reference to the mixed dipterocarp forests that form the core of the so-called tropical rain forest. Outstanding work on the mixed dipterocarp forests has already been published, and rather than gild the lily, I have chosen to concentrate on lesser-known fields.

Chapter 3 presents my earliest work, my doctoral dissertation submitted to the Faculty of Agriculture, Kyoto University. It is a survey of the structure and phenology of the forest of a Javanese volcano. Following the mountain gradient, it examines changes in the distribution of vegetation, particularly the stratification of the forest. For the phenological study, leaf fall collected in litter traps was analyzed by species to give a quantitative measure of their seasonal variations. This kind of work has seldom been seen since.

The first three chapters thus depict the tropical rain forests of Southeast Asia, from the swamps to the subalpine zones, focusing on forest structure, species composition, and phenology. While I have interspersed surveys of the literature, these all concern areas that I have visited on foot.

Chapter 4 concludes the book with a discussion of the problems of genetic resources, summarizing what I learned about this topic during my eight years at the Forest Tree Breeding Institute of the Ministry of Agriculture, Forestry and Fisheries. This was a period when, every year, about fifty research students visited from all over the world to learn about tree breeding and the treatment of tropical forests in Japan. I taught them what I had gained from my experiences, and at the same time I acquired much new knowledge from them. This I have again supplemented from the literature and presented in line with my own thinking.

Students coming to Japan from tropical countries often questioned whether Japanese-style techniques would be applicable in the tropics and to what extent the things they were taught rested on a firm grasp of the situation in tropical areas. Japan's international cooperation had only just begun in earnest, and the people who came to study seemed somewhat halfhearted, though my own lectures and practicals were favorably received. It was a pleasure, too, for me to learn the latest information from these students. This was also a time when I began to look at forestry areas across Japan from an administrative standpoint and when I studied in the field how forestry techniques were being established within the national framework.

The present state of forestry in Japan is far from ideal, but nevertheless much can be learned about the techniques practiced in the various forestry regions, some of which have a history of six centuries. These include basic philosophies that seem to be applicable anywhere in the world, and the long accumulation of substantial results. The most important thing for the students who came from overseas was to let them see these forests with their own eyes. And by walking the forests with them, I learned about viewpoints different from those of the Japanese.

I include in chapter 4, therefore, examples from various parts of the world, as well as Southeast Asia. I have not touched on the latest advanced technology; rather, I present a kind of guideline for practicable technological reform in the tropics.

The subjects and the methodology of each of the four chapters are different but linked by a common theme: the importance of discussion based on a close look at the situation in the field in each area. Today, when the problems of the tropical forests loom large, I believe it is the function of those researchers familiar with a particular location to provide as much pertinent information as possible about the situation on the ground in the forest, and from that to seek new directions.

Exploitation of the tropical rain forests has brought about a loss of diversity and an increase of uniformity. The multilayered forest has been felled and burned, and the land used for grazing or farming. Vast areas have been cleared to grow produce for humans, and single commercial crops, for example, rice, maize, cassava, rubber, or oil palm, now uniformly occupy land where complex forest formerly grew. Although such monoculture may be scenically uninteresting and biologically undesirable, it is by no means a bad thing. While not perfect, it is economically one of the most efficient methods presently available, and under today's economic system, it is probably inevitable.

The greatest problem is the conversion of forest to grassland, which results from repeated swiddening of land at short intervals. This becomes particularly severe if swiddening is followed by grazing. Such *alang-alang* grassland now extends over vast areas of tropical Southeast Asia. Here the soil condition has deteriorated, and in many places it is difficult even to dig planting holes for reforestation. In the swamps, the destruction of forest sometimes causes the appearance of such toxic soils as acid sulfate soils. This situation, of course, is disastrous for local residents. And for the hunter-gatherer peoples who live in the forest, the felling of the forest means utter deprivation of their livelihood.

Faced with this diversity of problems, how should we treat the tropical rain forests in the future? One method would be to leave intact as far as possible those areas of the remaining natural forest that show distinct regional features, and by exhaustive study of these representative forest ecosystems gain a basic knowledge of the structure and functions of tropical rain forest.

Where humans have already intervened, the situation must be examined from several aspects, depending on the degree of intervention. Monoculture is now at the review stage, and attempts to find solutions that accord with the conditions of tropical rain forest are getting under way. A comprehensive approach is required that incorporates the viewpoints not only of forestry but also of such basic industries as agriculture, stock raising, and fisheries, land use policies based on geology and soil, and even the field of social humanities. No arena is more suited to interdisciplinary study than the tropical rain forest, and such work seems likely to increase in the future.

The problems of the tropical forests have often tended to be treated crudely as part of the global environmental problem. In the absence of detailed data, they have brazenly been discussed in broad strokes. The problems of the tropical forests, however, concern not only the forests but a wide variety of people: people who inhabit the forests, people who log the forests, people who protect the forests, people in administration, and people in international environmental groups. Fundamental to the treatment of the tropical rain forests is the need for extremely detailed work that rests on a thorough grasp of the lives and thinking of these people and a solid recognition of the conditions in the areas concerned.

# 1.

# *Swamp Vegetation Landscape*

## 1. Mangrove

Research into mangrove has a long history, stretching back to the earliest record of this vegetation in the third century B.C. That mangrove has so long attracted the interest of researchers worldwide is probably because of the uniqueness of its growth environment and its external form.

Mangrove vegetation is unusual in being distributed along the border between sea and land. It consists, moreover, not of the kind of low vegetation found on sand dunes but trees rising to 30 m and forming large communities of a single species. With increasing distance from the sea, the constituent tree species of the communities change. The trees display remarkable root formations, including prop roots, pneumatophores, aerial roots, and buttress roots. In the trees are peculiar hanging fruits and a variety of flowers, and birds, moths, and insects gather there. The ground around is alive with crabs, prawns, and mud-skippers. Larger animals include snakes, lizards, crocodiles, tigers, and countless birds.

Records from the past leave no doubt about the deep impression that such mangrove scenery made on sea voyagers. Theophrastus (305 B.C.) first described the mangrove around the Persian Gulf, and by the time of Linné (1753), enough information had accumulated to allow the description of several species.

Scientific methods were first applied to the study of mangroves in the late nineteenth century. The early work of Schimper (1891), Karsten (1891), and others was furthered by those such as Mead (1912), Becking et al. (1922), Luytjes (1923), Watson (1928), and de Haan (1931). Surveys and research subsequently advanced in various fields, and from around the end of the 1950s, Ding Hou (1958), Macnae (1968) and others produced comprehensive surveys. The 1970s saw important works by Chapman (1976, 1977), and with international symposia being held in various countries mangrove research reached a kind of peak. This was the result of the weight attached to the three major primary industries of fishing, farming, and forestry for coastal development in tropical countries. For the foreseeable future, basic research and other works will likely continue to focus on applied aspects.

One practical problem that has been raised is the disappearance of mangrove in various localities. Development has finally reached the tropical swamps, where previously no one ventured. Swamps where large tracts of the primary forest have been felled and the land used for agriculture, fishponds, or settlements are a common sight in Southeast Asia. The rich mangroves that once delighted botanists from temperate climes and spawned countless studies are now threatened with extinction and have been divested of interest by the uncontrolled land reclamation that is in progress.

On entering the mangrove, a tranquillity reigns that is unimaginable from the noisy surroundings. All sound is absorbed by the mud and water and trees, and a world unfolds that has been forgotten in the modern age, accented only by the occasional birdcall or the beating of wings. Here the vegetation leads a rich and stable life.

It is quite an ordeal to enter a world that centers on a life form other than humans. Usually, you approach mangrove from the sea by small boat, aiming to enter a tidal creek at high tide. When alighting from the boat to walk in the forest, you may encounter various conditions. In some places, you may sink chest-deep in mud, while elsewhere walking may be surprisingly easy. Most troublesome are the mosquitoes. It is a myth that there are no malaria mosquitoes in the mangroves. Many people have contracted malaria after being bitten by the mosquitoes that swarm there in columns. Nevertheless, the mangroves hold sufficient attraction that people still venture forth.

Mangroves can be found in tropical and subtropical parts of the world, but their center is Southeast Asia, whether in terms of the number of species, the diversity of constituent species, the size of individual trees, or the scale of communities. The reports of early researchers overwhelmingly focused on Southeast Asia, and in discussing mangroves, one must look first at this region.

Here, I have attempted to position the mangrove of Southeast Asia in terms of the distribution of species, zonation and its causes, morphology, exploitation, and other aspects. Literature on mangrove is said to extend to almost seven thousand titles, of which I have been able to consult only a small number. My objective is to grasp the situation of Southeast Asia's mangroves by focusing on representative works.

## Mangrove Distribution

### Factors Governing Distribution

Walsh (1974) cites five factors governing the distribution of mangrove, while Chapman (1976) cites seven. I have rearranged them as follows:

- Factors governing regional distribution: geological factors, temperature, and ocean currents

- Factors governing local distribution: (a) terrestrial conditions—discharge of major rivers, shallow mud coastlines, and landforms offering protection from off-sea winds and waves; and (b) marine conditions—salinity and tidal range
- Factors determining zonation: tidal movements, soil type, soil salinity, light, and species characteristics

Mangrove is distributed in coastal zones of the world's tropics and subtropics, of which Southeast Asia is central, having the greatest number of species. The northern extremity of this sphere of distribution is marked by the *Kandelia candel* of southern Kyushu, Japan, at 35° N, the southern extremity by the *Avicennia marina* near Melbourne, Australia, at 38° S. In the east-west direction, with the exception of certain Pacific islands, mangroves circle the globe.

The distribution of vegetation is basically determined by temperature. Globally, the northern limit of mangrove corresponds with the isotherm of 16°C in January, while the southern limit in the Atlantic and Indian Oceans corresponds to 16°C in July (Chapman, 1975). Extensive mangrove occurs in areas with a monthly minimum average temperature of more than 20°C and a seasonal range not exceeding 5°C. There are, of course, exceptions; for example, in East Africa, where mangrove is found in places with a seasonal range higher than 10°C. The most prolific growth, however, is found in the range stated above. Rainfall, another important factor determining plant distribution, is not a major factor with mangrove, which is found widely from the humid tropics to the arid tropics.

Ocean currents influence the larger distribution of mangroves growing near coasts. Cold and warm currents to some extent determine distribution on the continental peripheries. The limits of distribution, though, are not determined simply by the temperatures of cold currents, which all flow northward, thus inhibiting the southerly drift of floating seedlings that might otherwise become established farther south. Also, no corroborated evidence exists of how far seedlings can travel by ocean currents. This is an important theme that should be examined over a wide area.

Mangroves are not generally found on rocky or sandy coasts. The primary terrestrial condition favoring mangroves is an extremely flat topography with a major river. Where sediments carried down over long periods have reduced river gradients, large volumes of mud raise the shallow seabed even further. Where, in addition, a landform such as an inlet, lagoon, bay, sandbar, island, or narrow strait provides protection from wave and tidal action, optimal mangrove conditions are established.

Marine conditions influencing mangroves are tides and salinity. Salinity varies with tidal range and with the flow of fresh water in a river, and vegetation can be found that is adapted to various conditions. Along a slow-moving river, the penetration of mangrove upstream is proportional to the tidal range. On the Fly River in Papua New Guinea, mangrove reaches 300 km upstream, while in South Sumatra it reaches approximately 100 km upstream as far as Palembang. Conversely, mangrove will not grow in areas constantly affected by strong winds and waves.

Under suitable conditions, mangrove forest will develop over a wide area. Although mangrove species are not all obligate halophytes, having been raised in fresh water in botanic gardens at Bogor and in Europe and India, they are under natural conditions generally restricted to areas of seawater or brackish water. Similarly, while mangroves sometimes grow on sand, peat, coral reefs, and other soil conditions, they do not form large communities.

Recently, human activities have added to the factors determining distribution. For example, the distributions of mangrove have shrunk in India because land has been converted to pasture and farmland, and in the Philippines and Java because fishponds have been created. There are also areas, as in Hawai'i, where mangrove has been introduced.

While the above factors are thought to operate in determining the distribution of mangroves, mangrove communities are generally structured in zones where the constituent species change successively with distance inland from the sea. Such zonation is regarded as a characteristic of mangrove distribution, and I shall examine this in more detail later.

## Origin of Distribution

Except for part of the Pacific Ocean, mangrove is now distributed along most of the coastlines in the tropics and subtropics. When and in what form it attained such a distribution is the subject of hypothesis advanced by Chapman (1975), based on the continental drift theory.

According to this theory, the original land mass of Pangaea existed in the Permian period about 200 million years ago. During the Mesozoic era, this began to break up, the northern continent of Laurasia separating from the southern continent of Gondwanaland, and the Tethys Sea opening up as far as Gibraltar. About 75 million years ago, from the end of the Cretaceous era into the beginning of the Cenozoic, the plate carrying India was juxtaposed with Madagascar and Africa. Around this time, the evolution of angiosperms was progressing, and the mangroves *Rhizophora* and *Avicennia* began to evolve in adaptation to a saline environment. Ten million years later, the Indian Ocean and remnants of the Tethys Sea were in evidence, as shown in figure 1-1. The region marked "M" in the figure shows the location where mangrove is thought to have originated.

As is clear from the figure, mangrove must have spread along the southern

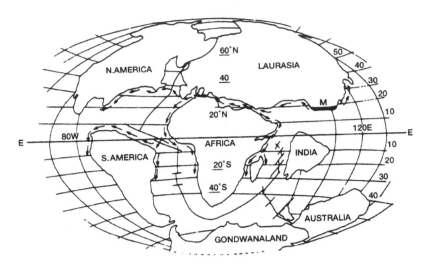

Figure 1-1. The world at the end of the Cretaceous. "M" indicates area of mangrove origin; arrows indicate the direction of dispersal. SOURCE: Chapman, 1977.

coast of Laurasia. If the entrance to the Mediterranean Sea had been open at that time, the ancestors of *Rhizophora* and *Avicennia* would have first entered the Mediterranean Sea and spread from there to the west coast of Africa, northward to Spain, and across the relatively narrow sea to North America, where they would have spread southward to Florida. In the colder winters of later times, both genera would have disappeared from the Mediterranean and northeastern America.

The situation of the Pacific Islands at that time is unclear, but it is thought that the ocean was too wide for seedlings to cross eastward from Laurasia.

Regarding the Indian Ocean, 35 million years ago India was separated from Laurasia by a narrowing sea (McKenzie and Sclater, 1973), so that the migration of *Rhizophora* and *Avicennia,* together with such genera as *Bruguiera, Ceriops, Xylocarpus,* and *Lumnitzera,* could have continued along these coasts. The entrance from the Tethys Sea to the Mediterranean Sea was closed off by this time, so these later genera could not have passed this way. Rather, they would have colonized the Maldives, Mascarenes, Chagos, and other islands and also spread southward along the east coast of Africa. The whole of the Indian Ocean at this time was warm and favorable to the growth of mangroves.

If *Rhizophora* and *Avicennia* did not pass through the Mediterranean Sea, they must have rounded Cape Horn. This seems highly unlikely, however, in view of the cold climate there at the time. If they made this passage later, it is difficult to explain why *Bruguiera* did not also succeed.

Collateral evidence that the climate was warm enough for mangroves is provided by the presence of *Nypa* pollen in London Clay of the late Eocene and similar strata in Brazil (Dolianiti, 1955). *Avicennia* pollen and *Acrostichum* fronds have been reported in Tertiary strata in England (Montford, 1970), while in West Africa *Nypa* pollen has been found in Upper Cretaceous to Upper Eocene deposits.

*Rhizophora* and *Avicennia* could also have crossed westward from Spain and Morocco to South America, where they crossed to the Pacific coast and spread southward before the isthmus of Panama closed.

In the Pacific Islands, eastward spread from the original center could have occurred before Australia approached Asia, and in the Northern Hemisphere, the oceanic distances to the Pacific Islands were too great for mangroves to have spread westward from the west coast of the Americas.

The above argument is the outline of Chapman's thesis that the ancestral mangroves *Rhizophora* and *Avicennia* evolved through a westward transmission. On the other hand, van Steenis (1962) and Ding Hou (1960) claim that *Rhizophora, Avicennia, Xylocarpus, Lumnitzera,* and *Laguncularia* originated in the Indo-Malaysia region and spread westward to East Africa, while the four species other than *Laguncularia* spread eastward to the west coast of America. These reached the Caribbean Sea between the Upper Cretaceous and Lower Miocene periods, when the isthmus of Panama was still open, and having established themselves on the east coast of America, their seedlings were carried by ocean currents across to the west coast of Africa. Ding Hou claims that there are no lands suitable for mangroves in the Pacific Islands, but Chapman (1975) states that this is not necessarily so.

To carry these arguments further, we need phylogenic research into the genera constituting mangrove, research into mangrove pollens in post-Cretaceous and later sediments in the Mediterranean Sea, and corroborative studies of the drift capacity of seedlings. These, together with progress in geologic studies, can be expected to narrow down further the debate on the origin of mangroves.

## Distribution of Species

The mangroves distributed across the globe have been divided into two types by Chapman (1975): the Indo-Pacific type and the New World–West Africa type.

The Indo-Pacific type stretches from East Africa to the Red Sea, India, Southeast Asia, China, southern Japan, the Philippines, Australia, New Zealand, and the Pacific Islands as far as Samoa (fig. 1-2). The New World–West Africa type is found on the Atlantic coasts of Africa and America, the Gulf of Mexico, the Pacific coast of tropical America, and the Galápagos Islands. In Hawai'i, *Rhizophora mangle, Bruguiera sexangula, Sonneratia caseolaris,* and *Conocarpus erectus* have been introduced, but the Hawaiian Islands have no indigenous mangroves.

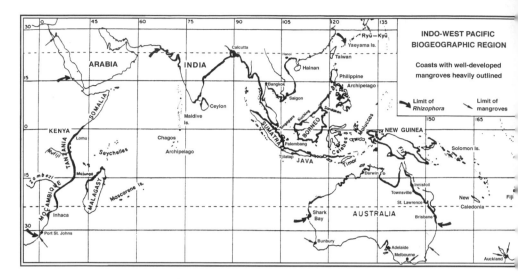

Figure 1-2. Distribution of mangrove from Africa to the Pacific Ocean.
SOURCE: Macnae, 1968.

Table 1-1 shows the worldwide distribution of mangrove genera according to this classification. The Indo-Pacific type (which includes the communities of East Africa) is shown in the column headed Indian Ocean and West Pacific; the New World–West Africa type is listed under three regions: Pacific U.S.A., Atlantic U.S.A., and West Africa. The main families include Rhizophoraceae, Avicennaceae, Meliaceae, Sonneratiaceae, Combretaceae, Bombacaceae, Plumbaginaceae, Palmae, and Myrsinaceae. Of these, the Rhizophoraceae is represented by the greatest number of species, with 16 species in 4 genera, of which 14 species are found in the Indo-Malaysia region. The fact that 63 of the total 90 species are found in this region demonstrates fully that this region is the center of distribution. In comparison, the New World type is impoverished, with no more than 17 species in the Atlantic U.S.A.

Table 1-2 shows the distribution of 50 typical mangrove species. Nine species are in East Africa, 28 in India, 28 in Malaysia, 31 in Borneo, and 25 in Australia. Thus the center of distribution is the tropical rain forest belt from the Malay Peninsula to Borneo.

## Zonation

Mangroves are distributed in a series of zones paralleling the coast, in which the species change successively with distance inland. Macnae (1968) summarized the zonation of the mangroves distributed from East Africa to Australia into the following six main types, from inland to the coast.

Table 1-1

World Distribution of Mangrove Types by Genus
(including some distributed in several regions)

| GENERA | INDIAN OCEAN WEST PACIFIC | PACIFIC U.S.A. | ATLANTIC U.S.A. | WEST AFRICA | TOTAL SPECIES |
|---|---|---|---|---|---|
| *Rhizophora* | 5 | 2 | 3 | 3 | 7 |
| *Bruguiera* | 6 | — | — | — | 6 |
| *Ceriops* | 2 | — | — | — | 2 |
| *Kandelia* | 1 | — | — | — | 1 |
| *Avicennia* | 6 | 3 | 3 | 1 | 11 |
| *Xylocarpus* | ?8 | ? | 2 | 1 | ?10 |
| *Laguncularia* | — | 1 | 1 | 1 | 1 |
| *Conocarpus* | — | 1 | 1 | 1 | 1 |
| *Lumnitzera* | 2 | | | | 2 |
| *Camptostemon* | 2 | | | | 2 |
| *Aegialitis* | 2 | | | | 2 |
| *Nypa* | 1 | | | | 1 |
| *Osbornia* | 1 | | | | 1 |
| *Sonneratia* | 5 | | | | 5 |
| *Scyphiphora* | 1 | | | | 1 |
| *Aegiceras* | 2 | | | | 2 |
| Others | 19 | 9 | 7 | 4 | 35 |
| Totals | ?63 | ?16 | 17 | 11 | ?90 |

SOURCE: Chapman, 1976.

1. The landward fringe
2. Zone of *Ceriops* thickets
3. Zone of *Bruguiera* forests
4. Zone of *Rhizophora* forests
5. The seaward *Avicennia* zone
6. The *Sonneratia* zone

For the Malaysia area, Chapman (1976) identified nine types.

1. *Avicennia* pioneer community
2. *Sonneratia alba* pioneer community
3. *Rhizophora mucronata* pioneer community
4. *Rhizophora apiculata* community

Table 1-2

Distribution of Fifty Typical Mangrove Species

| | East Africa | Mada-gascar | Ceylon | India | Burma | Ma-laysia | Su-matra | Java | Borneo | New Guinea | Aus-tralia | New Zealand | Philip-pines |
|---|---|---|---|---|---|---|---|---|---|---|---|---|---|
| *Rhizophora apiculata* | — | | — | | | | | | | | | | — |
| *R. mucronata* | — | | — | | | | | | | | | | — |
| *R. stylosa* | | | | | | — | | | | | — | | — |
| *R. harrisonii* | | | | | | | | | | | | | — |
| *R. mangle* | | | | | | | | | | | | | |
| *Bruguiera cylindrica* | | | — | | | | | | — | | | | — |
| *B. exaristata* | | | | | | | | | | | — | | |
| *B. gymnorrhiza* | — | | — | | — | | | | | | — | | — |
| *B. parviflora* | | | — | | — | | | | | | — | | — |
| *B. sexangula* | | | — | | | | | | | | — | | — |
| *Ceriops tagal* | — | | — | | | | | | | | — | | — |
| *C. decandra* | | | ? | | | | | | | | | | |
| *Kandelia candel* | | | ? | ? | — | | | | | | | | |
| *Sonneratia alba* | — | | — | | | | | | | | — | | — |
| *S. apetala* | | | — | | — | | | | | | | | |
| *S. caseolaris* | | | — | | | | | | | | — | | |
| *S. griffithii* | | | | | | — | | | | | | | |
| *S. ovata* | | | | | | | | | — | | | | |
| *S. spp.* | | | | | | | | | — | | | | |
| *Xylocarpus granatum* | — | | | | | | | | | | — | | — |
| *X. moluccensis* | — | | | | | | | | | | | | — |
| *X. mekongensis* | | | | | | | | | | — | | | — |
| *Lumnitzera littorea* | | | | | | | | | | | — | | — |
| *L. racemosa* | — | | | | | | | | | | — | | — |
| *Aegiceras corniculatum* | | | — | | | — | | | | | — | | — |
| *A. floridum* | | | | | | | — | | | | | | — |

TABLE 12 (continued)

| | EAST AFRICA | MADA-GASCAR | CEYLON | INDIA | BURMA | MA-LAYSIA | SU-MATRA | JAVA | BORNEO | NEW GUINEA | AUS-TRALIA | NEW ZEALAND | PHILIP-PINES |
|---|---|---|---|---|---|---|---|---|---|---|---|---|---|
| Avicennia alba | | ? | | ——————————— | | | | | | | | | — |
| A. marina | ——————————————— | | | | | | | | | ? | ——— | — | |
| A. resinifera | | | ——— | | | | | | | | | | |
| A. officinalis | | | | | | | | | | | | | |
| A. lanata | | | | | — | | | | | | | | |
| A. bicolor | | | | | — | | | | | | | | |
| A. germinans | | | | | | | | | | | | | |
| A. africana | | | | ——————————— | | | | | | | | | |
| Aegialitis rotundifolia | | | | | | ——— | | | | | | | |
| A. annulata | | | ——————————————————— | | | | | | | — | | | — |
| Scyphiphora hydrophyllacea | | | | | | | | | | | ——— | | — |
| Derris hexapetala | | | | ——— | | | | | | | | | |
| Laguncularia racemosa | | | | | | | | | | | | | |
| Conocarpus erectus | | | | | | | | | | | | | |
| Pelliciera rhizophorae | | | | | | | | | | | | | |
| Nypa fruticans | | | | ——————————————— | | | | | | | | | |
| Acanthus ilicifolius | | | | ——————————————— | | | | | | | | | |
| Acrostichum aureum | | | | ——————————————— | | | | | | | | | — |
| A. speciosum | | | | | | | | | | | ——— | | |
| Camptostemon philippinensis | | | | | | | | | | | | | — |
| C. schultzii | | | | | | | | | ——— | | | | |
| Heritiera minor | | | | — | | | | | | | | | |
| H. littoralis | ——— | | | | | | ——————— | | | | | | |
| Excoecaria agallocha | | | | ——————————————————————— | | | | | | | | | |

(continued on next page)

Table 1-2 (continued)

| | EAST THAILAND | VIETNAM | TAIWAN | HAINAN | RYUKYUS | SOUTH JAPAN | WEST AFRICA | BRAZIL | GULF OF MEXICO | FLORIDA | CENTRAL AMERICA | ECUADOR |
|---|---|---|---|---|---|---|---|---|---|---|---|---|
| *Rhizophora apiculata* | — | — | | — | | | | | | | | — |
| *R. mucronata* | — | — | | | | | | | | | | |
| *R. stylosa* | | | | | | | | — | | | | |
| *R. harrisonii* | | | | | | | | | | — | | |
| *R. mangle* | — | — | | | | | | | | | | |
| *Bruguiera cylindrica* | — | — | | | | | | | | | | |
| *B. exaristata* | | | | | | | | | | | | |
| *B. gymnorrhiza* | | | | | — | | | | | | | |
| *B. parviflora* | | | | — | | | | | | | | |
| *B. sexangula* | | | | | | | | | | | | |
| *Ceriops tagal* | | | | | | | | | | | | |
| *C. decandra* | | | | | | | | | | | | |
| *Kandelia candel* | — | — | — | | — | | | | | | | |
| *Sonneratia alba* | — | — | — | | — | | | | | | | |
| *S. apetala* | — | — | — | | | | | | | | | |
| *S. caseolaris* | — | — | — | | | | | | | | | |
| *S. griffithii* | — | — | | | | | | | | | | |
| *S. ovata* | — | — | | | | | | | | | | |
| *S.* spp. | — | — | — | | | | | | | | | |
| *Xylocarpus granatum* | — | — | — | | | | | | | | | |
| *X. moluccensis* | — | — | — | | | | | | | | | |
| *X. mekongensis* | — | | | | | | | | | | | |
| *Lumnitzera littorea* | ? | — | — | | | | | | | | | |
| *L. racemosa* | ? | — | | | | | | | | | | |
| *Aegiceras corniculatum* | — | — | | | | | | | | | | |
| *A. floridum* | — | | | | | | | | | | | |

(continued on next page)

TABLE 1-2 (continued)

| | EAST THAILAND | VIETNAM | TAIWAN | HAINAN | RYUKYUS | SOUTH JAPAN | WEST AFRICA | BRAZIL | GULF OF MEXICO | FLORIDA | CENTRAL AMERICA | ECUADOR |
|---|---|---|---|---|---|---|---|---|---|---|---|---|
| *Avicennia alba* | — | | | | | | | | | | | |
| *A. marina* | | | | | | | | | | | | |
| *A. resinifera* | | | | | | | | | | | | |
| *A. officinalis* | — | ? | ? | | | | | | | | — | — |
| *A. lanata* | | | | | | | | | | | | |
| *A. bicolor* | | | | | | | | | | | — | — |
| *A. germinans* | | | | | | | | | | | | |
| *A. africana* | | | | | | | — | — | | | | |
| *Aegialitis rotundifolia* | | | | | | | | | | | | |
| *A. annulata* | | | | | | | | | | | | |
| *Scyphiphora hydrophyllacea* | — | | | | | | | | | | | |
| *Derris hexapetala* | | | | | | | | | | | | |
| *Laguncularia racemosa* | | | | | | | — | | | | | — |
| *Conocarpus erectus* | | | | | | | — | | | | | — |
| *Pelliciera rhizophorae* | | | | | | | | | | | — | |
| *Nypa fruticans* | | | | | | | | | | | | |
| *Acanthus ilicifolius* | — | | | | | | | | | | | |
| *Acrostichum aureum* | | | | | | | | | | | | |
| *A. speciosum* | | | | | | | | | | | | |
| *Camptostemon philippinensis* | | | | | | | | | | | | |
| *C. schultzii* | | | | | | | | | | | | |
| *Heritiera minor* | | | | | | | | | | | | |
| *H. littoralis* | | | | | | | | | | | | |
| *Excoecaria agallocha* | | | | | | | | | | | | |

SOURCE: Chapman, 1975.

5. *Bruguiera cylindrica* (Berus) type
6. *Bruguiera parviflora* (Lenggadai) type
7. *Bruguiera gymnorrhiza* (Tumu) type
8. *Ceriops tagal* scrub
9. *Lumnitzera littorea*

Figure 1-3 summarizes the succession of these nine types in Malaysia. As Chapman notes, not all stages will necessarily be found in any one locality.

The classifications of Macnae and Chapman are not, however, complete, and I present my own interpretation in table 1-3. This takes the community as its unit and lists the thirteen major communities found in the India-Malaysia area in four divisions according to ecological environment. The first, saltwater pioneer type, consists of four communities that form stands as pioneer species on sea faces and land near the sea. The second, intermediate type, includes the four communities at the heart of mangrove, and these seldom face the sea. Inland from these is the third, inland transitional type, which includes two communities named after the dominant species and a mixed forest community of various transitional

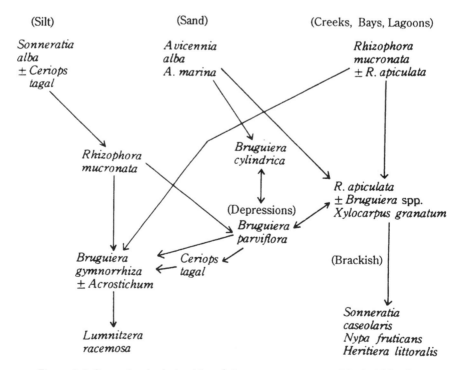

Figure 1-3. Successional relationships of nine mangrove communities in Malaysia.
SOURCE: Chapman, 1976.

## Table 1-3
### Main Groups of Mangroves Distributed in Southeast Asia and Characteristics of Their Growth Areas

| ECOLOGICAL DIVISION | COMMUNITY | SUBSTRATE AND POSITION |
|---|---|---|
| I. Saltwater pioneer | *Avidennia alba* | Fringes of large and small islets, muds mixed with sand or silt |
| | *Avicennia marina* | Sea faces, relatively hard basis; this group is followed by *B. cylindrica* |
| | *Sonneratia alba* | Extremely deep; soft basis of silt deposits near rivers |
| | *Rhizophora mucronata* | Hard, deep mud of small inlets and banks of creeks; prefers salinity of at least 20% |
| II. Intermediate | *Rhizophora apiculata* | Basis of dark, fertile humus mixed with fine sand; also grows in considerably dry soil |
| | *Bruguiera cylindrica* | Hard clay behind *Avicennia* and *B. parviflora*, near estuaries, shallow organic layer, places without canals |
| | *Bruguiera parviflora* | Called an opportunist, enters and leaves before and after *Rhizophora;* appears in locations similar to *B. cylindrica* or on wetter land |
| | *Bruguiera gymnorrhiza* | Landward side on dry, well-ventilated soil; a species of the final stage of mangrove |
| III. Inland transitional | *Ceriops tagal shrubs* | Appears on the landward fringe of mangrove, sometimes beneath *Bruguiera* or *Rhizophora,* height 1–6 m |
| | *Lumnitzera littorea* | Appears on soils near riverbanks; associated with *Scyphiphora hydrophyllacea* |
| | *Mixed forest* | Mixed inland species, including *Alstonia scholaris, Intsia bijuga,* and *Ficus retusa;* sometimes adjoins forest (annual rainfall over 2,000 mm); sometimes adjoins grassland or desert (500–1,500 mm) |
| IV. Brackish water | *Nypa fruticans* | Found in areas flooded by highest spring tides; grows on banks of rivers and creeks |
| | *Sonneratia caseolaris* | Grows along rivers, where salinity is diluted by abundant fresh water |

SOURCE: Prepared with reference to Watson, 1928; Macnae, 1968; and Chapman, 1976.

species. Here the variation with locality is most marked. The fourth, brackish
water pioneer type, contains the two major communities that are conspicuous
along predominantly freshwater river courses. This division, I believe, clarifies
the locations and characteristics of the various communities.

Table 1-4 shows the distribution of these communities from India to the
Pacific Ocean. Chapman (1977) discusses even more mangrove communities, but
the situation is easier understood if the above are regarded as the basic communities to which the other, smaller ones are appended.

## Factors Controlling Zonation

Examples in various localities make it clear that mangrove species are basically
distributed in zones. Although no corroborative answer has yet been advanced
concerning what factors control this zonation, the question has been examined
by several workers, including Walter and Steiner (1936), Macnae (1968), Baltzer
(1969), Chapman (1976), and Snedaker (1982). Macnae (1968) cites frequency of
tidal flooding, salinity tolerance, and waterlogging of the soil as controlling
factors, and the effects of waterways in mangrove and erosion of mangrove shores
as strong influences on these factors.

Chapman (1976) discusses ecological conditions controlling mangrove distribution rather than zonation itself, and cites the features of eight factors: (1) tidal
factors, (2) salinity, (3) soil composition, (4) water table and drainage, (5) aera-
tion, (6) soil chemistry other than sodium chloride, (7) climate, and (8) animal
biota. More recently, Snedaker (1982) has divided the existing theories pertain-
ing to the causes of zonation into four fields: (1) succession, (2) geomorphology,
(3) physiological ecology, and (4) population dynamics; and he has discussed
these in historical perspective, reasoning that in the future, research in geomorphology and physiological ecology will contribute to the elucidation of the zonation and succession of mangrove.

The factors controlling zonation can be examined from two aspects: the conditions of locations where species grow and the characteristics of species that
grow in these locations. Concerning the former, two major natural factors operate
in areas where mangroves grow: tidal action and the sedimentary action of rivers.
Concerning tides, basic classifications have been proposed by Watson (1928) and
de Haan (1931) (table 1-5), in which zonation is related to frequency of tides,
depth of flooding, and penetration inland. Sedimentation is determined by the
balance between tides from the sea and rivers from the land, leading to the development of the substrates that produce zonation. Salinity, water depth, and soil
conditions, specifically, are important.

Salinity has been measured extensively, and the salinity tolerances of most
species are known. The most salt-tolerant species are *Avicennia marina* and *Lumnitzera racemosa,* which can grow in salt concentrations of up to 90 per mille.

Table 1-4

Distribution of Thirteen Mangrove Groups from India
to the Pacific Ocean

| Ecological Division | Community | India | Burma | Malaysia | Papua New Guinea | Philippines | Pacific Islands |
|---|---|---|---|---|---|---|---|
| I. Saltwater area | *Avicennia alba* | + | + | + | + | + | + |
| | *Avicennia marina* | + | + | + | + | + | + |
| | *Sonneratia alba* | + | + | + | + | + | + |
| | *Rhizophora mucronata* | + | + | + | + | + | + |
| II. Intermediate | *Rhizophora apiculata* | + | + | + | + | + | + |
| | *Bruguiera cylindrica* | | + | + | + | + | |
| | *Bruguiera parviflora* | | + | + | + | + | |
| | *Bruguiera gymnorrhiza* | + | + | + | + | + | + |
| III. Inland transitional | *Ceriops tagal* shrubs | + | + | + | + | + | + |
| | *Lumnitzera littorea* | | + | + | + | + | + |
| | Mixed forest | + | + | + | + | + | + |
| IV. Brackish water | *Nypa fruticans* | + | + | + | + | + | + |
| | *Sonneratia caseolaris* | + | + | + | + | + | + |

SOURCE: Prepared from Chapman, 1977.

Table 1-5

Comparison of Mangrove Divisions by Watson, de Haan, and Macnae

| WATSON, 1928 | deHAAN, 1931 | DOMINANT PLANT AND SYSTEM USED IN MACNAE, 1968 |
|---|---|---|
| | A. Brackish to salt water—salinity at high tide 10–30‰ | |
| 1. Land flooded at all high tides | A1. Areas flooded once or twice daily on each of 20 days per month | Seaward fringe of *Sonneratia alba* or *S. griffithii* |
| 2. Areas flooded by "medium high tides" | | Zone of *Avicennia marina* |
| | A2. Areas flooded 10–19 times per month | Zone of *Rhizophora* forest |
| 3. Areas flooded by normal high tides | | Zone of *Bruguiera* forests |
| | A3. Areas flooded 9 times per month | Forests of the landward fringe |
| | | *Xylocarpus granatum* or |
| 4. Areas flooded by spring tides only | | *Lumnitzera littorea* or |
| | | *Bruguiera sexangula* or |
| | | *Samphire* association or |
| 5. Areas flooded by exceptional high tides | A4. Areas flooded on only a few days each month | *Barringtonia* association |
| | B. Fresh to brackish water—salinity 0–10‰ | |
| | B1. Areas more or less under the influence of the tide | *Nypa* association |
| | B2. Areas seasonally flooded either by fresh or brackish water | |

SOURCE: Macnae, 1968.

These are followed by *Ceriops tagal,* 60 per mille, and *Rhizophora mucronata* and *R. stylosa,* 55 per mille. *Rhizophora apiculata, Bruguiera gymnorrhiza,* and *B. parviflora* range from 10 to 30 per mille, and for the last-named species 20 per mille is the optimal concentration. *Xylocarpus* and *Nypa* range from 1 to 30 per mille, *Bruguiera sexangula* ranges from 1 to 10 per mille, and *Sonneratia* species favor about the usual salt concentration of seawater, except for *S. caseolaris,* which favors less than 10 per mille. Mangroves are facultative halophytes and in many cases can be raised artificially in fresh water, but since they grow optimally in nature, there is no doubt that their salt tolerances contribute to zonation along the environmental salinity gradient.

Depth and frequency of tidal inundation almost certainly influence zonation, as is illustrated by the inundation classes proposed by Watson, de Haan, and others. An early attempt to relate mangrove vegetation to topography was made by Davis (1940). In this regard, the tidal formation of a succession of flats at different levels, each with differing distributions of plants and animals, is important (Macnae, 1968). Further clarification of the origins, structures, and biota of these flat surfaces should provide clues to the explanation of zonation.

The final factor is substrate conditions, which many researchers believe to be more important than salinity. Mangrove soils are fine-grained, and often semifluid and poorly consolidated. The humus layer contains the remains of roots and other woody structures, and crab feces are an important constituent. The soils have a low oxygen content, with abundant hydrogen sulfide. In the mangroves of Sarawak, the average conductivity of clays in the *Avicennia* fringe was 5,000–8,000 mho, pH was more than 7, and weight loss on ignition was 5–15 percent. Similar values were found in a *Rhizophora* forest. In the muck soils of a *Nypa* belt, on the other hand, conductivity was as low as 0–1,000 mho, pH approached the acidic values of peat soil, and loss on ignition was 80–90 percent, rising above 95 percent in peat swamp (Macnae, 1968).

Soil properties are controlled by the basal rock in the upper reaches of the rivers. The soils carried down from areas of quartzite, old granite, or limestone are infertile, whereas young volcanic ash produces highly fertile soils, as can be seen, for example, in Java. In the Strait of Malacca, moreover, the marine sediments that built up in the river floodplains in the Pleistocene epoch are said to be eroded by tidal action and added to the mangrove soils.

These relatively young mangrove soils also contain calcareous materials originating from mollusks. Conchiolin, the protein basis of mollusk shells, is degraded by sulfur bacteria in the anoxic muds, and the shells break down and mix with the soil. Bacteria, blue-green algae, diatoms, green algae, and other organisms participate in making these soils usable by higher plants, among which nitrifying bacteria and sulfate-reducing bacteria are particularly important (Schuster, 1952).

Besides the nutritional quality of mangrove soils, water conditions are also important, and these are influenced by waterways in and around the mangroves. *Xylocarpus, Lumnitzera,* and *Avicennia marina* among others favor well-drained soils, while *Nypa* favors waterlogged soils.

The response of the various tree species to salinity, tidal inundation, and soil conditions by growing within their respective ranges of tolerance is manifested in the phenomenon of zonation. The requirements for growth of various species have been summarized simply by Kira (1967), and these are presented in slightly more detail in table 1-6.

The growth requirements of various species relate not only to substrate but

## Table 1-6

### Inundation Classes and Mangrove Species of the Malay Peninsula

| Name after Watson | Name after Chapman | Inundation class | Substrate and position |
|---|---|---|---|
| *Acanthus ebracteatus* | *Acanthus ebracteatus* Vahl. | 4, 5 | Loams or clays of riverbanks or cleared land |
| *A. ilicifolius* | *A. ilicifolius* L. | 4, 5 | Loams or clays of riverbanks or cleared land |
| *Acrostichum aureum* | *Acrostichum aureum* L. | 3, 4, 5 | Bright places |
| *Aegiceras majus* | *Aegiceras corniculatum* (L.) Blanco | 3, 4 | Loam of riverbanks not too distant from sea |
| *Avicennia alba* | *Avicennia alba* Blume | 2 | Deep mud within the influence of a river |
| *A. intermedia* | *A. marina* (Forsk.) Vierh. | 2, 3 | Stiff mud facing the sea |
| *A. lanata* | *A. lanata* Ridl. | 2, 3 | Sandy mud not too distant from sea. Singapore and the east coast |
| *A. officinalis* | *A. officinalis* L. | 3, 4 | Stiff mud on riverbanks |
| *Brownlowia lanceolata* | ——— | 4 | Riverbanks and rough open land |
| *B. riedelii* | ——— | 5 | Sandy soil on riverbanks |
| *Bruguiera caryophylloides* | *Bruguiera cylindrica* (L.) Bl. | 4 | Recent clay deposits facing sea |
| *B. eriopetala* | *B. sexangula* (Lour.) Poir. | 3, 4 | Loam. Often associated with *Br. cy.* and *Br. pa.* Usually sporadic |
| *B. gymnorrhiza* | *B. gymnorrhiza* (L.) Lamk. | 3, 4, 5 | Loam or sandy mud. Abundant in drier places |
| *B. parviflora* | *B. parviflora* (Roxb.) W. & A. ex Griff. | 3, 4 | Well-drained land |
| *Carapa moluccensis* | *Xylocarpus moluccensis* (Lamx.) Roem | 4, 5 | Loam near tidal limit |
| *C. obovata* | *X. granatum* Koenig | 3, 4, 5 | Sandy mud and riverbanks |
| *Cerbera lactalia* | *Cerbera manghas* | 4, 5 | Riverbanks near tidal limit |
| *C. odollam* | *C. odollam* Gaertn. | 5 | Sandy seashore |
| *Ceriops candolleana* | *Ceriops tagal* (Perr.) C. B. Rob. | 3, 4 | Loam near estuaries |
| *Cycas rumphii* | ——— | 5 | Coast and inland (east coast only) |
| *Daemonorops leptopus* | ——— | 5 | Riverbanks above mangrove |

(continued on next page)

Table 1-6  *(continued)*

| NAME AFTER WATSON | NAME AFTER CHAPMAN | INUNDATION CLASS | SUBSTRATE AND POSITION |
|---|---|---|---|
| *Derris uliginosa* | *Derris heterophylla* (Willd.) Back. | 4, 5 | Loam on riverbanks |
| *Excoecaria agallocha* | *Excoecaria agallocha* L. | 4, 5 | Clay. Often behind *Br. cy.* |
| *Heritiera littoralis* | *Heritiera littoralis* Drynand ex H. Ait. | 5 | Sandy loam. Riverbanks and inland fringe of mangrove |
| *Hibiscus tiliaceus* | ——— | 5 | Loam. Riverbanks inland |
| *Intsia retusa* | *Intsia bijuga* (Colebr.) O. Ktze | 5 | Loam above the tidal limit |
| *Kandelia rheedii* | *Kandelia candel* (L.) Druce | 4 | Riverbanks. Not seen in Perak. Sporadic |
| *Lumnitzera coccinea* | *Lumnitzera littorea* (Jack) Voigt | 4, 5 | Loam. Inland fringe of mangrove. Sometimes near sea |
| *L. racemosa* | *L. racemosa* var. *racemosa* van St. | 4, 5 | Clay. Sometimes behind *Br. cy.* |
| *Nipa fruticans* | *Nypa fruticans* Wurmb. | 3, 4, 5 | Riverbanks influenced by fresh water |
| *Oncosperma filamentosa* | *Oncosperma filamentosum* Blume | 5 | Clay or loam above the tidal limit. Also inland |
| *Pluchea indica* | ——— | 5 | Cleared land behind mangrove |
| *Podocarpus polystachyus* | ——— | 5 | Sandy mud or high riverbanks behind mangrove. East coast only |
| *Rhizophora conjugata* | *Rhizophora apiculata* Blume | 3, 4 | Any loam mud not facing the sea |
| *R. mucronata* | *R. mucronata* Lamk. | 2, 3 | Deep mud under the influence of a river, not facing the sea |
| *Scyphiphora hydrophyllacea* | *Scyphiphora hydrophyllacea* Gaertn | 3, 4 | Loam on riverbanks, sandy mud, or rough cleared land |
| *Sonneratia alba* | *Sonneratia ovata* Backer | 3, 4 | Loam. Often somewhat distant from riverbanks |
| *S. acida* | *S. caseolaris* (L.) Engler | 4, 5 | Loam. Within the range influenced by river water |
| *S. griffithii* | *S. alba* J. Sm. | 2, 3 | Fertile mud facing the sea, or riverbanks. Often associated with *R. mucronata* |
| *Thespesia populnea* | *Thespesia populnea* (L.) Soland ex Correa | 5 | Sandy shores |

SOURCE: Prepared from Watson, 1928, and Chapman, 1975.

also light conditions. Macnae (1968) has listed *Acanthus ilicifolius, Avicennia alba, Sonneratia alba, S. apetala,* and *S. griffithii* as requiring full sunlight, and *Aegialitis annulata, Aegiceras corniculatum, Ceriops decandra, Lumnitzera racemosa, Rhizophora stylosa, R. mucronata,* and *R. apiculata* as being basically light demanding but slightly shade tolerant. All others are shade tolerant, and *Bruguiera gymnorrhiza* is typical. The light-demanding species are the first to colonize open land created by the interaction of tidal action and riverine sedimentation. With their establishment and growth, the initial conditions are altered, and successive zones of species gradually develop. In the final stage of mangrove development, *Bruguiera gymnorrhiza* makes its appearance, after which mangrove gives way to inland forest.

Soils suited to mangroves thus are built up of sediments brought by tides and rivers, calcareous materials, litter, animal feces, and so on, and these are first colonized by the fully light-demanding species. As a community develops, its environmental conditions change, and species suited to the changed conditions invade. Repetition of this process leads to the formation of the basic zonal structure. Such zonation requires the distribution of homogeneous conditions over a certain range, and this requirement is most readily met on gently sloping coastal lowland. These conditions, however, are readily susceptible to change by such agents as waterways and creeks, wind, waves, and sea currents, and by felling of trees. This is borne out by the fact that comparatively well-defined zonation is found on seacoasts where no major river debouches. Conversely, around the estuary of a major river, a complexity of factors interact, and species become distributed in much narrower zones (fig. 1-4).

In a study of the propagules of American mangroves, Rabinobitz (1978) concluded that (1) the propagules of mangrove genera occupying lower land on sea faces are large and heavy, while those of genera inland are small; (2) the mortality of seedlings is inversely proportional to the weight of the propagules; and (3) seedlings do not grow well under an adult tree. While, as Snedaker (1982) points out, it is problematic to link this research into population dynamics directly to the factors of zonation, it may well be possible, given a more detailed follow-up, to explain zonal distribution to some extent in terms of the morphology, buoyancy, viability, mode of establishment, soil preference, and other features of propagules.

In a study of the metabolism of mangroves in southern Florida, Lugo and Snedaker (1975) found that the dominant species in any particular zone showed the most efficient metabolism among the species of the same community. This is interesting evidence that the properties possessed by a dominant species are employed to the maximum effect under the optimal conditions. Such research has yet to be conducted on the mangroves of Southeast Asia.

One final direction for research is the analysis of the relation between races of

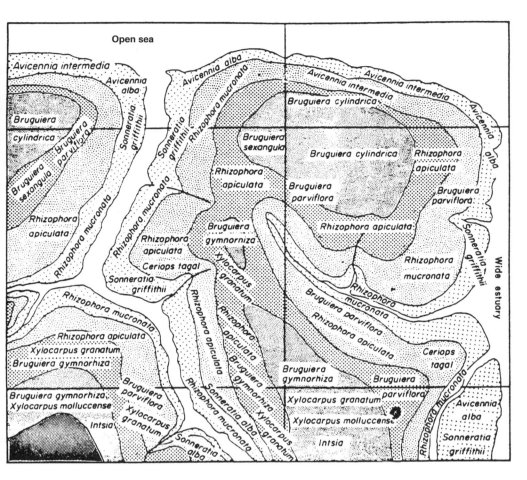

Figure 1-4. Zonal distribution of mangroves in Malaysia. SOURCE: Watson, 1928.

the species constituting mangrove. Since species are limited in number and widely distributed, and, moreover, trees do not attain the great size of lowland forest, the survey should be relatively simple. This may provide a clue to the relation between zonation and the wider area distribution.

## Biomass

Although many reports deal descriptively with mangrove, few give quantitative treatments.

First, let us look at basal area (table 1-7). In South Sumatra, the highest value is the 25 m²/ha of the fairly dense mixed forest of *Rhizophora apiculata* and *Avicennia alba*. All others show fairly low values. In the mixed forest of *Rhizophora apiculata* and *Bruguiera gymnorrhiza* lying farthest inland, it is only 13 m²/ha. This is

Table 1-7

Basal Area at Breast Height and Density of Various Mangroves

| REGION | MAIN TREE SPECIES | D ≥ 10 DENSITY (/ha) | BASAL AREA AT BREAST HEIGHT (m²/ha) | REFERENCE |
|---|---|---|---|---|
| South Sumatra | *Rhizophora apiculata* | 386 | 21.2 | Sukardjo and Kartawinata, 1978 |
|  | *Avicennia spp.* |  |  |  |
|  | As above | 426 | 12.9 | As above |
|  | As above | 452 | 15.5 | As above |
|  | As above | 366 | 14.5 | As above |
|  | *Rhizophora apiculata* | 596 | 25.0 | Yamada and Soekardjo, 1979 |
|  | *Avicennia alba* |  |  |  |
|  | *Rhizophora apiculata* | 190 | 13.0 | As above |
|  | *Bruguiera gymnorrhiza* |  |  |  |
| East Indonesia | *Rhizophora belt* | 761 (D ≥ 8) | 22.8 | Komiyama et al., 1988 |
|  | *Bruguiera belt* | 300 (D ≥ 8) | 36.2 | As above |
| Panama | *Rhizophora brevistyla* | 712 | 13.6 | Golley et al., 1975 |
|  | *Laguncularia racemosa* |  |  |  |
| South Florida | *Rhizophora mangle* | 5900 (2.5 ≥ D) | 20.3 | Lugo and Snedaker, 1975 |
|  | *Avicennia germinans* |  |  |  |

the aggregate for a 1-ha survey plot but is probably representative of the average, since although there will naturally be places where the figure is considerably higher, there will inevitably be others with *Nypa,* creeks, and so on. Density is not particularly high either, ranging from 190 to 596 stems per ha. In eastern Indonesia, the *Bruguiera* belt shows slightly higher values. Mangroves in the New World show similar values, with 13.6 m²/ha recorded in Panama and 20.3 m²/ha in Florida. This similarity despite the smaller individual trees in the New World is thought to be the result of surveying small areas of dense forest.

Records of biomass are even fewer, although new data recently have begun to emerge. In Southeast Asia, there are examples only for Thailand and Indonesia. For the *Rhizophora apiculata* of South Thailand, the dry weight of above-ground matter, including branches, trunks, and roots, was estimated at 159–281 ton/ha. For planted *Rhizophora apiculata* on the Gulf of Siam, the biomass was measured at 20.8 ton/ha in the third year and 188 ton/ha in the fourteenth year.

Table 1-8

Biomass of Mangrove and Tropical Forests in Various Regions

| Region | Main species, forest type | Survey area, tree number | Trunks | Branches | Prop roots | Leaves | Flowers, fruits, seeds | Above-ground biomass | Reference |
|---|---|---|---|---|---|---|---|---|---|
| South Thailand | Rhizophora apiculata | 25 m² | 74.4 | 15.8 | 61.2 | 7.4 | 0.3 | 159.1 | Christiansen, 1978 |
|  | Rhizophora belt | 26 trees | 214.2 | 58.2 | — | 8.1 | — | 281.2 | Tamai et al., 1986 |
| Gulf of Siam | 3-year R. apiculata | 1m × 1m, 30 trees | 4.7 | 3.7 | 6.2 | 6.2 | — | 20.8 | Aksornkoae, 1975 |
|  | 6-year R. apiculata | As above | 17.3 | 10.9 | 10.8 | 11.0 | — | 50.0 | As above |
|  | 9-year R. apiculata | As above | 41.1 | 17.9 | 17.8 | 16.3 | — | 93.1 | As above |
|  | 11-year R. apiculata | As above | 59.4 | 19.8 | 18.7 | 18.4 | — | 116.3 | As above |
|  | 12-year R. apiculata | As above | 86.3 | 22.5 | 20.7 | 19.8 | — | 149.3 | As above |
|  | 13-year R. apiculata | As above | 103.8 | 23.1 | 20.3 | 20.2 | — | 167.4 | As above |
|  | 14-year R. apiculata | As above | 135.9 | 15.1 | 22.9 | 14.0 | — | 187.9 | As above |
| Eastern Indonesia | Sonneratia belt | Sample trees felled | 130.0 | 32.2 | 0.1 | 6.5 | 0.17 | 169.1 | Komiyama et al., 1988 |
|  | Rhizophora belt | As above | 215.7 | 36.8 | 40.7 | 5.6 | 0.32 | 299.1 | As above |
|  | Bruguiera belt | As above | 333.3 | 62.8 | 29.3 | 10.6 | 0.39 | 436.4 | As above |
| Panama | R. brevistyla etc. | 625 m², 2 sites | 159.3 | | 116.9 | 3.6 | 0.02 | 279.3 | Golley et al., 1975 |
| Puerto Rico | R. mangle | 25 m², 2 sites | 28.0 | 12.7 | 14.4 | 7.8 | — | 62.9 | Golley et al., 1962 |

(continued on next page)

Table 1-8  (*continued*)

| Region | Main species, forest type | Survey area, tree number | Trunks | Branches | Prop roots | Leaves | Flowers, fruits, seeds | Above-ground biomass | Reference |
|---|---|---|---|---|---|---|---|---|---|
| Florida | *Laguncularia racemosa* | 25 m² | 4.0 | | 3.2 | 0.7 | — | 7.9 | Lugo and Snedaker, 1975 |
| | *Rhizophora mangle* | As above | | 70.4 | 52.0 | 7.3 | 0.02 | 129.7 | As above |
| | As above | As above | | 70.5 | 41.9 | 6.9 | 0.2 | 119.6 | As above |
| | *R. mangle, L. racemosa* | As above | 41.0 | 17.0 | 22.3 | 5.9 | 0.03 | 86.2 | As above |
| | As above | As above | 65.2 | 19.1 | 27.2 | 5.8 | 0.2 | 117.5 | As above |
| | As above | As above | 110.0 | 18.6 | 17.2 | 7.0 | 0.1 | 152.9 | As above |
| | *R. mangle, Avicennia germinans* | As above | 62.9 | 16.8 | 14.6 | 3.8 | 0.1 | 98.2 | As above |
| | As above | As above | 133.7 | 27.7 | 3.1 | 9.5 | 0.0004 | 174.0 | As above |
| | Tropical rain forest | 0.10ha | 528 | 126 | — | 8.8 | — | 663 | Ogawa, 1974 |
| | Tropical evergreen seasonal forest | 0.32ha | 230 | 93 | — | 7.8 | — | 331 | As above |
| | Monsoon forest | 0.16ha | 209 | 53 | — | 3.8 | — | 266 | As above |
| | Savanna forest | 0.16ha | 55 | 11 | — | 2.7 | — | 69 | As above |

Unit: ton-t/ha dry weight

In eastern Indonesia, a figure of 436 ton/ha has appeared for the *Bruguiera* belt (table 1-8).

Compared with other forest types, the values for mangroves rival those of tropical evergreen seasonal forest, if not tropical rain forest. In many cases they are higher than the values reported in Japan (Shidei and Kira, 1977) for thirty-nine-year-old larch (*Larix leptolepsis,* 207 ton/ha) and natural beech forest (*Fagus crenata,* 354 ton/ha) at Ashu. Also of interest is the extremely high root biomass of mangroves, of which the highest recorded value in South Thailand is 509.5 ton/ha (Komiyama et al., 1987). Fine roots of less than 2 mm in diameter account for up to 66.4 percent of this biomass. These data may involve considerable scatter arising from sampling methods and other causes, but they serve to illustrate the peculiarities of mangroves.

## *Mangrove Plant Morphology*
### Root Systems

The most remarkable feature of mangrove vegetation is its modified roots. The prop roots, buttress roots, aerial roots, and pneumatophores that protrude from belowground or descend from aboveground create a distinctive atmosphere. Also important is the negative geotropism of mangrove roots. It would be no exaggeration to say that this morphological characteristic has captivated many researchers. Research on roots has been summarized by Troll and Dragendorff (1931), Scholander et al. (1955), van Steenis (1958), Macnae (1968), and others.

Most mangrove species lack deep root systems or taproots, having shallow, spreading root systems. Rhizophoraceae have seedlings with a long radicle that appears suited to develop into a taproot, but ceases development once the seedling is established. In *Rhizophora,* the tree is anchored by bunches of roots that develop from the ends of the prop roots about 30 cm underground. The prop roots originate from the hypocotyl and later the stem, descending obliquely and branching repeatedly before striking the ground and rooting. It has been stated that normal prop roots do not branch, but that when root tips have been killed by heat, drought, insect damage, or other causes, secondary adventitious roots are subsequently produced. Repetition of this produces the sympodially branched prop roots (fig. 1-5).

In *Bruguiera,* the horizontal cable roots send up knee- or elbow-shaped projections above the ground, which bear lenticels and function as pneumatophores. *Ceriops* has similar roots. *Kandelia* has a thickened stem base with buttress roots and pneumatophores.

*Sonneratia, Avicennia,* and *Xylocarpus* have no taproots at any stage, but develop extensive systems of cable roots that lie 20–50 cm below the surface. In the former two genera, the cable roots give off anchor roots downward and aerial roots or pneumatophores upward. The pneumatophores in turn produce a net-

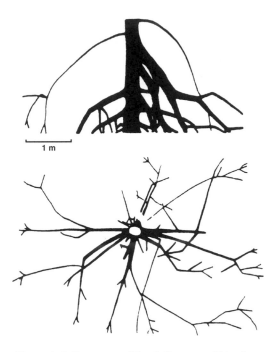

Figure 1-5. Prop roots of South Sumatran *Rhizophora*
*apiculata* (tree height 25.5 m; DBH 29 cm).
SOURCE: Yamada and Soekardjo, 1979.

work of fine nutritional roots in the uppermost soil layer. As silt accrues on the
surface, new nutritional roots are produced above the old ones, ensuring a continued supply of nutrients. The rich subsurface also contains oxygen, caused by the
activities of burrowing animals. Bivalve mollusks living there extend their siphons
to the surface and circulate seawater through their mantle cavities. This activity
oxygenates the surrounding mud, turning it paler in color.

Pneumatophores in *Avicennia* and *Lumnitzera racemosa* are slender, pencil-like
structures, while in *Sonneratia* and *Xylocarpus moluccensis* they are stout and knob-
bly. In anoxic soils, *Xylocarpus granatum* produces an upward-projecting flange
along its cable roots that lie just below the surface.

Scholander et al. (1955) showed that in *Avicennia nitida* and *Rhizophora mangle*
from Florida, the air capillary system is in communication with the lenticels of
pneumatophores or prop roots. With *Avicennia,* when the tide covered the lenticels on the air roots, the pressure in the root system continued to drop until the
tide began to recede, when air was rapidly absorbed. At low tide, oxygen concentration in the roots was 10–18 percent. This level fell when the lenticels were
submerged but returned when the tide ebbed and the lenticels were again exposed

to the air. When the lenticels were covered with grease, the oxygen concentration in the roots fell to 1 percent or less in one or two days, demonstrating that the pneumatophores function as ventilators of the root system in the anaerobic mud. In the case of *Rhizophora,* the lenticels of prop roots play the same role, and the oxygen concentration in the underground roots showed similar fluctuations to that in the cable roots of *Avicennia.*

### Vivipary

Like the modified root formations, vivipary is a feature of mangroves. It is most remarkable in the Rhizophoraceae but is seen in other families. Seeds have virtually no resting stage but germinate while in the fruit or on the branch, producing long hypocotyls and radicles. The endosperm plays only a brief role and is soon obsolete. Vivipary in various species has been summarized by van Steenis (1958), as follows.

In *Aegiceras corniculatum,* the embryo ruptures the testa and fills the pericarp, which then enlarges in proportion to the growth of the embryo. The embryo pierces the pericarp after the enlarged fruit has fallen. *Avicennia* develops similarly, but the embryo may pierce the pericarp while the fruit is still on the tree.

In *Bruguiera,* the apex of the hypocotyl pierces the fruit apex and grows cigarlike out of it. It drops together with the fruit while still attached. In *Bruguiera parviflora,* the fruit base is pierced by the growing plumule, and the pericarp persists as a cuff around the seedling. In *Bruguiera,* the cotyledons are connate only at the base, but they absorb nutrients from the pericarp through their glandular epidermal tissue and provide nutrients to the hypocotyl.

Viviparous seedlings of *Rhizophora* develop until the hypocotyl reaches a considerable length before ripening and dropping from the fruit. The plumule is entrenched in a clavate cotyledonary body, formed by the fleshy connate cotyledons, which absorbs nutrients from the pericarp for the growth of the hypocotyl and often protrudes from the fruit, being demarcated from the hypocotyl by an articulation. At maturity, this articulation ruptures, and the hypocotyl, crowned by the plumule, falls from the fruit. In *Ceriops,* the hypocotyl also separates and drops, but in *Kandelia,* as in *Bruguiera,* it falls together with the fruit (fig. 1-6).

It has been claimed that vivipary is an environmental adaptation of mangroves, but since this is not seen in such mangroves as *Sonneratia,* nor in species of the *Barringtonia* community growing in similar environments, nor in the trees of freshwater swamp forest and peat swamp forest, van Steenis (1958) rejected this claim. The advantage of vivipary is the shortening of the time between the fall and the attachment of the seedling. It should probably be regarded as a species character rather than an adaptation.

It is popularly held that when the hypocotyl drops, it sticks into the mud below and continues to grow; but in fact there are probably few people who have

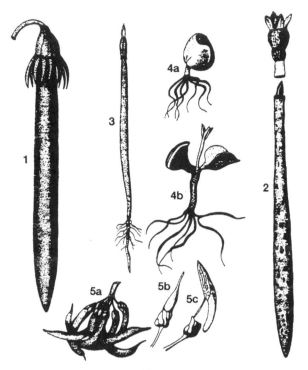

1   *Bruguiera gymnorrhiza*

2   *Rhizophora mucronata*

3   *Bruguiera parviflora*

4   *Avicennia marina*
     4a. Newly germinated. 4b. Plumule elongating

5   *Aegiceras corniculatum*
     5a. Bunch of fruits. 5b. Young fruit.

     5c. Germinating fruit.

Figure 1-6. Viviparous mangrove seedlings. SOURCE: MacNae, 1968.

observed a seedling at the moment it drops. For the seedling to penetrate, the mud would have to be suitably soft. Stiff mud or water underneath the tree would prevent this; and even if a seedling did stick into soft mud, it would be quite likely to topple later. In fact, the hypocotyl, which makes up the greater part of the fallen seedling, is dispersed by floating in the water, usually horizontally but occasionally vertically. Rootlets emerge from one end, and at low tide these attach the seedling horizontally to the mud. Once the anchorage is firm, the hypocotyl gradually turns itself upright (Egler, 1948; Lawrence, 1949). The prime time for establishment is during the spring ebb.

Mangrove seedlings and fruits are said to remain viable even after several months afloat (Guppy, 1906). In Malaysia, however, seedlings of *Rhizophora, Bruguiera,* and *Ceriops* species were observed to derive mainly from parent trees nearby. *Sonneratia* and *Avicennia* may regenerate through seeds that have drifted a considerable distance. The significance of drift capacity remains a task for the future.

Chapman (1940) has pointed out that successful rooting by seedlings requires extremely shallow sea with only a small rise in the tide or, in the case of a larger tidal range, that seedlings be exposed for several hours during the day, as they will soon die if the small shoot portion does not project above the water. Consequently, the length of the hypocotyl must play an important role, which is why species like *Rhizophora mucronata,* in which the hypocotyl reaches 60–90 cm, can grow in even the deepest water.

## *Utilization of Mangrove*

In the tropics, primary products are used in all aspects of daily life. Mangrove is no exception, and its uses are extensive. These have already been described in detail by Heyne (1950) and Burkill (1966) among others, and Watson, too, has summarized their information, so here I shall present the major uses in tabular form. Table 1-9 shows Watson's summary of Heyne's work, and I have simply updated the species names and divided the listings according to usage.

The basic uses can be divided into timber and others, of which the former can be subdivided into timber proper, firewood, and charcoal. Even today, you will encounter one or two charcoal kilns on entering the mangrove, and most tropical charcoal can be regarded as deriving from mangrove. Mangroves are important not only for charcoal but also as firewood, and table 1-10 shows an evaluation of mangrove species of Malaysia as firewood based on their specific gravity. Uses as timber also are manifold, ranging from the hard but durable woods of *Ceriops tagal, Lumnitzera littorea, Scyphiphora hydrophyllacea, Xylocarpus moluccensis,* and others that are used for fine domestic furniture, to the palm *Oncosperma filamentosum,* which has a wide range of uses, and those like *Rhizophora,* which are used for pilings.

The most characteristic usage other than as timber is as a source of the tannin contained in the bark. Table 1-11 shows the percentage of tannin in the bark of different mangrove species from different localities. These tannins are used in India in place of *Acacia* tannin; and in the Malaysia area they are widely used for tanning leather, for dying fishing nets and cloth, and other purposes. Abdul Razak et al. (1981) surveyed fifty tannin-containing species of plant in the Malay Peninsula, and among these the mangroves *Rhizophora mucronata, R. apiculata,* and *Bruguiera* occupied top positions. Species with higher contents than mangroves were limited to *Codiaeum variegatum* (Euphorbiaceae), *Calophyllum inophyl-*

Table 1-9
Uses of Mangrove

| | MAIN USES | | | | |
|---|---|---|---|---|---|
| SPECIES | TIMBER | FUEL | FOOD AND DRINK | MEDICINE | OTHERS |
| Acantbus ilicifolius | | | | Fruit pounded and used as blood purifier and dressing for boils; preparation of leaves for rheumatism; poultice of fruit or roots used for snakebite or arrow poisoning | |
| Acrostichum aureum | | | | | Litter for cattle; roofing |
| Avicennia alba A. lanata A. marina A officinalis | Inferior timber; better heartwood; resembles oak; rice mortars, cabinets, water pipes | Inferior firewood; smoking fish and rubber | | Ointment from seeds applied to tumors and smallpox ulcers; sap used for birth control | Ashes for washing cloth; bark as astringent and for tanning |
| Bruguiera cylindrica | Timber, pillars, rafters | Firewood | Young radicles boiled and eaten as vegetable | | |
| B. gymnorrhiza B. sexangula | Timber, durable, hard, heavy; cracks with seasoning | Firewood, charcoal | Radicles eaten as vegetable, fruits chewed | Medicine for sore eyes from fruit; scent from pneumatophores; condiment from bark | |
| B. parviflora | Timber, timber for mining | Firewood | | | |
| Cerbera mangbas C. odollam | | | Edible oil from seeds | Rubbing with fruit said to ease rheumatism; oil from seeds for scabies; bark and sap used as purgative and for manta stings | Oil from seeds as illuminant |

(continued on next page)

Table 1-9   (continued)

| SPECIES | MAIN USES | | | | OTHERS |
| --- | --- | --- | --- | --- | --- |
| | TIMBER | FUEL | FOOD AND DRINK | MEDICINE | |
| Ceriops tagal | Timber for mining, boat ribs, pillars, durable | Firewood | Calyx chewed | Dye and tan from bark; decoction of bark as hemostatic agent in obstetrical cases | |
| Cycas rumphii | | | Starch from seeds and pith | Ointment for ulcers from seeds | |
| Daemonorops leptopus | | | | | Tying |
| Derris heterophylla | | | | Weak fish-poison | |
| Excoecaria agallocha | | | | Skin diseases; purgative from sap and wood; fish-poison from sap | Incense wood |
| Heritiera littoralis | Good timber, pestles | Firewood | | Ground seeds for diarrhea | |
| Hibiscus tiliaceus | Cart wheels | | Young leaves eaten as vegetable | Leaves ground in water used as hair-restorer, expectorant, and to relieve retention of urine | Rope from blast fiber |
| Intsia bijuga | Good timber, but rare | | | | |
| Kandelia candel | | | | Bark used for making mordant and red dye; medicinally for diabetes | |
| Lumnitzera littorea L. racemosa | Timber, fence posts, most useful timber, bridge piers, durability comparable to rosewood | | | | |
| Nypa fruticans | | | Fruit for preserve; sap for sugar, alcohol | | Leaves for thatch; young leaves for cigarette wrappers |

(continued on next page)

Table 1-9  (continued)

| Species | Timber | Main Uses — Fuel | Food and Drink | Medicine | Others |
|---|---|---|---|---|---|
| Oncosperma filamentosum | Piles; houseposts, flooring, fishtraps, shuttles | | Cabbage-like young shoots for salads and pickles | | |
| Pluchea indica | | | Eaten as vegetable; leaves as tea | Mouthwash and various medicines; astringent ointment | |
| Rhizophora apiculata | Timber, piles | Firewood, charcoal | | Tanbark | Anchors for small boats |
| R. mucronata | Timber, piles | Firewood, charcoal | Light wine from juice of fruit | Tanbark; decoction of bark for hematuria | Anchors for small boats |
| Scyphiphora hydrophyllacea | Hard timber, tool handles, fence posts | | | | |
| Sonneratia alba | Hard but contains salt; piles, boat timber, bridges | Firewood | Fruit not edible | | Pneumatophores as fishing floats and substitute for cork |
| S. caseolaris | Not suitable as timber | | Fruit edible | Fruit used in poultices for sprains; fermented juice for hemorrhage | |
| S. ovata | Not suitable as timber | Firewood | Fruit edible | | |
| Thespesia populnea | Good quality timber | | | Heartwood for pleurisy and cholera; ointment from fruits for lice; decoction of leaves for headache and scabies | Rope from blast fiber |
| Xylocarpus granatum | Cabinets; pins for boatbuilding; use in quarrying | Firewood | | Decoction from bark for cholera and dysentery | |
| X. moluccensis | | | | | Oil from seeds as illuminant and hair oil |

source: Based on Watson, 1928, with species names modified after Chapman, 1976.

Table 1-10

Evaluation of Mangrove Species in Malaysia as Fuels,
Based on Specific Gravity

| CLASS | EVALUATION | SPECIFIC GRAVITY | SPECIES |
|-------|-----------|------------------|---------|
| I | Outstanding | > 90 | *Rhizophora spp.* |
| | | | *Bruguiera spp.* |
| | | | *Ceriops spp.* |
| | | | *Heritiera littoralis* |
| | | | *Cynometra ramiflora* |
| II | Excellent | 0.75–0.90 | *Lumnitzera spp.* |
| | | | *Sonneratia alba* |
| | | | *Pithecolobium umbellatum* |
| III | Good | 0.60–0.75 | *Xylocarpus spp.* |
| | | | *Aegiceras corniculatum* |
| IV | Moderate | 0.45–0.60 | *Dolichandrone longissima* |
| V | Poor | 0.30–0.45 | *Avicennia spp.* |
| | | | *Sonneratia caseolaris* |
| | | | *Excoecaria agallocha* |
| | | | *Cerbera manghas* |

SOURCE: Walsh, 1977; summarized from Cox, 1911, and Becking et al., 1922.

*lum* (Guttiferae), and the Leguminosae *Acacia auriculiformis* (exotic), *Cassia fistula,* and *C. suratensis* (exotic).

In Watson's day, the tan bark exported from Singapore originated from Sumatra rather than the Malay Peninsula, and in Penang it was entering from Thailand. The tanneries of Singapore employed mainly *Rhizophora mucronata* bark. This was kept moist by watering until use, when it was broken up by pounding and thrown into vats of water in which the hides were tanned. Fresh bark was estimated to tan about 125 percent of its own weight of leather. Thick bark from large trees was preferred; the best bark was not brittle and when fractured revealed a clean yellow color. The proportion of bark differs considerably from species to species and with the diameter of the tree, as shown in table 1-12.

Locally, the tan of *Ceriops tagal* was highly valued for dying and preservation of fishing nets, ropes, sails, and clothes, the life of which was said thereby to be prolonged at least three times. Where this species was unavailable, *Xylocarpus granatum* was sought. The quality and content of its tannin are superior, but its bark is thin.

## Table 1-11
### Percentage of Tannin in Mangrove Bark from Different Localities

| SPECIES | PHILIPPINES | | | INDONESIA | | | INDIA | MALAYSIA | AUSTRALIA |
|---|---|---|---|---|---|---|---|---|---|
| | MIN. | MAX. | AVE. | MIN. | MAX. | AVE. | AVERAGE | AVERAGE | AVERAGE |
| *Rhizophora mucronata* | 18 | 28 | 25 | — | — | 32 | 24–45 | 30–40 | 27–39 |
| *R. apiculata* | 18 | 39 | 27 | 24 | 36 | 30 | 25–36 | — | — |
| *Bruguiera gymnorrhiza* | 24 | 32 | 31 | 21 | 37 | 29 | 28–42 | — | 29–36 |
| *B. sexangula* | 27 | 32 | 32 | — | — | — | — | — | — |
| *B. parviflora* | 9 | 15 | 10 | — | — | — | — | — | — |
| *B. cylindrica* | — | — | — | 6 | 9 | 8 | — | — | — |
| *Ceriops tagal* | 17 | 31 | 27 | 25 | 30 | 27 | 29–41 | 24 | 21–34 |
| *Xylocarpus granatum* | 22 | 25 | 23 | 20 | 21 | 21 | — | — | — |
| *X. moluccensis* | 23 | 23 | 23 | 24 | 27 | 26 | — | — | — |
| *Sonneratia alba* | 12 | 12 | 12 | — | — | 9 | — | — | — |

Unit: Percent dry weight.
SOURCE: Chapman, 1976; summarized from Becking et al., 1922, and Howes, 1962.

Table 1-12
Bark Percentage by Volume of Mangrove

| SPECIES | GIRTH (INCHES) | | | | |
|---|---|---|---|---|---|
| | 19–21 | 22–24 | 25–27 | 28–30 | 31–33 |
| *Rhizophora apiculata* | — | 24.36 | 22.10 | 19.95 | 16.88 |
| *R. mucronata* | — | 23.42 | 29.50 | 26.56 | 24.03 |
| *Bruguiera gymnorrhiza* | 18.62 | 16.85 | 15.44 | 14.28 | 10.78 |

SOURCE: Watson, 1928.

Other uses of mangroves are in food and beverages and medicines, of which there are many examples. Such multiplicity of usage is not limited to mangroves but is common to all forest types in the tropics. These have been dealt with in such major works as those of Heyne and Burkill cited earlier, and minor uses are almost unlimited. In a recent work, Walsh (1977) presents new information on the uses of mangroves.

### Directions of Mangrove Research

Research on the mangroves of Southeast Asia patently lags behind that in the New World. Here I shall introduce part of the research in America, and then consider future research directions.

America's mangroves are small in comparison with Southeast Asia, but in terms of research, fairly detailed work has been done. In particular, research into metabolism flourished in the 1970s and produced many results, among which those of Golley et al. (1962, 1975), Lugo and Snedaker (1974, 1975), and Hicks and Burns (1975) are major. The last cited describes a survey of metabolism in the mangroves of southern Florida and shows the importance of supplementary nutrients carried in by fresh water flowing from surrounding swamp forest. Some of the potential ecological factors influencing mangrove ecosystem dynamics are summarized as follows:

I. TIDAL FACTORS
   A. Transport of oxygen to root systems
   B. Removal of toxic sulfides and reduction of total salt content of soil water
   C. Involvement in sediment deposition or erosion
   D. Transportation of nutrients to the root system through vertical movement of the water table

II. WATER CHEMISTRY FACTORS
  A. Total salt content governs the osmotic pressure gradient between the
     soil water and plant vascular system, thus affecting the transpiration
     rate of the leaves
  B. A high macronutrient content of the soil solution allows the main-
     tenance of high productivity in mangrove ecosystems despite the
     low transpiration rates caused by high salt concentrations in
     seawater
  C. Allochthonous macronutrients contained in wet-season surface runoff
     may dominate the macronutrient budgets of mangrove ecosystems

Together with the high metabolic efficiency of dominant species that was
mentioned in the section on zonation, these observations point the way to future
quantitative studies. Future developments relating to the uses of mangroves will
also have to be based on such surveys (Lugo and Cintron, 1975). Lugo and
Snedaker (1974) have produced extremely detailed material flow diagrams for
mangroves, and Lugo (1980) has reported on the directions and causes of
succession.

Pool et al. (1977) compared the structure of mangrove forests in Florida,
Puerto Rico, Mexico, and Costa Rica. Cintron et al. (1978) surveyed the mangroves of arid environments in Puerto Rico and surrounding islands and argued
that when soil salinity rises as a result of prolonged drought, the mortality of
mangroves rises and the area of salt flats increases, and that hurricanes reverse the
direction of succession.

Specific properties of mangroves were examined by Connor (1969), who found
that for artificially cultivated *Avicennia marina* the optimal sodium chloride concentration for growth was half that of seawater. Walsh et al. (1979) investigated
the tolerance of *Rhizophora mangle* for lead, cadmium, and mercury and reported
that only 500 micrograms of mercury per gram of soil was fatal. Johnstone
(1981) investigated insect damage to mangrove leaves.

Besides the above quantitative works, studies have been made involving
detailed morphological observation of species and their growth and phenology
(Gill, 1977; Gill and Tomlinson, 1969, 1971a, 1971b). Tomlinson et al. (1979)
have studied the floral biology of the Rhizophoraceae and clarified the features of
various genera. These works constitute a new field of study, which is a cross
between morphology and species ecology and is distinct from ecosystem ecology.
Tomlinson (1986) has reviewed these results in his general work on the botany of
mangroves.

For Malaysia, on the other hand, Srivastava (1980) has proposed the following
themes for future research.

I. SILVICULTURE
  A. Study the growth and yield of *Rhizophora* species and *Bruguiera parviflora*
  B. Study the natural regeneration of *Rhizophora* species
    1. Evolve an efficient and reliable procedure for sampling regeneration
    2. Study the effect of logging on natural regeneration and conditions suitable for optimal stocking
    3. Study the influence of slash on the presence and progress of regeneration
    4. Study the site conditions where *B. parviflora* seedlings outnumber *Rhizophora* seedlings after final felling
    5. Study conditions favorable for *Acrostichum aureum* invasion after felling and its effect on natural regeneration
    6. Study the role of water, mother trees, and coppicing ability of *Rhizophora* seedlings in regenerating residual stands
    7. Determine the extent of damage to natural regeneration by crabs and monkeys
  C. Study the artificial regeneration technique
II. ECOPHYSIOLOGICAL STUDIES
  A. Study the primary productivity, mineral cycling, and energy relations
  B. Study mangrove soils
  C. Effect of management practices
  D. Study the waters in and adjacent to mangrove ecosystems in terms of geomorphology and ecophysiology
III. OTHER
  A. Evaluate mangrove forest as an agent for prevention of coastal erosion
  B. Determine local labor skills in relation to the mangrove ecosystem
  C. Study the socioeconomic structure of the human community as it relates to the mangrove ecosystem
  D. Study land use patterns in mangrove areas

These topics include several practical applications, a tendency apparent in most of the projects currently in progress in Southeast Asia. Since basic research is necessary to conduct applied research, the future trend will probably be toward the accumulation of basic data over a longer term. For this, long-term surveys must be conducted in experimental plots that will not be buffeted by the waves of development. In the New World, research establishments are closely linked to survey areas, and the development of small-scale equipment is also progressing. In Southeast Asia, access is difficult, and many of the world's biggest mangrove communities remain unsurveyed. Establishment of research systems is urgently required, before they are obliterated.

## 2. Freshwater Swamp Forest

Compared with mangrove and the peat swamp forest, which will be described next, little work has been done on freshwater swamp forest. Difficulty of approach and the prevalence of insect-borne disease are probably the main deterrents to survey. In Southeast Asia, the freshwater swamp forest of Malaysia has been relatively well surveyed. Corner, who spent many years working on plant classification at the Singapore Botanic Garden, is among the foremost tropical botanists, and in the 1930s he surveyed the swamp forests of south Johore and Singapore. His account appeared in 1978, almost half a century later. The work was disrupted by the Second World War and scrub-typhus, and it was also an achievement of great patience that awaited the specific identification of many of the specimens he had collected. The book contains many plant descriptions and is somewhat difficult to summarize, but it is a detailed account that only someone who has spent long years in the field could have written.

Here, based mainly on this account, I shall summarize the features of freshwater swamp forest, one of the major types of lowland swamp forest.

### Distribution

Freshwater swamp forest is widely distributed across the globe in the tropical regions of Southeast Asia, Africa, and Central America and is found in the largest scale in the Amazon basin (Richards, 1952). In Southeast Asia, it is found in the Indochina Peninsula, Thailand, and Burma along the Mekong, Chaophraya, and Irrawaddy Rivers, and in the archipelago, it covers large areas on the lower reaches of the Fly, Sepik, and other rivers in New Guinea. Generally, it occupies the lower reaches of rivers where coastal topography is gently sloping. It is found under seasonal as well as perhumid climates. Because of its prevalence on alluvial coastal lowlands, much has now been cleared for use as rice fields and palm and rubber plantations, and only small pockets remain.

### Environment

Freshwater swamp forest is forest that is temporarily or permanently inundated by mineral-rich fresh water. If peat is present, it is no more than a few centimeters thick, rather than the several meters developed in tropical peat swamp forest. While peat swamps are supplied with water only by rainfall, and consequently are mineral deficient, freshwater swamp forest is also inundated by mineral-rich fresh water from rivers (Wyatt-Smith, 1961).

In the Malay Peninsula, these forests occur under varying degrees of inundation and show enormous variation in floristic composition and structure, from a sparse scattering of large trees to dense pole forest and forest resembling peat swamp forest. Often a single species is predominant.

Webber (1954) compared diatom flora obtained from freshwater swamp forest that somewhat resembled peat swamp forest with diatom flora from mangroves and found that, in common with mangroves, the diatoms of freshwater swamp forest were virtually all of marine origin. This shows that the area presently covered by freshwater swamp forest was formerly occupied by mangrove.

The transition from mangrove to freshwater swamp forest can also be observed in South Sumatra. With distance inland from the sea, *Rhizophora, Bruguiera,* and other mangroves gradually decrease, while *Oncosperma, Campnosperma,* and other elements of freshwater swamp forest begin to appear, giving way eventually to authentic swamp forest with large *pulai (Alstonia)* trees. Most of these transitional forests have now been cleared to make way for settlements and rice fields, though fragments remain at the peripheries.

A hydrosere follows this transition from open water through hydrophytes to tall forest, but where this is interrupted by burning, flooding, or other factors, various communities appear, presenting a variety of landscapes.

## *Freshwater Swamp Forests of the Southern Malay Peninsula*

Corner (1978) surveyed the swamp forests of the Sedili rivers and Jason Bay in Johore in the southern Malay Peninsula and of Singapore between 1929 and 1941. Although the ecological descriptions are incomplete in some aspects, this is inevitable for that time. Parts of this retrospective account clearly describe the features of freshwater swamp forest. Many interesting observations are interspersed; for example, after a big flood, insects and animals that could do so climbed into the trees and would shower down if the tree was touched, and scrub-typhus was responsible for many deserted villages. Most of these forests, however, are no longer extant. The following account, which derives from Corner (1978) unless otherwise specified, summarizes the botanical facts that can be found only in his work.

### Riverside Vegetation

The best, if not only, approach to tropical wetlands is along a river. Upstream from the estuary, the riverside vegetation gradually changes, forming belts that are controlled by the interactions of salinity, silting, tides, illumination, freshwater floods, and the course of the river. Basically, because the lower reach is gradually advancing because of sedimentation from upstream, the upstream end of a belt contains increasing numbers of moribund plants unable to adapt to the changing growth conditions. In some places, these riverside belts overlap in a complex manner, but the succession from the estuary inland can be summarized as follows (see fig. 1-7):

- Mangrove (Rhizophoraceae, *Avicennia, Carapa, Lumnitzera*) in the sea-water estuary
- *Nipa* belt on the mud in brackish water and quieter parts of the estuary
- *Putat* belt *(Barringtonia conoidea)* pioneering submerged mudbanks in flowing water, mainly in the tidal freshwater region
- *Rassau* belt *(Pandanus helicopus)* with *rengas (Gluta velutina)* following closely behind the *putat* belt as the mudbank widens
- *Mempisang* belt *(Polyalthia sclerophylla)* with *medang jankang (Elaeocarpus macrocerus)* and *pianggu (Horsfieldia irya),* following behind the *rassau* belt as the first forest-formation on the stabilized mudbank, limited to the freshwater tidal region
- *Jejawi* belt *(Ficus microcarpa)* growing over *rassau* and *mempisang,* chiefly in the lower and variably brackish region
- *Tristania* banks (*T. sumatrana,* or *pelawan*) on the firm raised riverbanks, especially in the freshwater tidal zone, either fronting the river or behind the *mempisang* belt, but extending far above the tidal region
- *Saraca* streams (*S. bijuga*) along the banks of the tributaries under the forest canopy, at the periphery of the swamp forest, accompanied by *pelong (Pentaspadon officinalis), Schoutenia glomerata,* and various dipterocarps

Apart from *Saraca,* all the tree species are light demanding.

## Flora

The alluvial plain around the Sedili rivers is influenced to some extent by tides and floods. Topographically, there is a confusion of tributaries around the main streams. The tides are subject to diurnal and seasonal variations, and the extent of brackish water also varies. Seasonal rains swell the rivers and cause flooding. Under these conditions, the forest differs from the riverside vegetation, being the so-called freshwater swamp forest.

Corner recorded 1,712 species for the Sedili region, of which 1,082 occur in freshwater and saltwater swamp. Table 1-13 shows the numbers of species in each family in the area as a whole and in the freshwater swamp forest. The families represented by many species in the swamp forest include: Orchidaceae, Rubiaceae, Euphorbiaceae, Annonaceae, Palmae, Lauraceae, Moraceae, Myrtaceae, Meliaceae, Clusiaceae, Melastomataceae, Myristicaceae, Dipterocarpaceae, and Leguminosae. The larger genera include: *Eugenia, Ficus, Bulbophyllum, Calophyllum, Dendrobium, Ardisia, Shorea, Diospyros, Garcinia, Panda-*

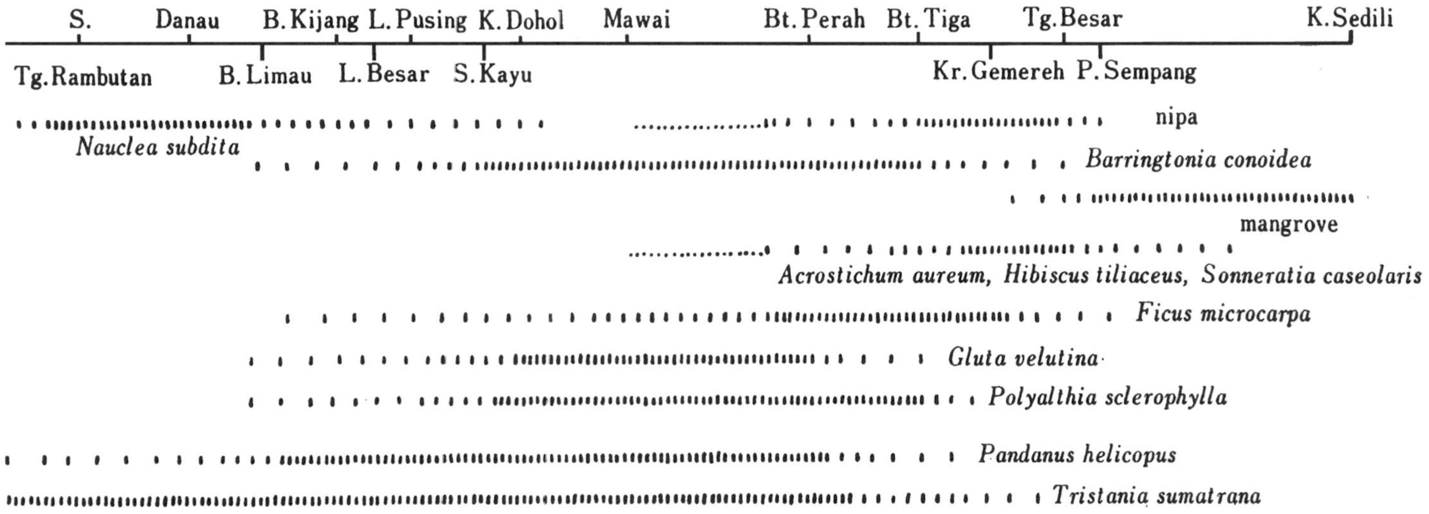

Figure 1-7. Vegetational change along the Sedili Besar River, South Johore. Tidal influence reaches Tg. Rambutan (35 miles from the bay).
SOURCE: Corner, 1978.

## Table 1-13
### Numbers of Species, by Family, in the Sedili Area, South Johore

| | TOTAL | | SWAMP FOREST | |
| | DICOT. | MONOCOT. | DICOT. | MONOCOT. |
|---|---|---|---|---|
| Acanthaceae | 12 | | 2 | |
| Aizoaceae | 1 | | 0 | |
| Amaranthaceae | 1 | | 0 | |
| Amaryllidaceae | | 2 | | 0 |
| Anacardiaceae | 27 | | 16 | |
| Annonaceae | 58 | | 49 | |
| Apocynaceae | 23 | | 15 | |
| Apostasiaceae | | 2 | | 2 |
| Aquifoliaceae | 3 | | 3 | |
| Araceae | | 23 | | 16 |
| Araliaceae | 5 | | 4 | |
| Asclepiadaceae | 21 | | 12 | |
| Bignoniaceae | 2 | | 2 | |
| Bombacaceae | 6 | | 4 | |
| Boraginaceae | 3 | | 1 | |
| Burseraceae | 12 | | 11 | |
| Capparidaceae | 1 | | 1 | |
| Casuarinaceae | 1 | | 0 | |
| Celastraceae | 12 | | 7 | |
| Clusiaceae | c. 40 | | 25 | |
| Combretaceae | 9 | | 3 | |
| Commelinaceae | | 5 | | 2 |
| Compositae | 6 | | 0 | |
| Connaraceae | 11 | | 8 | |
| Convolvulaceae | 10 | | 5 | |
| Cornaceae | 3 | | 3 | |
| Crassulaceae | 1 | | 0 | |
| Cucurbitaceae | 4 | | 3 | |
| Cyperaceae | | 31 | | 7 |
| Dichapetalaceae | 2 | | 1 | |
| Dilleniaceae | 12 | | 8 | |
| Dioscoreaceae | | 3 | | 3 |
| Dipterocarpaceae | 34 | | 22 | |
| Ebenaceae | ? 18 | | 11 | |
| Elaeocarpaceae | 11 | | 5 | |
| Ericaceae | 2 | | 1 | |
| Erythroxylaceae | 2 | | 1 | |
| Euphorbiaceae | 110 | | 71 | |
| Fagaceae | 20 | | 13 | |
| Flacourtiaceae | 15 | | 12 | |
| Flagellariaceae | | 2 | | 2 |

(continued on next page)

Table 1-13   (*continued*)

| | TOTAL | | SWAMP FOREST | |
|---|---|---|---|---|
| | DICOT. | MONOCOT. | DICOT. | MONOCOT. |
| Gesneriaceae | c. 15 | | 10 | |
| Gonystylaceae | 3 | | 3 | |
| Goodeniaceae | 1 | | 0 | |
| Gramineae | | 30 | | 3 |
| Hernandiaceae | 1 | | 0 | |
| Hypericaceae | 3 | | 2 | |
| Hypoxidaceae | | 2 | | 1 |
| Icacinaceae | 11 | | 7 | |
| Labiatae | 1 | | 0 | |
| Lauraceae | c. 60 | | 37 | |
| Lecythidaceae | 8 | | 5 | |
| Leguminosae | | | | |
|    Caesalpinioideae | 24 | | 12 | |
|    Mimosoideae | 10 | | 2 | |
|    Papilionatae | 20 | | 2 | |
| Lentibulariaceae | 1 | | 0 | |
| Liliaceae | | 14 | | 12 |
| Linaceae | 2 | | 2 | |
| Loganiaceae | 6 | | 4 | |
| Loranthaceae | 9 | | 7 | |
| Lowiaceae | | 1 | | 1 |
| Lythraceae | 4 | | 1 | |
| Magnoliaceae | 3 | | 1 | |
| Malpighiaceae | 2 | | 1 | |
| Malvaceae | 3 | | 0 | |
| Marantaceae | | 7 | | 2 |
| Melastomataceae | c. 35 | | 25 | |
| Meliaceae | c. 40 | | 26 | |
| Menispermaceae | 6 | | 4 | |
| Monimiaceae | 1 | | 1 | |
| Moraceae | 57 | | 32 | |
| Musaceae | | 2 | | ? |
| Myristicaceae | 30 | | 25 | |
| Myrsinaceae | c. 28 | | 16 | |
| Myrataceae | c. 65 | | 32 | |
| Nepenthaceae | 3 | | 3 | |
| Nymphaeaceae | 1 | | 1 | |
| Ochnaceae | 3 | | 3 | |
| Olacaceae | 3 | | 3 | |
| Oleaceae | 5 | | 5 | |
| Opiliaceae | 2 | | 1 | |
| Orchidaceae | | c. 115 | | 101 |

(*continued on next page*)

## Table 1-13  *(continued)*

| | TOTAL | | SWAMP FOREST | |
|---|---|---|---|---|
| | DICOT. | MONOCOT. | DICOT. | MONOCOT. |
| Oxalidaceae | 2 | | 2 | |
| Palmae | | c. 50 | | 37 |
| Pandanaceae | | 22 | | 18 |
| Passifloraceae | 1 | | 0 | |
| Piperaceae | c. 10 | | 7 | |
| Pittosporaceae | 1 | | | |
| Polygalaceae | 11 | | 8 | |
| Proteaceae | 5 | | 3 | |
| Rhamnaceae | 3 | | 2 | |
| Rhizophoraceae | 13 | | 4 | |
| Rosaceae | 6 | | 4 | |
| Rubiaceae | c. 110 | | 74 | |
| Rutaceae | 13 | | 7 | |
| Sabiaceae | 3 | | 0 | |
| Santalaceae | 1 | | 0 | |
| Sapindaceae | 18 | | 11 | |
| Sapotaceae | 23 | | 17 | |
| Saurauiaceae | 1 | | 0 | |
| Schisandraceae | 1 | | 1 | |
| Scrophulariaceae | 2 | | 0 | |
| Simaroubaceae | 2 | | 0 | |
| Solanaceae | 1 | | 1 | |
| Staphyleaceae | 1 | | 1 | |
| Stemonaceae | | 1 | | 1 |
| Sterculiaceae | 18 | | 11 | |
| Styracaceae | 1 | | 1 | |
| Symplocaceae | 4 | | 3 | |
| Taccaceae | | 4 | | 2 |
| Theaceae | 12 | | 9 | |
| Thymelaeaceae | 3 | | 3 | |
| Tiliaceae | c. 15 | | 9 | |
| Ulmaceae | 3 | | 1 | |
| Urticaceae | 7 | | 5 | |
| Verbenaceae | 18 | | 8 | |
| Violaceae | 4 | | 2 | |
| Vitaceae | 9 | | 6 | |
| Zingiberaceae | | 30 | | 20 |
| Totals | 1,285 | 346 | 794 | 230 |
| Total angiosperms | 1,631 | | 1,024 | |
| Gymnospermae | 9 | 6 | | |
| Pterydophyta | 72 | 52 | | |
| Total vascular flora | 1,712 | | 1,082 | |

SOURCE: Corner, 1978.

*nus, Lithocarpus, Macaranga, Lindera, Baccaurea, Memecylon, Elaeocarpus, Poly-althia, Calamus, Daemonorops, Horsfieldia, Psychotria, Xanthophyllum, Aglaia,* and *Dysoxylum.*

## Forest Physiognomy

Freshwater swamp forest is less mixed than lowland dipterocarp forest, also less tall, but it is not simple like mangrove. Its average height is about 35 m. Table 1-14 lists the main species found in the canopy. The largest among them include *Cratoxylum arborescens* (height 45 m, 20–28 m to the first branch, 70–80 cm in diameter), *Ganua motleyana* (height 50 m), and *Koompassia malaccensis* (height 50 m, 27 m to the first branch). In some areas, *Cratoxylum arborescens* and *Palaquium xanthochymum* show high degrees of dominance. Since lowland dipterocarp forest in Malaysia contains a considerable number of individuals in the 50–60 m class, freshwater swamp forest can be regarded as slightly inferior even in its tallest trees. Saplings of these large trees are abundant in the loose humus, many growing from the sides of fallen trunks. Many have lost the leading shoot, broken by falling limbs or palm leaves, or browsed by tapir.

Many of the trees of the swamp forest, as will be described later, display the remarkable morphological features of variant root formations (buttresses, stilt roots, and pneumatophores). Herbs are few, consisting mainly of monocotyledons, ferns, and the water lily *Barclaya.* Palms, including pandans, are abundant, as are lianes.

## Crown Types

Many trees in the swamp forest show characteristic crowns, and from the morphological changes accompanying the stages of growth, the following four types can be differentiated:

1. Persistently monopodial trees—Annonaceae, Myristicaceae, *Calophyllum,* and *Garcinia,* among which some species of *Calophyllum* and *Myristica* built large crowns

2. Those with a long phase of monopodial growth and finally a dense sympodial crown—*Shorea, Koompassia, Cratoxylum,* and *Parartocarpus*

3. Those with a shorter phase of monopodial growth, whose trunks divided at about 17 m high into two to four main ascending limbs that continued the sympodial growth—*Amoora rubiginosa, Ganua motleyana, Melanorrhoea wallichii, Mussaendopsis, Palaquium xanthochymum, Pentace triptera, Pometia, Terminalia phellocarpa,* and *Tristania merguiensis*

4. Small crowns on monopodial trunks that inserted themselves into gaps in the canopy—*Aromadendron nutans, Cyathocalyx ridleyi,* and *Tetractomia* (the smallest crown); all had few branches

# Table 1-14
## Canopy Trees of Freshwater Swamp Forest in Sedili, South Johore

**ANACARDIACEAE**
*Buchanania lucida*
*B. sessilifolia*
*Campnosperma squamata**
*Gluta malayana*
*Melanorrhoea wallichii*
*Pentaspadon officinalis*

**ANNONACEAE**
*Mezzettia leptopoda**
*Polyalthia hypoleuca**
*Xylopia fusca*

**APOCYNACEAE**
*Alstonia spathulata**

**BOMBACACEAE**
*Coelostegia griffithii*
*Kostermansia malayana*
*Neesia malayana**

**BURSERACEAE**
*Canarium littorale*
*Dacryodes macrocarpa**
*D. rostrata*
*Santiria laevigata**
*S. rubiginosa**
*S. tomentosa**

**CELASTRACEAE**
*Lophopetalum multinervium**

**CLUSIACEAE**
*Calophyllum incrassatum*
*C. inophylloide* var.
  *singapurense*
*C. macrocarpum*
*C. retusum**
*C. sclerophyllum**
*C. soulattri*
*C. wallichianum*

**COMBRETACEAE**
*Terminalia phellocarpa*

**DILLENIACEAE**
*Dillenia excelsa*
*D. grandifolia*
*D. pulchella**
*D. reticulata*

**DIPTEROCARPACEAE**
*Dryobalanops oblongifolia*
*Dipterocarpus lowii*
*Dipterosarpus sp.*
*Hopea mengarawan*
*Shorea bracteolata*
*S. exelliptica*
*S. lepidota*
*S. palembanica*

*S. platycarpa**
*S. singkawang*
*S. sumatrana*
*Vatica wallichii*

**ELAEOCARPACEAE**
*Elaeocarpus macrocerus*
*E. sphaericus*

**ERYTHROXYLACEAE**
*Ixonanthes reticulata*

**EUPHORBIACEAE**
*Blumeodendron tokbrai**
*Macaranga griffithiana*

**FAGACEAE**
*Lithocarpus bennettii*
*L. leptogyne*
*L. urcelolaris*

**HYPERICACEAE**
*Cratoxylum arborescens**
*C. formosum*

**ICACINACEAE**
*Platea latifolia**
*Stemonurus scorpioides**
*S. secundiflorus*

**LEGUMINOSAE**
*Dialium patens*
*D. playtysepalum*
*D. wallichii*
*Intsia palembanica*
*Koompassia malaccensis*
*Parkia speciosa*
*Sindora coriacea*
*S. wallichii**

**LYTHRACEAE**
*Lagerstroemia ovalifolia*

**MELIACEAE**
*Amoora rubiginosa**
*Dysoxylon marothyrsum*

**MORACEAE**
*Artocarpus elasticus*
*A. kemando*
*Ficus calophylla*
*F. crassiramea**
*F. consociata**
*F. delosyce*
*F. sumatrana**
*F. sundaica**
*Parartocarpus venenosus* ssp.
  *forbesii*

**MYRISTICACEAE**
*Gymnacranthera eugeniifolia*
  var. *griffithii**

*G. forbesii*
*Myristica crassa*
*M. iners*
*M. lowiana**

**MYRTACEAE**
*Eugenia cerina**
*E. leptostemon*
*E. nigricans*
*E. papillosa*
*E. pseudosubtilis*

**OLACACEAE**
*Ochanostachys amentacea*
*Scorodocarpus borneensis*
*Strombosia maingayi*

**ROSACEAE**
*Parinari costata*
*P. nannodes*
*P. oblongifolia*

**RUBIACEAE**
*Mussaendopsis beccariana**
*Nauclea maingayi*

**SAPINDACEAE**
*Nephelium glabrum*
*N. rubescens*
*Pometia pinnata* f. *alnifolia**
*Xerospermum muricatum**

**SAPOTACEAE**
*Ganua motleyana**
*Palaquium confertum*
*P. macrocarpum*
*P. obovatum*
*P. semaram*
*P. xanthochymum*
*Planchonella maingayi**

**STERCULIACEAE**
*Heritiera elata*
*H. simplicifolia*
*Pterospermum javanicum*
*Scaphium macropodum**
*Sterculia macrophylla**

**THEACEAE**
*Adinandra sarosanthera*
*Gordonia singapureana*

**THYMELAEACEAE**
*Aquilaria malaccensis*

**TILIACEAE**
*Pentace triptera*

**GYMNOSPERMAE**
*Podocarpus motleyi*
*P. neriifolius*
*P. wallichianus*

*Recorded from peat swamp forests of Sarawak and Brunei
SOURCE: Corner, 1978.

## Epiphytes

Corner made no detailed survey of epiphytes but states that in 1937 he recorded the following thirty-eight species on a 20-m tree of *Norrisia major* growing on the bank of the Sedili Besar River. From this the richness of epiphytic vegetation can well be imagined.

Asclepiadaceae: *Dischidia nummularia;* Gesneriaceae: *Aeschynanthus parvifolius;* Melastomataceae: *Medinilla hasseltii, M. maingayi, Pachycentria tuberosa;* Moraceae: *Ficus deltoidea var. deltoidea;* Orchidaceae: *Bulbophyllum pulchellum, B. purpurascens, B. sessile, B. vaginatum, Cymbidium finlaysonianum, Dendrobium acerosum, D. aloifolium, D. spurium, Eria pudica, E. vestica, Sarcostoma javanicum, Taeniophyllum sp.;* Rubiaceae: *Hydnophytum formicarum, Myrmecodia;* Ferns: *Asplenium nidus, Cyclophorus acrostichoides, Davallia solida, Drymoglossum piloselloides, Drynaria quercifolia, Humata repens, Hymenophyllum neesi, H. polyanthos, H. serrulatum, Lycopodium laxum, Microsorum punctatum, Ophioglossum pendulum, Phymatodes sinuosum, Platycerium coronarium, Pyrrosia longifolia, Selliguea heterocarpa, Thelypteris crassifolia, and Vittaria ensiformis.*

## Root Morphology

The roots found in freshwater swamp forest fall broadly into three morphological types: buttresses, stilt roots, and pneumatophores. These are not found in shrubs, climbers, and small trees in general, and it is not known exactly at what stage of growth they begin to develop in the large trees. In the swamp forest of the Sedili area, 151 species in 37 families showed one or more of these root structures, and some had all three (table 1-15). These are not characteristic of genus but are thought to be characters shown by the respective species under swamp conditions. Species with special root formations can be divided as follows:

1. Those with none of the three types—most species of Annonaceae, Lauraceae, *Calophyllum, Dipterocarpus,* and *Elaeocarpus,* and all species of *Cratoxylum, Mangifera, Sindora, Stemonurus,* and *Strombosia.*

2. Those with only buttresses—51 species have strong buttresses and 8 species have slight buttresses.

3. Those with only stilt roots—49 species have stilt roots and 1 has slight stilting.

4. Those with only pneumatophores—10 species have distinct pneumatophores and 2 more are uncertain.

5. Those with buttresses and stilt roots—*Dryobalanops oblongifolia, Eugenia nigricans, Lithocarpus ruceolaris, Myristica iners,* and *M. lowiana.*

6. Those with buttress roots and pneumatophores—*Alstonia spathulata, Ctenolophon,* and *Santiria tomentosa; Horsfieldia irya* also sometimes has small buttresses and pneumatophores.

## Table 1-15
### Trees of Freshwater Swamp Forests with Buttresses (B), Stilt Roots (SR), and Pneumatophores (P)

| | B | SR | P | | B | SR | P |
|---|---|---|---|---|---|---|---|
| **ANACARDIACEAE** | | | | **CORNACEAE** | | | |
| *Campnosperma macrophylla* | + | | | *Alangium ebenaceum* | + | | |
| *Melanochyla* sp. (Pontian) | | + | + | **DILLENIACEAE** | | | |
| *Melanorrhoea aptera* | + | | | *Dillenia albiflos* | | (+) | |
| *Gluta malayana* | + | | | *D. grandifolia* | | + | |
| *Parishia* spp. | + | | | *D. reticulata* | | + | |
| *Pentaspadon officinalis* | + | | | **DIPTEROCARPACEAE** | | | |
| **ANNONACEAE** | | | | *Dryobalanops oblongifolia* | + | + | |
| *Goniothalamus malayanus* | | + | | *Dipterocarpus grandiflorus* | (+) | | |
| *Polyalthia sclerophylla* | | | + | *D. sublamellatus* | + | | |
| *Xylopia ferruginea* | | + | | *Hopea mengarawan* | | + | |
| *X. fusca* | | + | + | *H. resinosa* | | + | |
| *X. malayana* | + | (+) | | *Shorea* spp. | + | | |
| **APOCYNACEAE** | | | | **EBENACEAE** | | | |
| *Alstonia spathulata* | + | | + | *Diospyros lanceifolia* | | + | |
| **BOMBACEAE** | | | | *D. maingayi* | + | | |
| *Coelostegia griffithii* | + | | | *D. siamang* | | + | + |
| *Kostermansia malayana* | + | | | **ELAEOCARPACEAE** | | | |
| **BURSERACEAE** | | | | *Elaeocarpus macrocerus* | | + | + |
| *Dacryodes macrocarpa* | | + | | *E. griffithii* | | + | |
| *Santiria laevigata* | + | | | *E. stipularis* | | + | + |
| *S. rubiginosa* | (+) | | + | *E. paniculatus* | | + | |
| *S. tomentosa* | + | | + | **ERYTHROXYLACEAE** | | | |
| **CELASTRACEAE** | | | | *Ixonanthes reticulata* | + | | |
| *Lophopetalum multinervium* | (+) | | + | **EUPHORBIACEAE** | | | |
| **CLUSIACEAE** | | | | *Baccaurea bracteata* | | + | |
| *Calophyllum curtisii* | + | | | *Blumeodendron tokbrai* | | + | |
| *C. inophylloide* | | + | + | *Bridelia pustulata* | | + | |
| *C. kunstleri* | | + | | *Macaranga amissa* | | + | |
| *C. retusum* | | + | | *M. puncticulata* | | + | |
| *C. sclerophyllum* | | + | + | *Neoscortechinia nicobarica* | | + | |
| *C. soulattri* | | + | | **FAGACEAE** | | | |
| *Calophyllum* sp. | | + | + | *Castanopsis inermis* | + | | |
| *Garcinia bancana* | | + | | *C. megacarpa* | + | | |
| *G. forbesii* | | + | | *Lithocarpus bennettii* | + | | |
| *G. maingayi* | | + | | *L. ? cyclophorus* | + | | |
| *G. nigrolineata* | | + | | *L. hystrix* | | + | |
| *G. ? rostrata* | | + | + | *L. ? javensis* | + | | |
| *Mesua lepidota* | | + | + | *L. leptogyne* | | + | + |

*(continued on next page)*

Table 1-15    *(continued)*

|  | B | SR | P |  | B | SR | P |
|---|---|---|---|---|---|---|---|
| *L. urcelolaris* | + | + |  | *M. iners* | + | + |  |
| *Lithocarpus* sp. (Pasania A) | + | (+) |  | *M. lowiana* | + | + | ? |
| GONYSTYLACEAE |  |  |  | *M. maingayi* | + |  |  |
| *Gonystylus bancanus* |  |  | + | MYRTACEAE |  |  |  |
| ICACINACEAE |  |  |  | *Eugenia atronervia* |  | + |  |
| *Platea excelsa* var. |  | + | + | *E. cerina* |  | + | + |
| *riedeliana* (Pontian) |  |  |  | *E. conglomerata* | + |  |  |
| *P. latifolia* |  | + |  | *E. cumingiana* |  | + |  |
| *Stemonurus scorpioides* |  |  | + | *E. garciniifolia* | + |  |  |
| LAURACEAE |  |  |  | *E. grata* |  | + |  |
| *Lindera* sp. | + |  |  | *E. kiahii* |  | + |  |
| *Litsea gracilipes* |  | + |  | *E. leptostemon* | + |  |  |
| *Notaphoebe coriacea* |  | + |  | *E. longiflora* |  | + |  |
| LEGUMINOSAE |  |  |  | *E. muelleri* |  | + |  |
| *Dialium* spp. | + |  |  | *E. nigricans* | + | + |  |
| *Intsia* spp. | + |  |  | *E. oblata* |  | + |  |
| *Koompassia malaccensis* | + |  |  | *E. oleina* | + | + | + |
| *Parkia speciosa* | + |  |  | *E. papillosa* |  | + | + |
| *Pterocarpus indicus* | + |  |  | *E. pauper* |  | + |  |
| LILIACEAE |  |  |  | *E. pseudocrenulata* |  | + |  |
| *Dracaena granulata* |  | + |  | *E. subhorizontalis* |  | + | ? |
| LINACEAE |  |  |  | *E. tumida* |  | + |  |
| *Ctenolophon parvifolius* | + |  | + | *Eugenia* sp. | + |  |  |
| MELIACEAE |  |  |  | *Pseudoeugenia singapurensis* |  | + |  |
| *Amoora rubiginosa* | + |  |  | PANDANACEAE |  |  |  |
| MORACEAE |  |  |  | *Pandanus atrocarpus* |  | + | + |
| *Artocarpus elasticus* | + |  |  | *P. helicopus* |  | + |  |
| *A. kemando* | + |  |  | *P. malayanus* |  | + |  |
| *A. maingayi* | + |  |  | *P. yvanii* |  | + |  |
| *Ficus* spp. (stranglers) |  | + |  | POLYGALACEAE |  |  |  |
| MYRISTICACEAE |  |  |  | *Xanthophyllum ? pulchrum* |  | + |  |
| *Gymnacranthera eugeniifolia* |  | + |  | RHIZOPHORACEAE |  |  |  |
| *Horsfieldia crassifolia* |  | + |  | *Pellacalyx axillaris* |  |  | + |
| *H. irya* | (+) |  | + | RUBIACEAE |  |  |  |
| *H. polyspherula* |  | + |  | *Anthocephalus cadamba* | + |  |  |
| *Horsfieldia* sp. |  | + |  | *Mussaendopsis beccariana* | + |  |  |
| *Knema glaucescens* |  | + | + | RUTACEAE |  |  |  |
| *K. intermedia* |  | + | + | *Tetractomia tetrandra* |  |  | + |
| *K. plumulosa* |  | + |  | SAPINDACEAE |  |  |  |
| *Myristica crassa* |  | + |  | *Pometia pinnata* f. *alnifolia* | + |  |  |
| *M. elliptica* |  | + | + |  |  |  |  |

*(continued on next page)*

## Table 1-15   *(continued)*

| | B | SR | P | | B | SR | P |
|---|---|---|---|---|---|---|---|
| SAPOTACEAE | | | | *Pterospermum javanicum* | + | | |
| *Ganua motleyana* | | | + | *Scaphium linearicarpum* | + | | |
| *Palaquium confertum* | + | | | *S. macropodum* | + | | |
| *P. hexandrum* | + | | | *Sterculia macrophylla* | (+) | | |
| *P. macrocarpum* | + | | | THEACEAE | | | |
| *P. rostratum* | + | | | *Adinandra sarosanthera* | + | | |
| *P. semaram* | + | | | *Ploiarium alternifolium* | | + | |
| *P. xanthochymum* | + | + | + | *Tetramerista glabra* | | | + |
| *Palaquium* sp. | | + | | TILIACEAE | | | |
| *Planchonella maingayi* | (+) | | | *Pentace triptera* | + | | |
| *Pouteria malaccensis* | + | (+) | | VERBENACEAE | | | |
| STAPHYLEACEAE | | | | *Vitex peralata* | (+) | | |
| *Turpinia sphaerocarpa* | + | | | *V. pubescens* | (+) | | |
| STERCULIACEAE | | | | | | | |
| *Heritiera elata* | + | | | | | | |
| *H. simplicifolia* | + | | | | | | |
| *H. sumatrana* | + | | | | | | |

SOURCE: Comer, 1978.

7. Those with buttresses, stilt roots, and pneumatophores—*Palaquium xanthochymum* and *Eugenia oleina*.

8. Those with stilt roots and pneumatophores—18 species.

A trunk bearing buttress roots has its greatest diameter immediately above the buttresses and, with the development of the buttresses, tapers downward to its insertion in the soil. The same effect can be seen on trunks with stilt roots, and sometimes the tapering base may rot away, leaving the trunk supported only by the stilts (fig. 1-8). The height of emergence of stilt roots shows the average height of flooding. Buttresses and stilts are often composed of extremely hard wood.

In connection with the soil, the buttresses of *Koompassia excelsa* became smaller in firm, dry soil, while those of *Kostermansia* were largest in upland areas. *Dillenia grandiflora, Xylopia ferruginea,* and some species of *Garcinia* and *Pandanus* were habitually stilted, whether in upland or swamp forest. Much remains unknown about the relation between root morphology and environment.

Pneumatophores, on the other hand, have soft, aerenchymatous, spongy wood and develop best under strong tidal influence. Corner distinguished five forms of pneumatophore in the freshwater swamp forest.

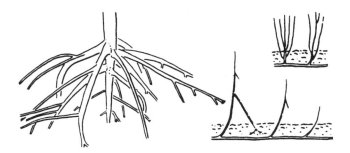

Figure 1-8. Prop roots, lambda roots, and clumps of erect roots of
*Garcinia? rostrata.* SOURCE: Corner, 1978.

1. Erect conical pegs—*Polyalthia sclerophylla* and *Sonneratia*

2. Erect planks—*Lophopetalum multinervium,* degenerating into short, thick knee roots in places with shallow tidal flooding

3. Slender loop roots growing upward, then curving down into the soil—*Xylopia fusca* (fig. 1-9) and *Calophyllum*

4. As in 3, but forming thick knee roots—*Alstonia spathulata, Ganua motleyana, Horsfieldia irya,* and *Tetractomia*

5. Lambda roots formed by more or less erect growth followed by a downward-growing lateral into the soil, the distal part often drying up—*Elaeocarpus macrocerus*

## Phenology

The wet season in Sedili was from October to December or January, followed by lower rainfall from February to September. The rivers were lowest in July and August. The weather of south Johore and Singapore, with two dry spells in February-March and July-August, also had some influence in Sedili, while rainfall in the surrounding mountains could cause river floods throughout the year. The annual average rainfall was 2,600–2,900 mm.

Flowering was most prolific from January to May, with each species flowering for three to five weeks. Fruits ripened in three to six months, and were most prolific from May to October.

Most species flowered once a year, but some flowered twice. For example, *Tristania sumatrana,* two species of *Buchanania,* and *Ixora grandifolia* flowered in January-February and again in July-August; *Kopsia singapurensis* and a species of *Ixora* flowered in January-February and September-October; and *Pentaspadon officinalis, Pterospermum javanicum, Mesua rosea,* and others flowered in April and October. *Nephelium* and *Xerospermum* also flowered twice.

Many trees flowered not annually but once every several years. Dipterocarps

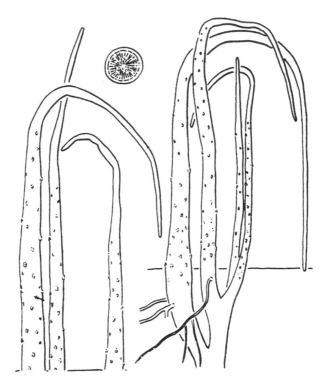

Figure 1-9. Looped absorption roots and bark of *Xylopia fusca*.
SOURCE: Corner, 1978.

are well known for their general flowering every several years after a severe dry spell, and in the Sedili area, Corner observed flowering in May 1935 following an exceptionally dry January. *Kostermansia* appears to have a period of eight years, since Corner observed fruits in September 1932 but not again until 1940, when flowering occurred in May. *Pentace triptera* seems to have a similar period. *Gluta renghas* flowered in two successive years in the period 1963–1969, while *Swintonia schenkii* flowered in three consecutive years in the same period (Medway, 1972). Corner also noted that several kinds of Anacardiaceae in the swamp forest seldom seemed to flower.

The phenology of plants in Sedili can be summarized as follows:

JANUARY The white flowers of *Buchanania sessilifolia* were conspicuous in the first half of the month, and the white, fragrant flowers of *Melanorrhoea wallichii* in the second half. *Tristania sumatrana* began to flower toward the end of the month, and the *Tristania* banks were whitened for three to four weeks. In the *mempisang* belt, leaf fall and the development of inflorescences began.

FEBRUARY The following events might overlap into March.

Gregarious flowering: *Buchanania lucida* (also flowers in August), *Campnosperma*, *Croton laevifolius* (fruits in May), *Eugenia* spp., *Ixora*, *Kopsia singapurensis*, various Lauraceae, *Lithocarpus* spp., *Macaranga baccaureifolia* (fruits in June), various Myristicaceae, *Neesia*, *Pometia*.

Fruiting: *Calophyllum* spp. (from flowering in October-November), *Connarus grandis*.

New leaves: *Shorea* spp., *Sindora coriacea* (deciduous), *Macaranga baccaureifolia* (deciduous).

MARCH Rain in March is uncertain. The main flowering begins in this month. Most striking were the leaf fall and flowering of *Cratoxylum formosum*. This deciduous tree is not felled for lumber and thus remains in abundance. The bare crowns were reddish pink with buds and flowers.

Gregarious flowering: *Castanopsis* spp., *Cerbera odollam*, *Crateva religiosa*, *Elaeocarpus* spp., *Entada phaseoloides*, *Eugenia spicata*, *Flagellaria*, *Gardenia tubifera*, *Gluta velutina* (flowering prolonged into June or July), *Goniothalamus* spp., *Horsfieldia irya*, *Poikilospermum* spp., *Vitex clarkeana*.

Fruiting: *Mesua ferruginea*, *Myristica elliptica*, *Neoscortechinia forbesii*, *Nephelium glabrum*, *Vatica wallichii*.

New Leaves: *Calophyllum sclerophyllum*, *Dalbergia beccarii*, *Koompassia malaccensis*, *Polyalthia sclerophylla* (olive-buff new leaves), *Pongamia pinnata* var. *xerocarpa*.

These events might begin in February or continue into March. Regarding *Koompassia malaccensis*, Medway (1972) noted the development of new leaves about every six months, but flowering was uncertain, with only one general flowering recorded in the period 1963–1969.

APRIL, MAY Gregarious flowering: *Barclaya* (aquatic plants), *Barringtonia filirachis*, *Chisocheton amabilis*, *Cratoxylum arborescens*, *Cryptocoryne* spp. (aquatic plants), *Dillenia excelsa*, *Elaeocarpus macrocerus*, *Eugenia spicata*, *Garcinia bancana*, *Gardenia tubifera*, *Horsfieldia irya*, *Ixora congesta*, *I. lobbii*, *Jackia*, *Lophopetalum multinervium*, *Mussaendopsis*, *Myristica elliptica*, *Palaquium xanthochymum*, *Pandanus helicopus* (prolonged into August or September), *Pentaspadon officinalis* (also October), *Polyalthia sclerophylla*, *Premna trichostoma*, *Pterospermum javanicum*, *Shorea* spp., *Sterculia bicolor*, *Vitex peralata*.

Fruiting: *Bhesa paniculata*, *Connarus grandis*, *Gnetum gnemon* var. *brunoniarum*.

New leaves: *Podocarpus neriifolius*.

JUNE, JULY Gregarious flowering: *Connarus grandis*, *Elaeocarpus* spp., *Eugenia cerina*, *Fagraea racemosa*, *Nephelium rubescens*, *Tristania sumatrana* (second flowering), *Vitex clarkeana* (second flowering).

Fruiting: *Buchanania, Dillenia excelsa, Elaeocarpus macrocerus, Gardenia tubifera, Grewia antidesmaefolia, Pandanus helicopus* (continues until October or November), *Pentace triptera, Pentaspadon officinalis, Rinorea anguifera, Schoutenia accrescens, Sterculia macrophylla, Vitex clarkeana, Xanthophyllum affine.*

New leaves: many kinds of trees. *Calophyllum sclerophyllum* changed leaves in March and July.

AUGUST General fruiting, and second flowering of *Buchanania lucida.*

SEPTEMBER General flowering of *Dysoxylon macrothyrsum* and second flowering of *Ixora* spp. and *Kopsia purensis.*

OCTOBER Gregarious flowering: *Dacryodes macrocarpa, Dryobalanops oblongifolia, Pentaspadon officinalis* (second flowering), *Pterospermum javanicum* (second flowering), *Xanthophyllum affine.*
Fruiting: *Garcinia penangiana.*
New leaves: *Melanorrhoea aptera* (deciduous).

NOVEMBER Many trees in fruit.

DECEMBER Several species of *Calophyllum* flowered in this month or in November. As the wettest time of year, collecting was often impossible.
Deciduous trees: The following were deciduous trees: *Cratoxylum formosum, Dillenia grandifolia, D. reticulata, Heritiera simplicifolia, Macaranga baccaureifolia, Melanorrhoea aptera, M. pubescens, Parkia speciosa (Pentace triptera), Pentaspadon officinalis, Polyalthia sclerophylla (Pterospermum javanicum), Scaphium macropodum, Sindora coriacea, Sterculia bicolor, S. macrophylla.*
Ever-flowering trees: *Dillenia suffruticosa, Hibicus tiliaceus,* possibly *Barringtonia racemosa.*

## Succession from the Coast to Freshwater Swamp Forest

The floristic succession inland from the coast of Jason Bay and the estuaries of the Sedili rivers shows various aspects, but broadly it is of two types: one beginning from sandy shores and one beginning from mangrove.

On sandy shores, the sea-facing beaches and sandbanks were colonized by a so-called *Ipomoea* community, where *Ipomoea* and *Canavalia* predominated, together with grasses and sedges. This community also included seedlings of *Casuarina equisetifolia* and of other coastal trees whose fruits were thrown up by the waves. *Casuarina* seedlings sprouted only on the open dunes, never in the shade of parent trees, and they rapidly formed a belt 10–50 m wide.

Inland, the region was invaded by *Eugenia grandis,* which grows wild in the Malay Peninsula only on rocky and sandy shores. It formed a community behind *Casuarina* that contained almost sixty species of medium and tall trees.

This in turn was succeeded by dry climax forest, in which *Eugenia grandis* and *Beilschmiedia tonkinensis* persisted up to 400 m from the coast. *Dipterocarpus hasseltii* also entered the *Eugenia grandis* forest, with some trees 25–30 m high standing only 60 m from the shore. *Cratoxylum,* a further three species of *Eugenia,* and *Pterospermum* were gregarious. In one patch about 30 × 50 m, for example, there were seven specimens of *Cratoxylum formosum* measuring 25–35 m high and 30–50 cm across at breast height. While *Cratoxylum* and *Pterospermum* were also common in swamp forest, all of the species of *Eugenia* were restricted to the coast. In the understory, *Tacca palmata* and *Petunga roxburghii* were present and extended into the *Eugenia* forest.

Farther inland was freshwater swamp forest, and farther again was lowland dipterocarp forest. Thus, in terms of forest types, the succession was from *Casuarina* to *Eugenia grandis,* to dry climax forest, to freshwater swamp forest, and to lowland dipterocarp forest. In this area, forest dominated by *Pterocarpus indicus* sometimes replaced the dry climax forest.

On mangrove coasts, the mangrove is gradually replaced by freshwater swamp forest. Since mangrove may extend a considerable way inland along a river, a transect through a meander where the river runs parallel to the coast may show mangrove following on from the *Pterocarpus indicus* forest and giving way to freshwater swamp forest. Although this will depend on how a transect is drawn, there are basically two types of succession.

## *Freshwater Swamp Forests of Southeast Asia*
### Burma

The lower reaches of the major rivers of continental Southeast Asia formerly must have been occupied by extensive swamp forests. However, the only recorded survey of such forest is that conducted by Stamp (1925) on the lower reaches of the Irrawaddy River, which can be summarized as follows.

The freshwater swamp forests of Burma are known as *myaing.* In Upper Burma, they occur in areas flooded for three to four months annually. They are also found in much of the area between the Rangoon-Prome Railway and the Irrawaddy River. This area, probably an old course of the Irrawaddy River, is flooded during the rainy season from July to October, and trees stand in 2–3 m of water. The forest is open and sparse, and the main components are *Albizia procera, Anogeissus* sp., *Barringtonia acutangula, Butea frondosa,* several species of *Lagerstroemia, Schleichera trijuga, Stephegyne parviflora,* and the bamboo *Bambusa arundinaceae.* In the wettest areas, which are dry for only a few weeks, *Xanthophyllum glaucum* and *Dalbergia reniformis* form almost pure stands, together with thickets of *Combretum* and *Glochidion.*

Another type of swamp forest is known as *kanazo (Heritiera fomes)* forest. It is found where surface water completely disappears for a few hours between tides, occupying land that is slightly higher and has lower salinity than that occupied by mangrove. It occurs in two types.

The first type grows where water is comparatively fresh. Growth is more vigorous and some *kanazo* exceed 30 m, but it is not a dense forest. Associated species include *Afzelia bijuga, Amoora cuculata, Barringtonia acutangula, Dysoxylum* sp., lianous *Combretum* sp., and, on the ground, *Acanthus ilicifolius* and others. In the highest part of this forest, the peglike pneumatophores become shorter, the trees become larger, and large *Pandanus foetidus, Litsea,* and *Cynometra ramiflora* appear.

The second type of *kanazo* forest grows in brackish water. The trees are smaller than in the first type but form denser stands. Associated species are few and also include mangroves: *Acrostichum aureum, Xylocarpus moluccensis, X. granatum, Cerbera odollam, Cynometra ramiflora, Bruguiera parviflora, Excoecaria agallocha, Sonneratia griffithii,* and others. On slight elevations where inundation is less and salinity higher because of evaporation, the forest becomes impoverished. *Kanazo* occurs with an understory of *Cynometra ramiflora* and sometimes *Ceriops roxburghiana,* while *Acrostichum aureum* covers the ground.

*Kanazo* forest can be killed by submergence throughout the whole or part of the rainy season, and it cannot exist above the level of the spring tides. Between these extremes, it appears in areas of almost fresh or brackish water, occupying several hundred square miles of the Irrawaddy delta and large areas of South Arakan and Mergui. Some of these forests contain large scattered trees of *Sonneratia apetala.*

On clay soils above the upper limits of the *kanazo* forest are low mixed scrub forests, known locally as *byaik* or *kon-byaik,* where tree growth is scattered and poor. The main tree species include *Amoora cuculata, Calophyllum* sp., *Diospyros* sp., *Elaeocarpus hygrophilus, Eugenia, Lagerstroemia flosreginae* (Myristicaceae), *Litsea,* and *Mangifera caloneura.* The understory is thick and difficult to penetrate, and contains *Calamus arborescens, C. erectus, Clinogyne dichotoma, Flagellaria indica, Hibiscus tiliaceus, Phragmites* sp., and *Pinanga* sp. Gaps are common and are covered by grasses and *Hibiscus tiliaceus.*

Another type of scrub forest is characterized by *Phoenix paludosa,* and a further type with almost pure stands of *Cynometra ramiflora* is widespread in the Irrawaddy delta.

## Singapore

Singapore, in what is today the most developed part of Southeast Asia, also once had swamp forest. In the 1930s, when development had already begun, Corner surveyed the remaining forests before they were felled. He has described three sites, Mandai Road and Jurong Road in Singapore, and Pontian in Johore.

## MANDAI ROAD

The canopy height was 32 m, the tallest tree being *Palaquium xanthochymum*. At one plot this tree comprised nearly half of the canopy, and *Xylopia fusca* was the second most abundant. Of the small trees, *Talauma singapurensis* was abundant, and the undergrowth consisted mainly of Pandanaceae, Palmae, and Araceae.

At other plots on the same site, *Xylopia fusca* and *Cratoxylum arborescens* were dominant in one area and *Palaquium xanthochymum* at another. Under dryer conditions, *Melanorrhoea wallichii, Shorea macroptera,* and *S. platycarpa* were dominant, and the dominant species of the other plots were unrepresented.

These findings showed that *Palaquium xanthochymum* grew under the most swampy conditions, with *Xylopia fusca* on slightly higher ground. On even higher, drier land, in a transitional zone to lowland dipterocarp forest, the species peculiar to swamp forest no longer appeared.

## JURONG ROAD

The canopy height was again 30–32 m, with *Cratoxylum arborescens* the highest tree and reaching 70 cm in diameter. Thirty-one species of major trees and 10 species of small trees were recorded. Climbers included *Daemonorops leptopus, Dapania, Embelia, Flagellaria indica, Gnetum, Nitrella kentii, Morinda rigida, Pothos latifolia,* and *Uncaria.* The undergrowth included *Ardisia tuberculata, Cyrtostachys, Hanguana, Labisia, Nenga, Salacca,* and *Sterculia coccinea.* The presence here of species not found in other swamp forests, and the absence of other species that were found elsewhere, suggest that Jurong represents a distinct phytogeographical area. The state of regeneration, reaching 20,000 small stems per ha, can be regarded as favorable.

## PONTIAN

By 1939, the surrounding forest had all been converted to pineapple estates. The remaining swamp forest contained fewer species than the two described above. Some species were common to all, but species typical of freshwater swamp forest were fewer at Pontian. *Cratoxylum, Gonystylus,* and *Tristania* were abundant, but *Melanorrhoea* and *Mussaendopsis* were absent. While I shall not list here all species present or absent, Corner concludes that the Pontian forest was a freshwater swamp forest on the way to becoming peat forest.

From consideration of the swamp forests at these three sites and along the Sedili rivers, together with data on peat swamp forest, Corner presented the scheme shown in figure 1-10 for the sequence of the swamp forests.

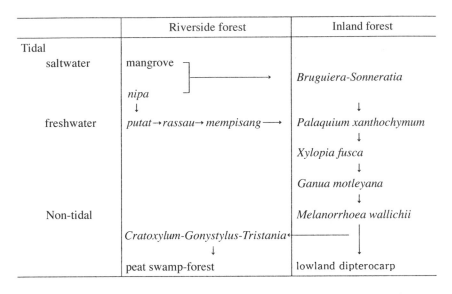

|  | Riverside forest | Inland forest |
|---|---|---|
| Tidal<br>  saltwater | mangrove<br>nipa | *Bruguiera-Sonneratia* |
| freshwater | putat→rassau→mempisang → | *Palaquium xanthochymum*<br>↓<br>*Xylopia fusca*<br>↓<br>*Ganua motleyana*<br>↓ |
| Non-tidal |  | *Melanorrhoea wallichii* |
|  | *Cratoxylum-Gonystylus-Tristania*<br>↓ |  |
|  | peat swamp-forest | lowland dipterocarp |

Figure 1-10. Transition from mangrove to freshwater swamp forest, peat swamp forest, and lowland dipterocarp forest. SOURCE: Corner, 1978.

## Swamp Forest of the Malay Peninsula

Wyatt-Smith (1961) gathered information not just on the south but the whole of the Malay Peninsula. Having earlier discussed the environment of freshwater swamp forest, here I shall record briefly the areas and tree species not mentioned by Corner.

The Utan Melintang forest reserve in Perak is characterized by *Sindora intermedia, Melanorrhoea torquata,* and *Lophopetalum* spp. In the Rompin district on the east coast are large areas dominated by sedges (*Scirpodendron ostatum* in deeper parts), *Licuala,* and others, including the palms *Livistona, Pinanga,* and *Salacca;* and this type also can be found in considerable areas of swamp forest. The *Livistona-Pandanus-Stemonurus* community appears on the Tinggi River in Selangore and the Tasek Bera area of southwest Pahang. *Fagraea crenulata* often appears as a narrow belt a short distance from the sea. *Hopea griffithii* appears in seasonal swamp forest in north Kelantan.

In old mining areas where drainage is impeded, *Ploiarium alternifolium* and *Alstonia spathulata* form almost pure communities, often with *Scleria* spp., *Lycopodium cernuum,* and *Gleichenia linearis. Macaranga maingayi* is common on abandoned farmland where freshwater swamp forest formerly grew.

*Oncosperma filamentosum,* a typical palm of freshwater swamp forest, forms almost pure stands in a narrow belt a short distance from the sea. It is widely distributed along the east coast, and fairly rich areas of this species have developed in central Johore where shallow peat swamp forest has been cleared. Its gray-black stems rise close to 30 m and bear sharp thorns protruding at right angles. It is an important palm used for pillars and flooring of houses in the coastal zone and swamps.

No description of freshwater swamp forest would be complete without mention of *Melaleuca leucadendron*. Found in Malacca, Kelantan, Trengganu, Kedah, Perlis, and elsewhere, it was estimated to occupy an area of 200,000 ha in 1966 (Sandrasegaran, 1966), though this has subsequently shrunk dramatically. It grows in lowland swamp, coastal alluvium, and even scrubland, under a variety of conditions, including sandy land behind mangrove, acidic soil, saline soil, and waterlogged land. Its morphology is also diverse, with tall shrubs, bushy coppice forest (particularly after fire), and tall trees 15–25 m. The species is readily distinguishable by its small, gray-tinged crown and the constant flaking of its pale bark from the slightly twisted trunk. It is often associated with *Ploiarium alternifolium*. It is highly fire resistant and regenerates rapidly through coppices and suckers. Fire created almost pure forests of *Melaleuca*, often associated with the climbing fern *Stenochlaena palustris* and *Scleria* spp.

*Melaleuca leucadendron* attains a girth of 23 cm in six to seven years, 30 cm in fifteen years, and thereafter grows at a rate of 0.8 cm/yr. Growth is most vigorous in the seventh year, and the tree takes nine years to attain a height of 9 m. It is used as firewood and for fishing stakes. Medicinal *kayu puteh* oil is extracted from its leaves. The dried fruits are sold as black pepper *(mercha bolong)*. The bark is suitable for making torches. It is not suitable for pulp, however, having short fibers and being difficult to bleach and macerate.

The freshwater swamp forest of the Malay Peninsula described above is a thing of the past. In an exhaustive survey of the coastal regions of the peninsula in 1978, I found no large areas of freshwater swamp forest remaining. The majority had been converted to rice fields, pineapple estates, and the like. It was also rare to see *kayu puteh* forest, which remained only as single trees or small clumps. Wyatt-Smith (1961) lists the following tree species as characteristic of the freshwater swamp forests formerly found throughout the peninsula.

*Alstonia spathulata, Artocarpus penduncularis, Calophyllum* spp., *Campnosperma* spp., *Coccoceras muticum, Cratoxylum arborescens, Dialium* spp., *Dillenia* spp., *Dipterocarpus costulatus, Dryobalanops oblongifolia, Eugenia* spp., *Hopea griffithii, H. mengarawan, Ilex cymosa, Intsia palembanica, Koompassia malaccensis, Lophopetalum* spp., *Madhuca* spp., *Melanorrhoea* spp., *Mezzettia leptopoda, Nauclea maingayi, Palaquium* spp., *Parastemon urophyllum, Shorea bentongensis, S. hemsleyana, S. macrantha, S. platycarpa, Sindora intermedia, Vatica flavida, V. lobata,* and *V. wallichii.*

South Sumatra

In insular Southeast Asia, swamp forests occupy large areas along the east coast of Sumatra. Mangrove occupies the most seaward areas of this region, giving way inland to freshwater swamp forest and peat swamp forest. Yamada and Soekardjo (1979) surveyed the mangrove and swamp forest of the Palembang coast.

In 1978, South Sumatra was one of the target areas of the Indonesian government's transmigration policy. Swamp forest was cleared, canals were dug, rice fields were made, and large numbers of migrants arrived from densely populated Java. Along the courses of the Musi River running northward from Palembang toward the sea, the transmigrants engaged in a variety of settlement activities, which I shall touch on later.

The entire east coast of Sumatra facing the Strait of Malacca is covered by mangroves. This is in the center of Southeast Asia's tropical rain forest climate, and the stock of biomass is immense. *Avicennia alba* of 85 cm DBH (diameter at breast height) can be seen, and *Rhizophora apiculata, Bruguiera gymnorrhiza,* and other mangroves reach heights of 36 m and more. Freshwater swamp forest occupies river courses inland of the mangrove-dominated coastal strip, invades the tributaries where *Sonneratia alba* dominates, and appears on the margins of areas where *nipa* constricts the black-water rivers.

Our survey was conducted in the latter part of September, when the forest was flooded to a depth of about 30 cm with black water, below which was a layer of about 20 cm of humus. Knee-like pneumatophores protruded from the forest floor, the spiny *Salacca conferta* grew everywhere, and old spines floated in the water, making the survey difficult. The forest was light and far more open than mangrove forest, lowland dipterocarp forest, or montane forest. The canopy contained *Alstonia angustiloba,* called *pulai,* reaching a height of 50 m. Below this was *Campnosperma auriculata,* which formed the middle layer with *Macaranga trichocarpa* and species of *Palaquium.* This layer also contained stands of the *nibong* palm, *Oncosperma filamentosum,* some of which reached 25 m. Low trees included *Polyalthia glauca, Neonauclea obtusa,* and *Baccaurea* spp., while the forest-floor layer included *Salacca conferta* and seedlings of the upper-story trees.

Table 1-16 shows the composition of species with DBH higher than 10 cm. On a per ha basis, the figures are equivalent to a total of 290 stems with a total basal area of 16 m². These are appreciably smaller than the 596 stems and 25 m² obtained for the mangrove forest dominated by *Rhizophora apiculata* in the most seaward area of the same region. However, they are larger than the 190 stems and 13 m² of the mixed forest of *Bruguiera gymnorrhiza* and *Rhizophora apiculata* at the landward side of the same mangrove. The latter figures were obtained for a transect of 20 × 500 m containing considerable areas occupied by creeks and

## Table 1-16
### Composition of Species with DBH Higher than 10 cm in Freshwater Swamp Forest, Banjuasin, South Sumatra

| SPECIES | NUMBER | BASAL AREA (CM$^2$) |
|---|---|---|
| Polyalthia glauca | 6 | 1,259.5 |
| Campnosperma auriculata | 5 | 3,812.2 |
| Ochanostachys amentacea | 2 | 215.7 |
| Alstonia angustiloba | 1 | 5,407.9 |
| Macaranga trichocarpa | 1 | 1,224.8 |
| Palaquium sp. | 1 | 779.0 |
| Eugenia sp. | 1 | 498.5 |
| Elaeocarpus obtusus | 1 | 359.5 |
| Brochidia sp. | 1 | 326.7 |
| Artocarpus anisophyllus | 1 | 277.4 |
| Memecylon sp. | 1 | 277.4 |
| Litsea sp. | 1 | 254.3 |
| Baccaurea sp. 1 | 1 | 224.2 |
| Neonauclea sp. | 1 | 158.3 |
| Dendrocnide stimulans | 1 | 141.0 |
| Neonauclea calycina | 1 | 126.6 |
| Macaranga ipolica | 1 | 102.1 |
| Neonauclea obtusa | 1 | 84.9 |
| Baccaurea sp. 2 | 1 | 80.1 |
| Total | 29 | 15,610.1 |

(20 × 50 m$^2$ plot)
SOURCE: Yamada and Soekardjo, 1979.

clumps of *nipa*, but although fairly small they probably represent the average accumulated biomass for this forest type. Even in freshwater swamp forest, the total biomass can be altered appreciably by the presence of a single large specimen of *pulai*, but the above figures do provide some indication.

Table 1-17 shows the composition of species with DBH of 2–10 cm. Saplings of the same species also appear in the forest-floor layer. The *Salacca* palm is not included in the table, but it constitutes an important element. Figure 1-11 shows the crown projection of the survey plot, revealing the high degree of *pulai* dominance, with its large buttress roots, and *nibong*.

Table 1-17
Composition of Species with DBH of 2–10 cm in Freshwater Swamp
Forest, Banjuasin, South Sumatra

| | |
|---|---|
| *Euonymus javanicus* | *Polyalthia sumatrana* |
| *Syzygium polycephalum* | *Campnosperma auriculata* |
| *Dendrocnide stimulans* | *Leea indica* |
| *Canarium* sp. | *Elaeocarpus glaber* |
| *Eugenia* sp. 1 | Species of Annonaceae |
| *Macaranga* sp. | *Euodia* sp. |
| *Elaeocarpus odontopetalus* | *Aglaia* sp. |
| *Neonauclea calycina* | *Timonius flavescens* |
| *Elaeocarpus obtusus* | *Anthocephalus chinensis* |
| *Glochidion zeylanicum* | *Artocarpus* sp. |
| *Eugenia reinwardtiana* | *Ficus glaberrima* |
| *Litsea* sp. | *Parastemon urophyllum* |

(20 × 50 m² plot)
SOURCE: Yamada and Soekardjo, 1979.

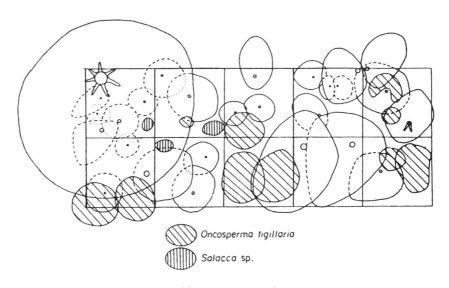

Oncosperma tigillaria

Salacca sp.

Figure 1-11. Crown projection of freshwater swamp forest, Banjuasin, South Sumatra.
The tallest tree with buttress roots is *Alstonia angustiloba* (height 50 m).
SOURCE: Yamada and Soekardjo, 1979.

## Present State of Forest Exploitation

The exploitation of freshwater swamp forest in Singapore began, as indicated earlier, in the 1930s, and the Malay Peninsula forest was felled at about the same time. The forest remaining in South Sumatra is presently being exploited. Pulau Rimau, an island in the estuary of the Banjuasin River where freshwater swamp forest once grew, was formerly inhabited not by humans but, as the word *rimau* indicates, by tigers. The large *pulai* that grew there attracted the attention of the headman of Sunsang, a town on the Musi River estuary. Beginning in 1970, *nipa* was cleared along streams to allow extraction of lumber. Where the streams grew too narrow, canals were dug. The felled *pulai* were cut into lengths of 4.4 m, maneuvered onto rollers by several men, and taken to the canals. At collection points, where the canals had been widened, the logs arriving from upstream were joined into two lines of ten logs each and taken by one man to the junction with the main stream. There the logs were realigned sideways into rafts of twenty or thirty logs, which were towed to the outskirts of Palembang, where they were sorted for export or domestic use.

The land left behind, accessible by canal, was cleared of large trees and was therefore easy to reclaim. It attracted migrants, mainly Bugis from Sulawesi, and each family received 2 ha. They cleared the remaining medium and small trees and burned them in September, the month of lowest rainfall. Then they prepared the land, starting with small plots of about 20 × 20 m where there were fewer felled trees and drainage was good.

At the same time, they prepared rice nurseries on hardened earth roads dug out of the canals, on artificially raised beds, on their roofs, and elsewhere. First they spread *nipa* leaves, which they covered with about 3 cm of mud before broadcasting their seed. This was then enclosed on four sides with banana or *nibong* stems, covered with *nipa* or banana leaves, and watered every morning and evening. After about one week the seedlings reached 10 cm high, and after two weeks they were ready for the first transplanting. This was done with a planting stick called a *tugal*, the holes penetrating only into the humus layer to prevent the roots being cut when they were transplanted the second time. After the first transplanting, surrounding land was prepared for the second transplanting, carried out after about forty days. This time, the holes reached deep into the soil.

The crop was harvested in April. Yields the first year were about 1 ton/ha of unhulled paddy, and this was expected to increase every year by about 1 ton/ha up to 8 ton/ha in the future. In the neighboring Jambi Province, yields of 8 ton/ha were obtained in the seventh year under similar conditions. The factors in the rising yields are the nutrients provided by the rotting of the unburned wood and the progress in reclamation of surrounding lands. Here the fields are rain fed, and the canals provide drainage.

The reclamation of wetland currently in progress in Sumatra is of three types. The first type involves heavy investment of capital by the government, large-scale engineering works, and mass migration. It is highly planned and engineered, employing heavy machinery to clear the natural forest, dig canals, create large paddy areas, and establish residential areas. It can be seen in the vicinity of Upang district. The second type is the cooperative operation at the village level, such as that described above. And the third is totally individual reclamation. Migrants dig canals navigable by small boat and create small fields for themselves. Many peasants engaged in such small-scale reclamation are congregated along the small rivers running through the swamp between Palembang and the sea. Because of the proliferation of weeds, many of them move to new locations after a few years.

Of these three types, the second probably shows the most stable tendency because of the conditions of the wetlands. The first type, large in conception, involves much waste and failure. Migrants tend to know little of local conditions and, worst of all, evince little desire to develop the new land. The third type is limited by the abilities of the individual and only spreads secondary forest to no purpose. What makes the second type the most desirable are the intimacy of local residents with local environmental conditions and the organized nature of their communal efforts.

When forest is cleared, the usable lumber must be put to the most significant use possible. *Pulai* provides timber for export, *Campnosperma* provides timber for domestic use, and *nibong* provides building materials for local people. The small trees of no use are cut and burned. Wood left after the burning is not wastefully cleared away but left to decompose and become fertilizer. Standing water is drained, and crops are planted selectively, depending on the nature of the land, sometimes on raised beds. House building, widening and extension of canals, land preparation, and other tasks requiring considerable labor are performed communally. The essential feature of such exploitation is that it involves no waste. The obstacles presented by the natural conditions are overcome gradually, and the forest's fertilizing capacity is converted only as it is needed.

The exploitation of freshwater swamp forest, however, with the exception of New Guinea, is proceeding rapidly. For future forest management, and for future environmental conservation, it is most desirable that a conservation area with a certain continuity be established that includes remaining natural forest. The Alangantan district on the border of South Sumatra and Jambi Provinces is a strong candidate highly suited to this purpose.

## 3. Peat Swamp Forest

Peat swamp forest also has been little studied, for a variety of reasons. First, being located landward of mangrove, it is difficult to approach. And once access has been gained, peat underfoot hampers progress. While it occupies the same tropical wetland, it has few of the morphological and ecological features that have attracted attention in the mangroves, such as the variant root formations and viviparous fruits. A person is always haunted, moreover, by unaccountable anxiety about insect-borne disease. Even from my own modest experience, I would prefer not to conduct a survey in peat swamp forest.

Nevertheless, felling began in the 1940s to obtain swamp forest timber. Shortly after this, Anderson began his survey of swamp forest in Sarawak, and his work remains the only substantial study of Southeast Asia's peat swamp forest. Here I shall give an outline of tropical peat swamp forest, focusing mainly on his work published in the 1960s.

### *Distribution and Environment*
### Distribution

Peat swamp forest in Southeast Asia can be found in the lowlands of eastern Sumatra, Sarawak, Brunei, the Malay Peninsula, southwestern New Guinea, and the southern Philippines. Estimates for peat swamp forest in these areas are 17 million ha for Indonesia (Coulter, 1957), 1.5 million ha for Sarawak (Anderson,

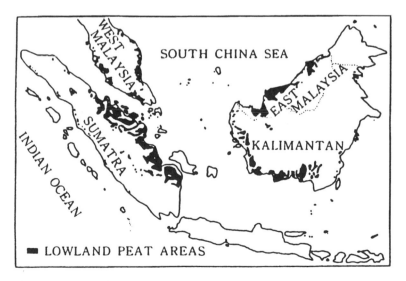

Figure 1-12. Distribution of peat swamp forest in the Malay Peninsula, Borneo, and Sumatra. SOURCE: Andriesse, 1974.

1963), and 0.5 million ha for the Malay Peninsula (Wyatt-Smith, 1963). Substantial portions of these areas have now been felled, and the present areas must be far smaller.

Globally, areas have been estimated at 7 million ha in Central and South America, including the Amazon lowlands, the Caribbean Islands, and the lowlands of Guyana, and 3 million ha in tropical Africa (Driessen, 1978). Research data are even fewer in these areas than in Southeast Asia, and the economic value of the forests is also lower. Like mangrove, tropical peat swamp forest can be seen on the largest scale under the rain forest climate of Southeast Asia, centered on Borneo and Sumatra.

## Environment

The peat swamp on which tropical peat swamp forest grows has a domed structure. This is true in Sumatra, Java, New Guinea, and elsewhere as surveyed by Polak (1933), and also in Sarawak and Brunei. The oldest part of the swamp, located farthest from the sea, is thickest and shows a clear dome. Peat accumulation is greatest at the center of the swamp, but the depth of the peat base is variable. The base of the deepest peat in the Rejang Delta lies below mean sea level. Below the peat is a thin layer of black mud, and below this is bluish gray or bleached yellow clay. The groundwater level is high, and in Brunei it lies about 10 cm below the surface in April. River water never inundates the surface peat. Consequently, rainfall is the only water supply.

Peat is highly acidic, with a pH of 3.85–4.15, and extremely oligotrophic, with a loss on ignition of 77–94 percent. Its phosphorus content overall is low, and its nutrient content tends to decrease toward the center of the swamp, with phosphorus and calcium being particularly low (Anderson, 1964b; Muller, 1972). In Sarawak, the Baram swamp has the best-developed peat, with a nutrient content much lower than in other regions; loss on ignition is 98–100 percent.

### *Peat Swamp Forests of Sarawak and Brunei*

In 1954, Anderson was employed by Sarawak's Forest Department, and early in 1960, he completed an almost perfect monograph on the peat swamp forests of Sarawak and Brunei. At the same time, he conducted an ecological survey. In five years, he assembled complete specimens of 95 percent of all tree species distributed in the peat swamp forest. He lived in Singapore for many years, conducting surveys of Southeast Asia's forests as a forestry consultant, before retiring to live in Edinburgh.

### Overview of Peat Swamp Forest

Peat swamp forests cover 8.4 percent of the national area of Sarawak and 13.6 percent in Brunei. In Sarawak, the largest forests are in the Rejang Delta and along

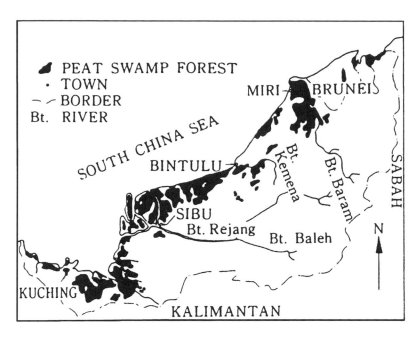

Figure 1-13. Distribution of peat swamp forest in Sarawak. SOURCE: Lee, 1979.

the Baram River. Peat swamp forest is particularly extensive on the Maludam Peninsula, extending 64 km inland from the mangrove and covering 1,070 km² (fig. 1-13). In Brunei, it can be seen in the Belait, Tutong, and Temburong areas.

The vegetation that develops on the dome-shaped peat swamp changes successively from the margin to the center. Anderson (1961b) has divided it into six phases. The results of pollen analysis in relatively well-developed peat swamps showed a peat depth of up to 13 m on a basement of hard clay. On this was a *Campnosperma coriacea–Crytostachys lakka–Salacca conferta* association in which the six phases appeared successively (Anderson, 1964a; Muller, 1965, 1972). This process of succession has been estimated by Wilford (1960) to require 4,500 years. Further, in a different boring, mangrove appeared in the oldest layer, indicating a history of peat swamp forest developed after mangrove.

The six phasic communities, or forest types, are not found in all swamps; generally, one or another is missing. In particular, the sixth phase, which is the final stage, is only found in the best-developed peat swamps.

While these peat swamp forests contain a variety of tree species, as will be described later, one species, *Shorea albida,* is characteristic of Sarawak and Brunei. Found from the estuary of the Kapuas River to Tutong in Brunei, it forms pure stands, which are rare in the tropics, and can be considered specific to peat swamp forest.

## Physiognomy

The six successive phasic communities that accompany the development of peat swamp can be described as follows:

*Phasic community 1:* Gonystylus-Dactylocladus-Neoscortechinia *association (mixed swamp forest).* This community appears at the margins of swamps and covers large areas of relatively undeveloped coastal wetland. The forest crown is not uniform but resembles lowland dipterocarp forest, with the highest layer reaching 40–45 m and every layer being exceedingly mixed. The principal dominants are *Gonystylus bancanus, Dactylocladus stenostachys,* and three species of *Shorea* other than *Shorea albida* (*S. platycarpa, S. scabrida,* and *S. uliginosa*). In the middle and lower stories many species appear, among which *Alangium havilandii* and *Neoscortechinia kingii* are most widely distributed. Large trees, more than 30 cm in girth, number 60 to 70 species in an area of 4,046 m². The water table frequently rises above ground level, and aroids and the sedge *Thorachostachyum bancanum* are abundant. The palm *Salacca conferta* forms dense thickets on shallow peat.

*Phasic community 2:* Shorea albida-Gonystylus-Stemonurus *association* (alan *forest).* This community is transitional between communities 1 and 3. The forest crown is not uniform and is dominated by enormous *Shorea albida* trees, which often exceed 3.7 m in girth. Such large trees are usually moribund, having hollow stems and large, stag-headed crowns. Smaller trees and saplings of this species are virtually absent. The middle and lower stories are composed of virtually the same species as phasic community 1 and have similar physiognomy. *Stemonurus umbellatus,* a small tree 30–60 cm in girth, is most characteristic and is also found in heath forest.

*Phasic community 3:* Shorea albida *consociation (*alan bunga *forest).* This community covers extensive areas of the second and fourth divisions in Sarawak and the Badas area of Brunei, but its distribution in the Rejang Delta is very limited. The upper story is a pure, even canopy of *Shorea albida* 50–60 m tall, with a density of 88 to 125 stems per ha. The middle story is absent, but the lower story is moderately dense and often dominated by a single species, *Tetractomia holttumii, Cephalomappa paludicola,* or *Ganua curtisii.* Herbaceous vegetation is sparse. *Pandanus andersonii* sometimes forms a dense shrub layer.

*Phasic community 4:* Shorea albida–Litsea–Parastemon *association (*padang alan *or* padang medang *forest).* This community is found in the center of swamps. It constitutes a transitional zone in the Rejang Delta and part of the Baram swamp

and has an even, continuous canopy at 30–37 m. The constituent trees are relatively small in girth and rarely exceed 1.8 m. The outstanding features of the
forest are the polelike nature of the trees and the xerophytic aspect of the vegetation. The principal dominants are *Shorea albida* (with *padang alan*) and *Litsea
palustris* (with *padang medang*), the former reaching a density of 350 stems per ha.
Other characteristic species include *Parastemon spicatum, Combretocarpus rotundatus,* and *Calophyllum obliquinervum.*

*Phasic community 5:* Tristania–Parastemon–Palaquium *association.* This community appears as a narrow band between communities 4 and 6. The canopy is
dense, with an average height of 15–18 m and occasional emergents. Trees
exceeding 30 cm in girth number 1,000 to 1,250 stems per ha, but individual
trees are slender and none exceed 90 cm. The most abundant species include
*Tristania obovata, T.* aff. *maingayi,* Parastemon spicatum, Palaquium cochleariifolium, and *Dactylocladus stenostachys.* Herbaceous vegetation is virtually absent.

*Phasic community 6:* Combretocarpus-Dactylocladus *association* (padang
keruntum). This community appears in the final stage of development of the
peat dome. It is distributed over a wide area of deep peat on the Baram River
upstream from Marudi on the middle reach. The aspect of the vegetation is xerophytic and open, with stunted trees. *Combretocarpus rotundatus* is the only tree
species with a girth exceeding 90 cm, though it is rarely more than 12 m high.
*Dactylocladus stenostachys, Litsea palustris, Garcinia rostrata,* and others are common
but attain the height only of a shrub. Myrmecophytes and *Nepenthes* are particularly common. *Thorachostachyum bancanum* and *Pandanus ridleyi* are abundant on
the ground surface, and in such places sphagnum moss *(Sphagnum junghuh-
nianum)* is found.

The only species common to all six phasic communities is *Dactylocladus stenostachys,* which has a girth of 2.7–3.0 m in community 1 and appears as a small
tree about 3 m tall in community 6. In general, the number of species in the
swamp forest tends to decline with the development of the peat swamp, while
individuals tend to become smaller and grow more densely.

## Main Species and Genera and Species Numbers

Anderson (1963) collected specimens of 1,706 species, including 1,528 dicotyledons, 106 monocotyledons, 6 conifers, and 66 pteridophytes. The record includes 242 tree species, of which 38 are small trees of the understory with girths
of less than 30 cm. Sewandono (1938) estimated that the swamps of east Sumatra
contain fewer than 100 species of trees, whereas in one forest type alone in Sarawak and Brunei as many as 75 species were found in an area of 4,046 m². Even

## Table 1-18
### Major Families of the Peat Swamp Forests in Sarawak and Brunei, Showing Numbers of Genera and Species, and Vegetation Types

| FAMILY | GENERA (species) | TREES GIRTH (inches) >12 | TREES GIRTH (inches) <12 | SHRUBS | HERBS | CLIMBERS | EPIPHYTES |
|---|---|---|---|---|---|---|---|
| **Gymnosperms** | | | | | | | |
| Coniferae | 2 (2) | 2 (2) | | | | | |
| Gnetaceae | 1 (1) | | | | | 1 (1) | |
| **Angiosperms** | | | | | | | |
| **Dicotyledons** | | | | | | | |
| Dilleniaceae | 2 (2) | 1 (1) | | | | 1 (1) | |
| Magnoliaceae | 1 (1) | 1 (1) | | | | | |
| Annonaceae | 12 (17) | 5 (9) | 3 (3) | | | 5 (5) | |
| Menispermaceae | 2 (2) | | | | | 1 (1) | 1 (1) |
| Polygalaceae | 2 (4) | 2 (4) | | | | | |
| Hypericacea | 1 (2) | 1 (2) | | | | | |
| Flacouriaceae | 2 (2) | 1 (1) | 1 (1) | | | | |
| Guttiferae | 3 (16) | 3 (16) | | | | | |
| Ternstroemiaceae | 2 (4) | 2 (4) | | | | | |
| Dipterocarpaceae | 7 (15) | 7 (15) | | | | | |
| Malvaceae | 2 (2) | 2 (2) | | | | | |
| Sterculiaceae | 2 (4) | 2 (4) | | | | | |
| Tiliaceae | 2 (5) | 1 (2) | 2 (3) | | | | |
| Linaceae | 2 (2) | 2 (2) | | | | | |
| Rutaceae | 2 (3) | 1 (2) | 1 (1) | | | | |
| Simaroubaceae | 2 (2) | 1 (1) | 1 (1) | | | | |
| Ochnaceae | 3 (4) | 2 (2) | | 1 (2) | | | |
| Burseraceae | 2 (6) | 2 (6) | | | | | |
| Meliaceae | 3 (3) | 2 (2) | 1 (1) | | | | |
| Olacaceca | 4 (5) | 4 (5) | | | | | |
| Ilicinaceae | 1 (2) | 1 (2) | | | | | |
| Celastraceae | 3 (4) | 3 (4) | | | | | |
| Rhamnacecae | 1 (1) | | | | | 1 (1) | |
| Ampelidaceae | 3 (3) | | | | | 3 (3) | |
| Sapindaceae | 3 (4) | 3 (3) | 1 (1) | | | | |
| Anacardiaceae | 7 (11) | 7 (11) | | | | | |
| Connaraceae | 3 (3) | | 1 (1) | | | 2 (2) | |
| Leguminosae | 8 (8) | 6 (6) | 1 (1) | | | 1 (1) | |

*(continued on next page)*

Table 1-18   *(continued)*

| Family | Genera (species) | Trees Girth (inches) >12 | Trees Girth (inches) <12 | Shrubs | Herbs | Climbers | Epiphytes |
|---|---|---|---|---|---|---|---|
| Rosaceae | 2 (3) | 2 (3) | | | | | |
| Rhizophoraceae | 2 (2) | 2 (2) | | | | | |
| Myrtaceae | 3 (16) | 3 (15) | 1 (1) | | | | |
| Melastomataceae | 4 (8) | 2 (2) | | 1 (1) | | 1 (2) | 1 (3) |
| Curcurbitaceae | 1 (1) | | | | | 1 (1) | |
| Araliaceae | 2 (4) | 1 (1) | 1 (1) | 1 (1) | | | 1 (1) |
| Cornaceae | 1 (1) | 1 (1) | | | | | |
| Rubiaceae | 18 (22) | 6 (6) | 5 (5) | 1 (1) | 1 (1) | 5 (7) | 2 (2) |
| Ericaceae | 1 (1) | | | 1 (1) | | | |
| Myrsinaceae | 5 (11) | 1 (1) | 2 (3) | 3 (4) | | 2 (3) | |
| Sapotaceae | 4 (12) | 4 (12) | | | | | |
| Ebenaceae | 1 (4) | 1 (4) | | | | | |
| Oleaceae | 2 (4) | 1 (3) | | | | 1 (1) | |
| Apocynaceae | 5 (6) | 2 (2) | 1 (1) | | | 2 (3) | |
| Aslepiadaceae | 2 (6) | | | | | | 2 (6) |
| Loganiaceae | 1 (2) | | | 1 (2) | | | |
| Convolvulaceae | 1 (2) | | | | | 1 (2) | |
| Gesneriaceae | 1 (1) | | | | | 1 (1) | |
| Verbenaceae | 3 (3) | 1 (1) | | | 1 (1) | 1 (1) | |
| Nepenthaceae | 1 (5) | | | 1 (4) | | 1 (1) | |
| Piperaceae | 1 (4) | | | | 1 (1) | 1 (3) | |
| Myristicaceae | 4 (7) | 4 (5) | 2 (2) | | | | |
| Lauraceae | 9 (18) | 6 (14) | 3 (4) | | | | |
| Thymelaceae | 2 (4) | 1 (3) | | | | 1 (1) | |
| Loranthaceae | 3 (3) | | | | | | 3 (3) |
| Santalaceae | 1 (2) | | | 1 (1) | | | 1 (1) |
| Euphorbiaceae | 11 (17) | 7 (12) | 4 (4) | | 1 (1) | | |
| Urticaceae | 1 (2) | | | | | | 1 (2) |
| Moraceae | 3 (27) | 2 (3) | | 1 (1) | | 1 (19) | 1 (4) |
| Casuarinaceae | 1 (1) | 1 (1) | | | | | |
| Fagaceae | 2 (5) | 2 (5) | | | | | |
| **Monocotyledons** | | | | | | | |
| Orchidaceae | 12 (17) | | | | 3 (3) | | 11 (14) |
| Zingiberaceae | 2 (2) | | | | 1 (1) | | 1 (1) |
| Dioscoreaceae | 1 (1) | | | | | 1 (1) | |

*(continued on next page)*

Table 1-18    *(continued)*

| FAMILY | GENERA (species) | TREES GIRTH (inches) >12 | <12 | SHRUBS | HERBS | CLIMBERS | EPIPHYTES |
|---|---|---|---|---|---|---|---|
| Liliaceae | 1 (1) | | | 1 (1) | | | |
| Flagellariaceae | 2 (2) | | | | 1 (1) | 1 (1) | |
| Palmae | 7 (7) | 1 (1) | | 3 (3) | | 3 (3) | |
| Pandanaceae | 1 (4) | | | 1 (3) | | 1 (1) | |
| Araceae | 9 (9) | | | | 7 (7) | 2 (2) | |
| Cyperaceae | 2 (2) | | | | 2 (2) | | |

SOURCE: Anderson, 1963.

under peat swamp forest conditions, the species composition of the forest can be described as considerably complex.

Table 1-18 shows the numbers of genera and species in each of the families found in the peat swamp forests. With the exception of four families (Combretaceae, Lythraceae, Styracaceae, and Proteaceae), the predominant families of lowland dipterocarp forest are also represented in peat swamp forest. These include Dipterocarpaceae, Anacardiaceae, Annonaceae, Euphorbiaceae, Guttiferae, Lauraceae, Leguminosae, Myrtaceae, Rubiaceae, and Sapotaceae.

In the herbaceous vegetation, such aquatic dicotyledons as Nymphaeaceae, *Limnanthemum, Jussiaea,* and *Ludwigia* are absent. A possible reason is the extremely acidic and anaerobic conditions of the peat swamp soils. Also absent are the calciphyllous families such as Balsaminaceae *(Impatiens),* Acanthaceae, Scrophulariaceae, Begoniaceae, and Gesneriaceae. The only recorded example is of *Aeschynanthus hians* (Gesneriaceae). Araceae and Cyperaceae predominate.

Climbing plants are represented by many families, of which Annonaceae, Rubiaceae, and Moraceae are notable. Species of Capparidaceae, Malpighiaceae, Solanaceae, and Acanthaceae have not been recorded. Convolvulaceae is represented by only two species, neither of which is common.

## Stratification and Species

Stratification, as we have seen, is most developed in mixed swamp forest and becomes increasingly simplified as succession progresses. Strictly speaking, a more detailed examination is required, but table 1-19 lists the more widely distributed and characteristic peat swamp species according to canopy classes.

## Table 1-19
## Main Components of Peat Swamp Forests in Sarawak and Brunei

Tree species

(i) Upper story (60 in. girth and over):

| | |
|---|---|
| *Alstonia spathulata* | *Gonystylus maingayi* |
| *Anisoptera marginata* | *Koompassia malaccensis* |
| *Artocarpus rigidus* | *Litsea palustris* |
| *Calophyllum retusum* | *Lophopetalum multinervium* |
| *Campnosperma coriacea* | *Melanorrhoea beccarii* |
| *Casuarina* sp. nov. | *Melanorrhoea tricolor* |
| *Combretocarpus rotundatus* | *Mezzettia leptopoda* |
| *Copaifera palustris* | *Parartocarpus venenosus* ssp. *forbesii* |
| *Cratoxylum arborescens* | *Parastemon urophyllum* |
| *Cratoxylum glaucum* | *Parishia sericea* |
| *Dacrydium beccarii* var. *subelatum* | *Planchonella maingayi* |
| *Dactylocladus stenostachys* | *Shorea albida* |
| *Dillenia pulchella* | *Shorea inaequilateralis* |
| *Dipterocarpus coriaceus* | *Shorea platycarpa* |
| *Dryobalanops rappa* | *Shorea rugosa* var. *uliginosa* |
| *Durio carinatus* | *Shorea scabrida* |
| *Dyera lowii* | *Shorea teysmanniana* |
| *Ganua motleyana* | *Swintonia glauca* |
| *Gonystylus bancanus* | *Tetramerista glabra* |

(ii) Middle story (26–60 in. girth):

| | |
|---|---|
| *Alangium havilandii* | *Cotylelobium fuscum* |
| *Alseodaphne insignis* | *Ctenolophon parvifolius* |
| *Alseodaphne rigida* | *Dacryodes incurvata* |
| *Amoora rubiginosa* | *Dacryodes macrocarpum* var. |
| *Aromadendron nutans* | *maacrocarpum* |
| *Arthrophyllum rubiginosum* | *Dialium laurinum* |
| *Bhesa paniculata* | *Diospyros evena* |
| *Blumeodendron subrotundifolium* | *Diospyros maingayi* |
| *Blumeodendron tokbrai* | *Diospyros pseudomalabarica* |
| *Calophyllum fragrans* | *Elaeocarpus obtusifolius* |
| *Calophyllum obliquinervum* | *Eugenia christmannii* |
| *Calophyllum sclerophyllum* | *Eugenia havilandii* |

*(continued on next page)*

Table 1-19   *(continued)*

| | |
|---|---|
| *Eugenia incarnata* | *Nepheliium maingayi* |
| *Eugenia nemestrina* | *Palaquium cochleariifolium* |
| *Eugenia spicata* | *Palaquium pseucocuneatum* |
| *Ganua coriacea* | *Palaquium ridleyi* |
| *Ganua pierrei* | *Palaquium walsuraefolium* |
| *Garcinia havilandii* | *Parastemon spicatum* |
| *Garcinia vidua* | *Parkia singularis* |
| *Gardenia pterocalyx* | *Platea excelsa* |
| *Goniothalamus andersonii* | *Polyalthia glauca* |
| *Gonystylus forbesii* | *Pometia pinnata* f. *acuminata* |
| *Horsfieldia crassifolia* | *Santiria laeviata* |
| *Jackia ornata* | *Santiria rubiginosa* var. *rubiginosa* |
| *Kokoona ovato-lanceolata* | *Santiria tomentosa* |
| *Lithocarpus wenzigianus* | *Sindora leiocarpa* |
| *Litsea cylindrocarpa* | *Stemonurus scorpioides* |
| *Litsea gracilipes* | *Tristania* aff. *maingayi* |
| *Litsea grandis* | *Tristania grandifolia* |
| *Litsea nidularis* | *Tristania obovata* |
| *Longetia malayana* | *Vatica mangachapoi* |
| *Macaranga caladifolia* | *Xerospermum muricatum* |
| *Mussaendopsis beccariana* | *Xylopia coriifolia* |

(iii) Lower story (12–24 in. girth):

| | |
|---|---|
| *Antidesma coriaceum* | *Gomphandra comosa* |
| *Baccaurea bracteata* | *Goniothalamus malayanus* |
| *Brackenridgea hookeri* | *Gymnacranthera eugeniifolia* |
| *Canthium didymum* | var. *griffithii* |
| *Carallia brachiata* | *Ilex hypoglauca* |
| *Cephalomappa paludicola* | *Ilex sclerophylloides* |
| *Cryptocarya griffithiana* | *Knema kunstleri* var. *kunstleri* |
| *Cyathocalyx biovulatus* | *Lithocarpus* sp. |
| *Dillenia pulchella* var. | *Lithocarpus sundaicus* |
| *Diospyros elliptifolia* | *Litsea resinosa* |
| *Eugenia cerina* | *Mezzettia umbellata* |
| *Garcinia rostrata* | *Nauclea parva* |
| *Garcinia tetrandra* | *Neoscortechinia kingii* |
| *Glochidion lucidum* | *Pithecellobium borneense* |

*(continued on next page)*

Table 1-19 *(continued)*

*Polyalthia hypoleuca*
*Pygeum parviflorum*
*Samadera indica*
*Stemonurus umbellatus*
*Sterculia rhoidifolia*

*Ternstroemia hosei*
*Ternstroemia magnifica*
*Tetractomia holttumii*
*Timonius peduncularis*
*Xanthophyllum* aff. *citrifolium*

(iv) Understory (less than 12 in. girth):

*Antidesma phanerophlebium*
*Ardisia copelandii*
*Canthium umbellatum*
*Chisochetum brachyanthus*
*Cyrtostachys lacca*
*Dehaasia* sp.

*Disepalum anomalum*
*Gaertnera borneensis*
*Ixora pyrantha*
*Polyalthia* sp.
*Tarenna fragans*

Shrubs

*Daemonorops longipes*
*Euthemis leucocarpa*
*Euthemis obtusifolius*
*Fagraea litoralis*
*Fagraea racemosa*
*Ficus deltoidea* var. *motleyana*
*Hanguana malayana*
*Iguanura* sp.
*Labisia punctata* f. *pumila*
*Labisia punctata* f. *punctata*
*Medinilla hasseltii*

*Nepenthes bicalcarata*
*Nepenthes gracilis*
*Nepenthes rafflesiana*
*Pandanus andersonii*
*Pandanus brevifolius*
*Pandanus ridleyi*
*Pinanga* sp.
*Plemele cantleyi*
*Schefflera subulata*
*Zalacca conferta*

Herbs

*Aglaonema pictum*
*Alocasia beccarii*
*Alocasia longiloba*
*Argostemma psychotrioides*
*Bromheadia finlaysoniana*
*Clerodendron fistulosum*
*Cryptocoryne pallidinervia*
*Crytosperma lasiodes*

*Cystorchis variegata*
*Globba panicoides*
*Homalomena rostrata*
*Piper muricatum*
*Podolasia stipitata*
*Thorachostachyum bancanum*
*Zeuxine violascens*

SOURCE: Anderson, 1963.

UPPER STORY

In the upper story, Dipterocarpaceae species predominate. Of the six species of *Shorea,* the most remarkable is *Shorea albida,* whether in size, density, or area of distribution. Other than *Shorea,* the timber-producing species *Gonystylus bancanus, Dactylocladus stenostachys,* and *Copaifera palustris* are important. These species account for 80 percent of the upper story in mixed swamp forest and in the peripheral zones of swamps.

MIDDLE STORY

The middle and lower stories contain many species of Lauraceae, Euphorbiaceae, Guttiferae, Burseraceae, Ebenaceae, Fagaceae, and Annonaceae. Some of the more abundant species in the middle story are *Alangium havilandii, Blumeodendron tokbrai, Ctenolophon parvifolius, Diospyros evena, D. pseudomalabarica, Kokoona ovatolanceolata, Palaquium cochleariifolium, Parastemon spicatum,* and *Xylopia coriifolia.*

LOWER STORY

The most widely distributed species in the lower story of mixed swamp forest are *Neoscortechinia kingii, Cyathocalyx biovulatus,* and *Stemonurus umbellatus. Tetractomia holttumii* and *Cephalomappa paludicola* grow under pure stands of *Shorea albida.*

UNDERSTORY

*Ixora pyrantha* is abundant in *Shorea albida* forest, and *Tarenna fragrans* in mixed swamp forest.

**Shrubs.** Shrubby dicotyledons are rare, but in the open forest in the center of swamps may be found *Ficus deltoidea* var. *motleyana, Euthemis obtusifolius, Labisia punctata* f. *punctata,* and *Medinilla hasseltii.* Species of *Nepenthes* exhibit a climbing habit only when the forest canopy is open and may be more properly regarded as shrubs. Of the palms and pandans, the stemless spiny palm *Salacca conferta* is found particularly in mixed forest on shallow peat, while *Pandanus andersonii* forms large, dense thickets in *Shorea albida* forest. Other pandans include *Pandanus brevifolius* in mixed swamp forest and *P. ridleyi* in stunted forest in the center of swamps on the Baram River.

**Herbs.** Herbaceous dicotyledons are virtually absent in the understory. *Argostemma psychotrioides* is frequent in mixed swamp forest, and the myrmecophyte *Clerodendron fistulosum,* though not common, has a wide distribution. Aroids and sedges predominate, of which the most abundant and widespread species is *Thorachostachyum bancanum.* Most of the Araceae, including *Aglaonema pictum, Homalomena rostrata, Alocasia longiloba,* and *A. beccarii,* occur in perennially damp localities where the water table is high, and the aquatic aroid *Cryptocoryne pallidinervia* is abundant near streams and in permanently wet land. Conversely, this family is not found in the drier central areas of swamps. The large herb *Hanguana mala-yana* also favors the damp and is often found in holes where trees have been uprooted by the wind. The terrestrial orchids *Zeuxine violascens* and *Cystorchis variegata,* with attractive variegated leaves, occur on leaf litter in mixed swamp forest.

## Epiphytes and Climbers

In the tropics, plants often change their habit according to conditions. In the peat swamp forest, too, some species, such as *Ficus* spp., *Poikilospermum* spp., and *Pycnarrhena borneensis,* start life as epiphytes but develop a climbing habit once their roots have reached the swamp surface. *Fagraea littoralis* and *Randia* sp. begin as epiphytes but later become independent shrubs. *Bulbophyllum beccarii,* which is epiphytic on *Shorea albida,* is said to start as a climber from the ground surface and later to become an epiphyte spiraling around the upper boles of trees. In addition, *Ficus deltoidea* var. *borneensis* and *Dischidia nummularia* are normally epiphytic in tree crowns, but in the open forest in the center of swamps they are terrestrial. Such phenomena are not confined to the swamp forest; a species of *Vaccinium* in the montane forest of Java, for example, is a climber at low altitude but a small tree at higher altitude. Species other than those mentioned above can almost all be classified according to habit, as shown in table 1-20.

### SUN EPIPHYTES

These plants are particularly abundant in mixed forest in the crowns of *Gony-stylus bancanus* and *Dactylocladus stenostachys,* both of which have soft, fibrous bark to which they readily attach, and orchids are common among them, including *Bulbophyllum vaginatum, Dendrobium merrillii, Eria pannea,* and *E.* aff. *pulchella.* They are scarce in the crowns of *Shorea albida.* The *Dischidia* spp., *Hydnophytum formicarum,* and *Myrmecodia tuberosa* are largely confined to stunted forest where conditions are similar to open heath forest, in which these species also occur.

### SHADE EPIPHYTES

The representative species are *Medinilla laxiflora* and the orchid *Liparis lace-rata.* Both occur as epiphytes on small trees in dense forest at heights of 0.5–3.0 m aboveground.

### LARGE CLIMBERS

Lianes and climbers are generally less abundant in peat swamp forest than in lowland dipterocarp forest, with fewer species and smaller sizes. Lianes are rare in

## Table 1-20
## Epiphytes and Climbers in Peat Swamp Forests of Sarawak and Brunei

Epiphytes

  (i) Sun epiphytes:

| | |
|---|---|
| *Adenoncos sumatrana* | *Eria pannea* |
| *Bulbophyllum vaginatum* | *Eria pulchella* |
| *Dendrobium* aff. *merrillii* | *Ficus deltoidea* var. *deltoidea* |
| *Dischidia hirsuta* | *Grammatophyllum speciosum* |
| *Dischidia nummularia* | *Hoya coronaria* |
| *Dischidia rafflesiana* | *Hydnophytum formicarum* |
| *Eria obliqua* | *Myrmecodia tuberosa* |

  (ii) Shade epiphytes:

| | |
|---|---|
| *Appendicula pendula* | *Liparis lacerata* |
| *Bulbophyllum beccarii* | *Medililla laxiflora* |
| *Dendrobium cumulatum* | *Pogonanthera pulverulenta* |
| *Dipodium pictum* | *Pycnarrhena borneensis* |
| *Eulophia squalida* | |

Climbers

  (i) Large climbers, frequently attaining crowns of upper-canopy trees, including ground-rooting epiphytic and strangling figs:

| | |
|---|---|
| *Calamus* sp. | *Grenacharia beccariana* |
| *Erycibe impressa* | *Korthalsia rigida* |
| *Fibraurea chloroleuca* | *Mitrella dielsii* |
| *Ficus acamptophylla* | *Piper arborescens* |
| *Ficus consociata* | *Plectocomiopsis wrayi* |
| *Ficus crassiramea* | *Rourea mimosoides* f. *mimosoides* |
| *Ficus spathulifolia* | *Tetrastigma* sp. |
| *Ficus sundaica* | *Uncaria ovalifolia* |
| *Ficus sundaica* var. *beccariana* | *Willughbeia glaucina* |
| *Ficus xylophylla* | *Zizyphus suluensis* |

*(continued on next page)*

## Table 1-20    *(continued)*

(ii) Small climbers, usually confined to understory:

| | |
|---|---|
| *Aeschynanthus hians* | *Linostoma longiflorum* |
| *Ampeslocissus thyrsiflora* | *Lucinaea morinda* |
| *Connarus semidecandrus* | *Medinilla scandens* |
| *Epipremnopsis media* | *Nepenthes albomarginata* |
| *Ficus callicarpides* | *Nepenthes ampullaria* |
| *Ficus villosa* | *Pandanus* sp. |
| *Flagellaria indica* | *Psychotria sarmentosa* |
| *Gnetum neglectum* | *Rhaphidophora lobbii* |
| *Lecananthus erubescens* | |

SOURCE: Anderson, 1963.

*Shorea albida* forest and the central forest types. Rattan palms, though not rare, are small and of no commercial value. Of the three recorded species, *Plectocomiopsis wrayi* and *Korthalsia rigida* are common, the latter often in association with *Shorea albida.*

The commonest of the larger lianes is *Uncaria ovalifolia,* the sap of which is drunk. It is found in young secondary forest following exploitation, where it tends to form thickets. Other common lianes are *Willughbeia glaucina, Fibraurea chloroleuca,* and *Mitrella dielsii.*

*Ficus* species are particularly abundant in peat swamp forest, with 24 recorded species. Except for *Ficus crassiramea,* which is confined to shallow peat near the coast, true strangling figs are rare. Others start life as epiphytes and are partial stranglers that on occasion kill the host tree, and may be regarded as ground-rooting epiphytes. Common species include *Ficus acamptophylla, F. consociata, F. spathulifolia, F. xylophylla,* and *F. sundaica,* and these are particularly common on *Gonystylus* and *Dactylocladus* in mixed swamp forest. The small *Ficus callicarpides* is abundant on buttresses of *Shorea albida.*

### Small Climbers

The most common include *Lecananthus erubescens, Lucinaea morinda, Medinilla scandens, Aeschynanthus hians,* and *Rhaphidophora lobbii.* In peat swamp forest, *Nepenthes ampullaria* has a climbing habit, but in heath forest, it is terrestrial. The small twining climber *Gnetum neglectum,* rarely higher than 4.5 m, is found throughout the Rejang Delta and Maludam Peninsula but is never common.

MYRMECOPHYTES

These include two trees, *Macaranga caladifolia* and *M. puncticulata*. The former is found in primary forest, the latter in secondary forest, particularly along railway lines and on shallow peat cleared for cultivation, where it often forms almost pure stands. The herb *Clerodendron fistulosum* occurs in natural forest. The epiphytes *Dischidia nummularia, D. rafflesiana, Myrmecodia tuberosa,* and *Hydnophytum formicarum* occur in open stunted forest.

PARASITES

Parasitic vegetation is generally rare in peat swamp forest. *Lepidaria oviceps* is conspicuous in the crowns of upper-story trees, whereas *Macrosolen beccarii* occurs as a parasite of the middle- or lower-story trees in the shade and can be quite abundant locally. The parasitic shrub *Henslowia varians* is found in open stunted forest.

## Ferns

TERRESTRIAL FERNS

Terrestrial ferns are few. *Vittaria elongata, Schizoloma coriaceum,* and *Syngramma lobbiana* are typical, occurring on roots and pneumatophores of mixed swamp forest. The stemless tree fern *Cyathea glabra* is locally abundant in *Shorea albida* forest. *Schizaea malaccana,* usually a montane species, occurs in stunted forest in association with *Lycopodium cernuum. Ophioglossum intermedium* has been reported from only one locality in the Rejang Delta. *Nephrolepis biserrata* is rare in primary forest but forms dense thickets in completely open secondary forest, especially where the slash has been burned.

EPIPHYTIC FERNS

Because epiphytic ferns have stronger tolerance to light conditions than flowering plants, they are more difficult to classify into crown and shaded epiphytes. Among those found in gardens and rubber plantations, as well as in peat swamp, the light-demanding species include *Asplenium nidus* and *A. phyllitidis,* while the shade-bearing species include *Pyrrosia longifolia* and *Paragramma longifolia. Humata angustata* and *H. parvula* show a variety of habits, being found as either crown or shade epiphytes, and occasionally they are terrestrial even in the central forest types. The most typical shade epiphyte is *Lycopodium phlegmaria* var. *divaricatum,* which occupies a similar habitat to *Medinilla laxiflora* and *Liparis lacerata.* Only two climbing ferns are recorded: *Stenochlaena palustris,* found on the lower trunks of many trees on the shallow peat near the coast but becomes progressively rarer inland, and *Teratophyllum ludens,* seen in relatively dense shade in the understory (table 1-21).

## Table 1-21
## Ferns in the Peat Swamp Forests of Sarawak and Brunei

Aboveground ferns

| | |
|---|---|
| *Asplenium longissimum* | *Nephrolepsis biserrata* |
| *Cyathea glabra* | *Schizaea malaccana* |
| *Lindsaya scandens* var. | *Schizoloma coriaceum* |
|    terrestris | *Syngramma lobbiana* |
| *Lycopodium cernuum* | *Vittaria elongata* |

Epiphytic ferns

(i) Positive ferns

| | |
|---|---|
| *Asplenium nidus* | *Lecanopteris sinuosa* |
| *Asplenium phyllitidis* | *Photinopteris speciosa* |
| *Crypsinus albidopaleatus* | *Phymatodes crustacea* |
| *Drynaria involuta* | *Platycerium coronarium* |

(ii) Negative ferns

| | |
|---|---|
| *Asplenium glaucophyllum* | *Lycopodium pinifolium* |
| *Asplenium tenerum* | *Paragramma longifolia* |
| *Crypsinus albidopaleatus* | *Polypodium verrucosum* |
| *Humata angustata* | *Pyrrosia longifolia* |
| *Humata parvula* | *Selliguea heterocarpa* |
| *Lycopodium phlegmaria* var. | *Vittaria ensiformis* |
|    divaricatum | *Vittaria hirta* |

Climbing ferns

| | |
|---|---|
| *Stenochlaena palustris* | *Teratophyllum ludens* |

SOURCE: Anderson, 1963.

## New Species

Anderson's study represented the first full-scale survey of the peat swamp forest, and it uncovered many new species. The most significant discovery was *Litsea palustris* Kostermans. This species was found in abundance in the center of swamps in Sarawak and Brunei, and in the center of the Rejang Delta a pure stand with 300 stems per ha covered a wide area. The fact that such a new species had not until then been collected shows how limited the surveying of peat swamp had been.

Other new species include *Cephalomappa paludicola* Airy-Shaw, *Pandanus andersonii* H. St. John, *Knema uliginosa* Sinclair, *Goniothalamus* sp., *Polyalthia* sp., *Ficus callicarpides*, *F. spathulifolia*, *F. supperforata*, *Parishia*, *Xanthophyllum*, *Piper*, and *Cinnamomum*. The new genus *Jarandersonia* in the family Tiliaceae was also discovered.

## Shorea albida

As has been mentioned several times, *Shorea albida* is the characteristic species of the peat swamp forests of Sarawak and Brunei. Its distribution extends from the vicinity of Pontianak to Brunei, and besides growing mainly in peat swamps, it also occurs on overlying a white sand plateau at 60 m in Sarawak, at elevations of 1,200 m in northeast Sarawak, and in *kerangas* forest (Ashton, 1964b; Browne, 1955). For a tropical tree species, it is extremely rare in forming pure, even-aged stands covering several hundred square kilometers.

The tree occurs in various sizes, from small to extremely large. It is generally taller at the margins of swamps and smaller toward the center. The largest specimens reach 70 m, with a diameter of 2 m. It has a straight, cylindrical trunk, large, thick buttresses 5 m high and 3.5 m long, and surface roots spreading over the ground surface. The crown is a large, open dome, flat on top. The bark is a purplish brown, often whitening to a grayish pink. Deep fissures reach 2 m long; the surrounding wood is pink, and the heartwood is reddish brown.

*Shorea albida* has been reported to be susceptible to insects and lightning (Anderson, 1961a, 1964b; Brunig, 1964). The first report of insect damage concerned forest along the Baram River in 1953. This was clearly discernible from the air and covered an estimated 75,000 ha. This was followed in 1955 by an area of damage along the Puran River, estimated to have occurred ten to fifteen years earlier, where fallen and decaying trunks and roots were found but no sign whatsoever of the regeneration of *Shorea albida*. In the understory, *Tetractomia holttumii* and *Quercus sundaicus* of the 30–60-cm girth class were predominant. In a different area where damage was estimated to have occurred twenty-five to thirty years earlier, again no regeneration of *Shorea albida* was found. The understory contained many trees of the 90–120-cm and 120–150-cm girth classes, of which *Diospyros ferrea* var. *buxifolia* was dominant. Tree heights were in the range of 25–30 m, lower than the surrounding *Shorea albida* forest by 20 m.

In a recently damaged section of forest in the Badas district, all species had survived except for *Shorea albida,* which again showed no signs of regeneration. In a forest in the Maludam Peninsula known to have suffered damage for six months during 1948, the damage is estimated to have extended to 1,542,000 trees in a total area of 62,500 ha. Here, again, other species were not defoliated, but partial regeneration of *Shorea albida* was found. The undamaged sections of forest were dominated by *Gonystylus bancanus* and *Tetractomia holttumii*.

This defoliation was at first thought to be due to oil-related pollution, but the fact that damage did not extend over all swamps and that other species were unaffected pointed to a biological cause. The most likely possibility was an infestation of the hairy caterpillar known as *ulat bulu.* These caterpillars are 2.5–4.0 cm long and the thickness of a pencil, with black bodies covered with reddish brown, nonpoisonous, nonirritant hairs. The infestation was at its height during the rice-harvesting season, in March and April, when millions of larvae are reported to have covered the ground.

Although the insect responsible has not been identified, the fact that infestations have increased since 1950 suggest that it may not be indigenous. Defoliation has been observed in other species, but the trees have not died.

In the tropics, thunderstorms are more frequent than in the temperate zones. In Kuching, for example, between 1955 and 1961 thunderstorms occurred an average of 109 days per year, and thunder was heard on 163 days. Areas damaged by thunderstorms are circular, the largest being 80–100 m in diameter, about 6,000 m². The damage spreads in concentric circles, and in the center where lightning has struck, it is not limited to the canopy layer but extends to the smaller trees, which are also killed. Slightly away from the center, the canopy trees are killed but the smaller ones suffer lesser damage. Farther away, damage is limited to the canopy trees but is not fatal.

By forest type, the *Shorea albida* consociation sustains the greatest damage. This is thought to be because the forest canopy is flat, with an even distribution of a single species. A single bolt of lightning can kill up to fifty trees. *Rhizophora* mangrove forest is susceptible to the same kind of damage. Common features are a flat canopy of a single species of even age, level ground, and a high water table; but any causal relationship between these factors and lightning strikes is unclear.

In mixed forest, on the other hand, damage is limited to giant trees and a small number of middle- and lower-story trees around them. Though it tends to be overlooked, lightning damage is also common in *kerangas* forest and lowland dipterocarp forest. Abundant regeneration is often seen in areas struck by lightning in *Shorea albida* forest, with 11.1 saplings per 4 m² having been recorded.

A further example of climatic damage is by wind. The *Shorea albida–Litsea–*

*Parastemon* association is particularly susceptible, and damage covering 625 ha has been recorded. Large numbers of trees with small crowns and slender trunks were flattened. *Shorea albida,* however, can recover from wind damage, and evidence of wind damage has been found on close examination of old trees. In one locality on the Baram River, seven *Shorea albida* trees had been windthrown in a swathe 170 m long and 15 m wide. Abundant regeneration was evident, with 3,000 saplings in an area of 2,000 m². Seedlings were also seen growing on the trunks of the fallen trees. There appear to be multiple examples of successful regeneration when a seed year follows wind damage.

Seen from the air, the results of the insect damage described above are huge circular areas in the midst of an even forest canopy (fig. 1-14), while areas damaged by lightning are readily distinguishable as patches of dead trunks and branches. *Shorea albida* is the most important species of dark red *meranti* in Brunei, but the problem of its regeneration remains unsolved. As mentioned, regeneration in damaged areas was patchy; and while conclusive evidence must await future surveys, it has been demonstrated that under conditions of perfect sunlight, regeneration proceeds extremely favorably and growth is extremely

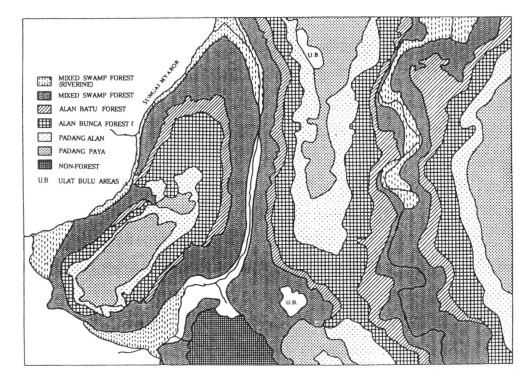

Figure 1-14. Forest types at Beluru Forest Reserve, Sarawak, and areas defoliated by *ulat bulu.* SOURCE: Lee, 1979.

rapid. Heavy fruiting is said to occur once every twenty-five years; and in 1955, 81,600 saplings per ha were observed in the *Shorea albida* consociation of phasic community 3, and 64,800 in the forest of phasic community 2. In phasic community 4, where regeneration is by vegetative propagation, the volume of timber per unit area of saplings and poles of *Shorea albida* is extremely high. Leo and Lee (1971) estimated the timber volume of *Shorea albida* in Sarawak to be $8 \times 10^6$ metric tons.

In its timber quality, *Shorea albida* has various shortcomings: It may show cavities or brittle heart, it has a tendency to split, and it does not readily take up preservative. It is easy to process and dry, though, and is widely used for beams, pillars, floorboards, and other heavy structural timbers. Above all, it is attractive for the size of its stock. This species should be exploited following the establishment of a method of regeneration based on further experimentation and scientific study of its properties in the field.

## *Peat Swamp Forests of Southeast Asia*
### The Malay Peninsula

Peat swamps are found in the coastal zones of the Malay Peninsula from the center to the south. Peat lies immediately inland from the coast and is at most about 6 m deep. As in Sarawak and Brunei, it exists under anaerobic conditions and is oligotrophic and highly acidic. At first sight, the water appears black, but when held up to the light, it displays the dark reddish brown color characteristic of peat swamp. On part of the west coast it overlies clay, on the east coast, sand.

The forest has a canopy height of 30 m and its density varies, but in everything it is smaller than lowland dipterocarp forest. Often it forms an extremely open, discontinuous canopy. No trees are seen projecting above it. The middle layer reaches 6–18 m and is often tightly closed. The shrub layer varies in density, the densest parts containing abundant stemless palms. Ground vegetation is comparatively poor (Whitmore, 1975).

A forest structure of concentric zones can be found only in Hutan Melintang in South Perak, where *Shorea uliginosa* is represented by 4.8 stems per ha at the periphery and 30–51 stems toward the center (Wyatt-Smith, 1963). The number of species is limited, comprising mainly the following:

*Trees: Amoora rubiginosa, Anisoptera marginata, Blumeodendron tokbrai, Calophyllum retusum, C. scriblififolium, Cratoxylum arborescens, Ctenolophon parvifolius, Dialium patens, Durio carinatus, Eugenia* spp., *Ganua motleyana,* Gonystylus bancanus, Koompassia malaccensis, Litsea grandis, Myristica lowiana, Neesia altissima, Palaquium ridleyi, Parastemon urophyllum, Polyalthia glauca, Santiria nana, Shorea platycarpa, S. uliginosa, S. teysmanniana, Stemonurus capitatus, Tetramerista glabra, and *Xylopia fusca.*

Others: *Cyrtostachys lakka, Mapania palustris, Nepenthes* spp., *Pandanus arto-carpoides, Salacca conferta,* and *Stenochlaena palustris.* (Wyatt-Smith, 1961)

The flora of the east and west coasts differ, and much remains unknown about the details of species distribution. Commercially important timbers include *Gonystylus bancanus, Calophyllum, Tetramerista, Cratoxylum arborescens, Koompassia malaccensis,* and *Durio carinatus,* and members of the mixed swamp forest type of phasic community 1 of Sarawak and Brunei are good producers. Natural regeneration is generally good, but regeneration of *Gonystylus bancanus* and *Tetramerista glabra* is not seen in natural forest. Regenerated forest is dominated by fast-growing species of *Shorea* and by *Cratoxylum arborescens. Koompassia malaccensis* is a large tree that grows under a variety of conditions, from deep swamp to the margins of peat swamp and infertile lowland.

Where peat swamp forest has been cleared for cultivation and subsequently abandoned, *Macaranga maingayi* develops almost pure communities. Other common species are *Pellacalyx axillaris* and *Scleria* spp. In Johore, shallow peat swamp has been converted to pineapple and rubber plantations. Although deep peat like that found in Sarawak and Brunei does not develop in the Malay Peninsula, it is still best to leave the forest intact where peat depths exceed 3 m. Soil properties have been discussed by Coulter (1950, 1957).

## Indonesia

The above discussion of peat swamp derives from research conducted in the twentieth century. Research on Indonesia's peat is much older: Hewitt (1967) mentions a description by Junghuhn (1854) of the "floating island" phenomenon of Java. Many reports followed on the peat and the floating islands of Java, and with the discovery of peat in Sumatra, tropical peat further attracted the attention of academic society, which until then had believed that peat developed only in temperate and cold climates.

Many reports focused on Sumatra. The swamp forests of South Sumatra and Riau were treated by Endert (1920), van Bodegon (1929), Sewandono (1938) and others; Mohr (1922) reported on the peat of Sumatra's east coast; and in the Malay Peninsula there is the study of Cooke (1930). E. Polak (1933) published a study of the peat of Sumatra, Kalimantan, Java, and Sulawesi. Later, B. Polak (1950) presented an overview of peat in Indonesia, and in a description of Java's swamps she pursued the origin and construction of the floating islands (B. Polak, 1951). More recently, she has presented a summary of peat in the Malaysian tropics (B. Polak, 1975). Anderson (1976) has reported a summary of his surveys in Sumatra and Kalimantan.

Research into the peat swamps of Indonesia thus has a long history, but regrettably, no comprehensive description exists of the peat swamp forest. Here, again, I rely on the work of Anderson (1976).

Anderson selected three locations in Riau Province, Sumatra, and one each in West and Central Kalimantan, which he surveyed by the transect method. He found that the Sumatran forests had slightly fewer standing trees per unit area than those of Kalimantan, though the timber volumes were virtually equal, indicating that there were more large individuals in Sumatra.

In stands of the mixed swamp forest type, many of the species in the canopy were common to Sumatra and Kalimantan. But in the open forest type, the dominants in Kalimantan were reported to be absent in Sumatra.

In Riau, the constituent tree species in the canopy of the mixed swamp forest were *Durio carinatus, Palaquium burckii, Dyera lowii, Shorea platycarpa, S. teysmanniana,* and *S. uliginosa,* while in the middle and lower layers *Knema intermedia, Mangifera havilandii,* and *Neoscortechinia kingii* were seen. In the open forest type, 150 stems per ha of *Campnosperma coriacea* were found, and *Parastemon urophyllum* and *Alstonia pneumatophora* were abundant in the middle and lower layers and sometimes attained the upper story.

In a second plot, the dominants of mixed swamp forest were *Artocarpus rigidus, Gonystylus bancanus,* and *Palaquium burckii.* In the open forest, *Eugenia elliptilimba, Shorea teysmanniana, Mangifera havilandii,* and other elements of the mixed swamp forest appeared as small trees. *Combretocarpus rotundatus,* a typical component of open forest in Borneo, was also recorded.

In a third plot, *Strombosia javanica, Mezzettia leptopoda, Palaquium walsuraefolium,* and *Koompassia malaccensis* were dominant in the transitional zone. In the open forest, 200 stems per ha of *Palaquium burckii* were found, together with *Blumeodendron kurzii, Palaquium walsuraefolium,* and *Campnosperma coriacea.*

In West Kalimantan, dominants in the mixed swamp forest included *Parastemon urophyllum, Diospyros pseudomalabarica, Dyera lowii,* and *Gonystylus bancanus.* In the middle layer, *Blumeodendron tokbrai, Alangium havilandii,* and *Stemonurus scorpioides* were found. In the open forest, *Palaquium cochleariifolium* and *Diospyros evena* were present with extremely high densities of, respectively, 440 and 185 stems per ha.

In Central Kalimantan, *Calophyllum* was abundant in all plots, being represented by five species (*C. retusum, C. sclerophyllum, C. lowii, C. soulattri,* and *C. fragrans*). In the open forest along river courses, on shallow peat subject to frequent inundation, the dominants were *Combretocarpus rotundatu*s, Diospyros evena, and *Dactylocladus stenostachys.* In the mixed swamp forest, *Gonystylus bancanus* was dominant together with *Calophyllum;* and in the transitional zone to open forest, *Palaquium burckii* was dominant. In the middle and lower layers, *Neoscortechinia kurzii* and *Stemonurus secundiflorus* were abundant. In the open forest, the most abundant species were *Calophyllum retusum* (395 stems per ha), *Palaquium cochleariifolium* (225 stems per ha), and *Diospyros evena*.

From the above findings, the dominants of mixed swamp forest common both to Sumatra and Kalimantan are *Alstonia pneumatophora, Campnosperma coriacea, Durio carinatus, Dyera lowii, Gonystylus bancanus, Koompassia malaccensis, Lophopetalum multinervium, Mezzettia leptopoda, Palaquium burckii, Parastemon urophyllum, Shorea platycarpa, S. teysmanniana, S. uliginosa,* and *Tetramerista glabra.*

In the middle story and understory, species common to both islands and fairly characteristic on them are *Blumeodendron kurzii, B. tokbrai, Diospyros pseudomalabarica, D. siamang, Engelhardtia serrata, Eugenia elliptilimba, Gymnacranthera eugeniifolia, Horsfieldia crassifolia, Ilex hypoglauca, I. sclerophylloides, Mangifera havilandii, Neoscortechinia kingii, Santiria laevigata,* and *S. rubiginosa.*

In the open type of peat swamp forest, species abundant in Kalimantan but absent in Sumatra include *Dactylocladus stenostachys, Diospyros evena, Garcinia cuneifolia, Lithocarpus dasystachyus, Palaquium cochleariifolium, Parastemon spicatum, Sterculia rhoidifolia,* and *Xylopia coriifolia.*

Standing tree densities were higher in Kalimantan than Sumatra, with 190 and 251 stems per ha, respectively, in mixed swamp forest and open forest in the former as compared with 99 and 154 stems per ha, respectively, in the latter (table 1-22). The average basal areas, however, showed no difference, which means that, as stated earlier, individual trees are larger in Sumatra. In species numbers, the mixed swamp forest of Kalimantan is richest, with 55 species per plot, while the open forest had 33 species. In Sumatra, the corresponding figures were similar, at 37 and 39 species. Differences in nutritional conditions of soil are thought to have a major influence.

Much remains unclear about the peat swamp forests of Kalimantan and Sumatra—concerning distinction of forest types, for example—and it would be

### Table 1-22
### Comparison of Peat Swamp Forests of Kalimantan and Sumatra

|  | SUMATRA | | KALIMANTAN | |
|---|---|---|---|---|
|  | MIXED FOREST | OPEN FOREST | MIXED FOREST | OPEN FOREST |
| Average species number | 37 | 39 | 55 | 33 |
| Average density | 99 | 154 | 190 | 251 |
| Total basal area (m²) | 7.206 | 5.868 | 7.195 | 6.309 |
| Average girth (cm) | 86.743 | 69.1 | 69.7 | 56.1 |

Figures are per 0.2 ha.
SOURCE: Anderson, 1976.

unwise at present to compare them with those of Sarawak and Brunei. A quantitative survey is required in the future.

## Present Situation and Problems of Exploitation

Exploitation of the peat swamp forest in Sarawak and Brunei began about 1945 with the logging of *Gonystylus bancanus*. Full-scale development of the timber industry in Sarawak began in the 1950s (Lee 1972). Swamp forest was divided into five forest types, of which the total area reached 1,455,000 ha. Of this, a 689,000-ha section was placed under the jurisdiction of the forestry department (see table 1-23). Tree species of the mixed swamp forest type naturally were most important, of which *ramin (Gonystylus bancanus), jongkong (Dactylocladus stenostachys), sepetir paya (Copaifera palustris), kapor paya (Dryobalanops rappa),* and three species of swamp *meranti (Shorea platycarpa, S. scabrida,* and *S. uliginosa)* are notable. Other desirable species that attracted the loggers were sixteen with light, soft wood, seven medium species, and one heavy, hard species.

For the extraction of peat swamp forest lumber, Yap (1966) advocated switching from the traditional *kuda-kuda* (wooden sledge) to a mechanized method, but at present this does not seem to have spread greatly. The current practice in Brunei is the combination of *kuda-kuda* and light railway. For this, the concessionaires lay a light railway with a semidiesel engine to the designated logging area. Trees are felled by teams of four men using chain saws, then cut into logs and taken on wooden sledges by manpower to the railhead; from there, they are transported by rail to a sawmill on a riverbank. *Shorea albida* is used for the railroad sleepers, which are changed every three years. The felled trees have a mini-

## Table 1-23
### Present Distribution of Forest in Sarawak

| FOREST TYPE | FOREST RESERVES AND PROTECTED FORESTS | STATELAND FOREST | TOTAL |
|---|---|---|---|
| Mixed swamp forest | 435 | 739 | 1,174 |
| *Alan batu* forest | 114 | 13 | 127 |
| *Alan bunga* forest | 68 | 8 | 76 |
| *Padang alan* forest | 37 | 4 | 41 |
| *Padang paya* forest | 35 | 2 | 37 |
| Total | 689 | 766 | 1,455 |

(Unit: 1,000 ha.)
SOURCE: Lee, 1979.

mum diameter of 45 cm and are cut into lengths of 3.7, 4.3, or 4.9 m, depending on the height of the first branch. A tree normally yields three or four logs. The manual work involves extremely heavy labor, and in Sarawak it is said to be possible only for young, healthy workers.

No particular regulations exist governing management of land after felling, but areas subjected to virtual clear-felling show little regeneration of *Shorea albida*. In Brunei, the question of how to promote such regeneration is an important task for the future. The abundance of oil has obviated the need to export timber, and as a result the country's logging industry is underdeveloped.

Various trials have been conducted concerning the management of Sarawak's swamps (Lai, 1976, 1978; Lee, 1972, 1976, 1977, 1979), but it is still too early for results to be conclusive. Establishment and growth of *Shorea albida* are favored by dry soil rather than swamp; and in the case of line planting, growth is accelerated by appropriate thinning of the canopy trees, although too drastic thinning allows other species to invade and hamper the growth of *S. albida*.

Under swamp conditions, the problem of regeneration is not simple; and with deep peat, the lack of any alternative but to maintain the land as forest means that trials and planning are urgently required. Regeneration of the useful species of mixed swamp forest is vigorous, but *Shorea albida* is problematic. Together with long-term surveys of regeneration at logging sites, in areas damaged by insects or wind, and in the natural forest, growth experiments in seedling beds and the laboratory using seeds, seedlings, and saplings are required.

Also important is the preservation of genetic resources. The six phasic communities that Anderson distinguished in Sarawak and Brunei are unique in the world. Rather than small pockets, areas as large as possible should be enclosed as permanent conservation areas, where the total ecosystem can be preserved.

# 2.

# *Tropical Rain Forests of Brunei*

The kingdom of Brunei, which faces the South China Sea at 114°23′ E to 115°23′ E and 4° N to 5°5′ N on the northwest coast of Borneo, has a population of 220,000 and an area of 5,765 km². Through the production of oil and natural gas, discovered in the 1920s, it is Asia's richest country and is closely linked to Japan through natural gas exports.

Since its independence from the United Kingdom in 1984, the fifth five-year plan has been announced, which includes promotion of industries other than oil. The fact that the Forestry Department has been set a goal of producing 6 percent of the national income in the future indicates the importance accorded to forest resources.

Although oil and natural gas have tended to overshadow Brunei's forest resources, forestry surveys and research by the British beginning in the early twentieth century produced some outstanding results, including those of Browne, Anderson, and Ashton. However, there was no organized institution for research, such as a forestry research center.

On the occasion of independence, the Forestry Department erected a building to house a forestry research center at Sungai Liang, located at the entrance to the forest zone. To give substance to the future forest research center and promote basic research with potential applicability, technical cooperation with Japan was planned, and in October 1985 the Brunei Forest Research Center project between Brunei and Japan was born.

I lived in Brunei for two years around the time this research project was launched, and I participated in establishing the basic concept of the research center project and in conducting survey research after its establishment. Here I shall introduce part of the Brunei forest survey.

## 1. Brunei Forest Types

In Brunei, representative forest types of tropical rain forest are well conserved. I surveyed species composition and forest structure of typical forest types and gathered basic data for the purpose of continuing periodic surveys in the future as a permanent project.

Anderson and Marsden (1984) have classified Brunei's forest types as follows:

1. Mangrove
2. Freshwater swamp forest
3. Peat swamp forest
3.1 Mixed peat swamp
3.2 *Alan* forest
3.3 *Alan bunga* forest
3.4 *Padang alan* forest
3.5 *Padang* forest
3.6 *Padang keruntum*
4. *Kerangas*
5. Mixed dipterocarp forest
6. Limestone forest (not found in Brunei)
7. Montane forest
8. Secondary forest
9. Urban, cleared land, and cultivation

At the request of the Forestry Department, permanent survey plots were established to obtain basic information on the forest types regarded as relatively important (see fig. 2-1). At this time the mangrove, freshwater swamp forest, and montane forest types were omitted because of difficult access, problems of distance, or relative unimportance, but in the future, permanent survey areas should be established.

In establishing the permanent plots, the larger framework was grasped from Anderson and Marsden's map of forest types, and after repeated observation surveys on the ground over as wide an area as possible, forest sites were chosen that were thought to be the most representative. The basic area was taken as 50 × 50 m; using compass and measuring tape, these were divided into smaller plots of 10 × 10 m and the boundaries marked with poles and tapes. Plot sizes were set at from 50 × 50 m to 100 × 100 m, taking into consideration the height of the tallest trees and the diversity of tree species.

Within the plots, the diameters of standing trees with DBH (1.3 m) of 10 cm and more were measured with a diameter tape, and tree heights and heights to lowest branches were measured with a Weisse height meter. To prepare for future measurements, a zinc plate stamped with one of consecutive numbers was nailed to each tree, and a white band was painted around the trunk at the position of the breast-height diameter. Maps were prepared showing the trunk position and crown projection of each tree. Leaf samples were taken for identification of unknown species.

The *alan bunga* and *Agathis* forest plots were established in a month's expedition in April 1983, while the remainder were established between 1984 and

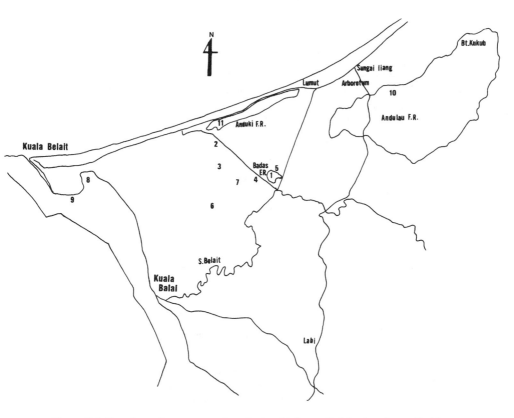

Figure 2-1. Locations of permanent plots: 1. *Agathis* forest; 2. *Alan batu* forest; 3. *Alan bunga* forest; 4. *Alan* forest; 5. *Padang alan* forest; 6. *Alan padang* forest; 7. *Ulat bulu* forest; 8. Mixed peat swamp forest (1); 9. Mixed peat swamp forest (2); 10. Mixed dipterocarp forest; 11. *Kapor paya* forest.

1986. This work was done by myself and three to five local staff. In the shortest cases, it was accomplished in a few days; in the longest, it took several months, including the collection of specimens. The most difficult plots were in *alan bunga* and *alan batu* forest, followed by mixed peat swamp forest. All were peat swamp forest plots, with very difficult footing, even more troublesome than I had previously experienced in surveying mangrove in Sumatra. Particularly difficult obstacles were *Pandanus andersonii* and the high buttress roots of *alan* (*Shorea albida*), which markedly decreased the efficiency of the survey.

The local staff, lead by Niga, were outstanding field surveyors, and however difficult the circumstances were, they worked steadily at all tasks, from plot establishment to measurement and collection of samples for identification. Their progress was as good or even better than what I have experienced in fieldwork in Indonesia, Thailand, Malaysia, and elsewhere.

## *Agathis Forest*

### Location

About ten minutes' ride by jeep from Sungai Liang toward Seria, turning left at the liquefied natural gas plant and following the gravel road through the *alan* forest alongside the water pipe to the factory, one comes to the intake on the Belait River. Immediately before the intake, you cross the pipe by a wooden bridge and enter the forest, which is a conservation area of *kerangas* forest. After a further ten minutes' ride, you come to a sign announcing VJR (Virgin Jungle Reserve), and about three minutes' walk beyond, we established a plot.

Today, it takes about forty minutes to reach the plot from Sungai Liang, but then, when road conditions were poor and the roadside ditches were blocked, puddles appeared at every turn through which the jeep could barely pass, and the journey took between one and a half and two hours. With the subsequent building of gravel pits and other facilities, the road was improved. The road beside the water pipe is flanked on both sides by pure *alan* forest, and the roadside ditches dug into the peat carry the black water that is characteristic of a peat area.

On observing the *alan* forest from the road, the almost gray-green crowns of the canopy are at first sparse, but on entering the forest, they become smaller and closer together. The trunks in many cases are bent just below the lowest branches, rather than straight, and above this sits a small crown. The yellow trunks of *Calophyllum* spp. and the reddish purple trunks of *Melanorrhoea beccarii* and *Gonystylus bancanus* are conspicuous. The cleared areas along the road show the start of regeneration, mainly of *Combretocarpus rotundatus*. Where the forest extends to the roadside, large *Nepenthes* overhang the road, the red stems of the palm *Cyrtostachys lacca* are conspicuous, and the large flowers of *Campnosperma* bloom. This road alongside the pipe passes straight through the *alan* belt to the Belait River, and near its terminus and at one point midway it passes through *kerangas* belts on slight elevations of white sand, at the periphery of which are dense stands of slender trees. *Tristania* sp., which readily sheds its bark, is conspicuous, and the trees become notably less tall. In *alan* forest, trees attain 50 m or even more, while in *kerangas* forest they are markedly shorter, at around 25 m.

Entering the forest across a wooden bridge, the road runs on top of white sand; sphagnum moss grows on the white sand, and many kinds of moss also grow on the lower trunks of the surrounding trees. Epiphytic orchids and ferns are abundant, and the green of the moss on the tree trunks presents an attractive contrast with the white sand of the forest floor. Conspicuous in the midst of this is *Agathis,* whose tapered, blackish red trunks with scaly bark rise commandingly to the canopy. Many secrete a white sap from the lower trunk, from which the resin dammar, formerly an important commercial product, is derived. *Agathis* also yields good timber, straight and unflawed and favored for its white grain. In Brunei it has almost all been cut, remaining in substantial areas only here and in the Temburong area on lowland and in the montane areas.

The footpath to the plot passes through *kerangas* forest, and where *Agathis* predominates, the ground underfoot is cushioned by a thick layer of 20–30 cm of litter and roots overlying the white sand. To the sides of the path, various species of orchid and *Agathis* seedlings are scattered over the deep litter, and epiphytic orchids and ferns are abundant on fallen trees. The forest is dark. The proportion of *Agathis* gradually increases, and eventually you reach the plot.

Beyond the plot, the proportion of *Agathis* again declines, the ground level drops by several meters, and suddenly you are in a damp peat swamp forest. The transition from *kerangas* forest to peat swamp forest is drastic and worth observing.

PHYSIOGNOMY

The plot was established in the location where *Agathis* grew most nearly as a pure stand. The tallest trees reached 45 m, and the largest diameters were 90 cm. These constituted the canopy, which was continuous with the second story below. There was no clear stratification, however, and *Agathis* trees of various stages appeared in various layers. Regeneration, too, was good, with abundant seedlings. The soil was white podzolic soil covered by a thick root hair layer of about 20 cm, and on top of that was a litter layer composed mainly of *Agathis* twigs and leaf fragments. Fallen trees were fewer than in peat swamp forest, but included a high proportion of medium and small trees. The tree density was lower than in the surrounding *kerangas* forest where there was no *Agathis,* and walking was easy. Epiphytes were limited to a small number of orchids and pandans, fewer than in the general *kerangas* forest. Many unripe female cones of *Agathis* lay on the forest floor, and particularly in March and April, sprouts were also found where ripe cones had fallen and scattered their seeds. The land was flat. Subsequently, a second 50 × 50 m plot was established, at the edge of which the land sloped down and the transition to peat swamp forest was apparent.

Modified root forms, such as buttresses and stilt roots, were rare; only the occasional flying buttress roots of *Casuarina* sp. were conspicuous.

A soil survey by Furukawa (1988) showed that in this locality fine roots were abundant to a depth of 35 cm.

The crown projection diagram in figure 2-2 reveals a fairly dense overlapping of tree crowns. Figure 2-3 depicts the stratification. From measurements of crown area, tree height, and height to first branch, the crown areas are plotted at 1-m intervals of height and show clearly the stratification of *Agathis* forest.

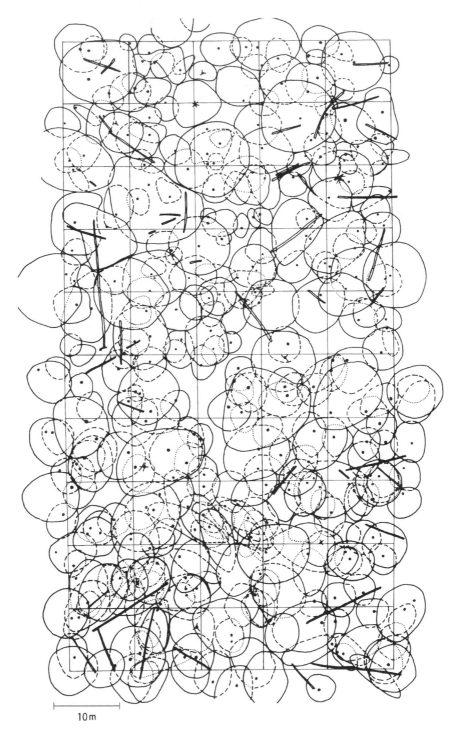

10 m

Figure 2-2. Crown projection of *Agathis* forest (50 × 100 m).

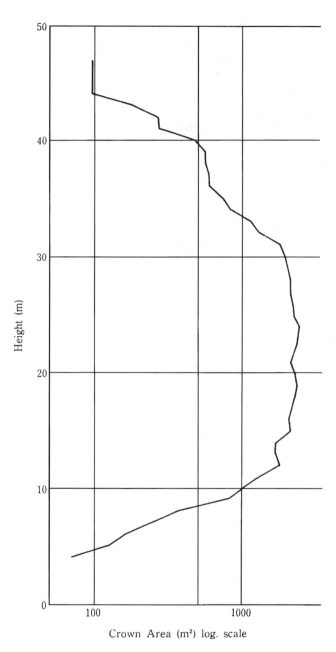

Figure 2-3. Stratification of *Agathis* forest.

TREE SPECIES COMPOSITION

Table 2-1 summarizes the survey results on the species composition of *Agathis* forest. In the first *Agathis* plot chosen, plot SP-1, *Agathis* was overwhelmingly predominant, with 83 out of 148 individuals. Next most common was *Cotylelobium burckii*. These two species can be regarded as dominant here. The first story also contained isolated individuals of *Calophyllum, Hopea pentanervia, Casuarina,* and others. In the second story and below, *Agathis* was uniformly dominant, while *Eugenia, Parastemon,* and others appeared, and in the third story, *Eugenia bankensis* was conspicuous. The fourth story was made up of low trees and the fifth of ground vegetation. On the ground, *Agathis* seedlings were abundant, not where other trees grew densely but on the comparatively open forest floor under the mother trees.

Plot SP-2 was established after SP-1, and the appearance of *alan* near its edge indicated the proximity of the boundary with peat swamp forest. The number of *Agathis* trees declined considerably to a value similar to those of *Cotylelobium burckii* and *Palaquium* (1). At the same time, the numbers of *Diospyros, Shorea albida, Melanorrhoea, Polyalthia* (1), and others increased. Comparison of the two subplots showed slightly more individuals in SP-2 but a greater total basal area in SP-1. These differences are thought to be because of *Agathis.* The combined figures for the two on a per hectare basis give considerably high values of 610 stems and a basal area of 38.4 m$^2$.

The total number of species, 38, was somewhat limited, in part by the soil conditions known as *kerangas.*

The distribution of diameters at breast height followed a gentle L-shaped curve to the 90-cm class, with the 10-cm class accounting for 56 percent.

## *Alan Batu Forest*

LOCATION

From Sungai Liang, you travel through Seria, turning inland at the center of the town, and on nearing the forest, you reach the new settlement of Kota Baru, at the southeast edge of which is the Hiap Hong lumber mill, where *alan* is processed. From here a light railway runs to the felling site. The trucks were of two types, larger ones for transporting logs, and smaller, hand-made ones for carrying people, on which we generally rode. Crossing a highway under construction and running through the *alan batu* forest, the trucks travel at considerable speed on an incredibly torturous track, which resulted in frequent derailments and spilling of the occupants. After about fifteen minutes, you encounter a wooden walkway built by the Shell Company for oil exploration. In this area, the forest consists of aging *alan* known as *alan batu,* the largest of the *alan*. Fallen *alan* trees, projecting buttress roots, and ground vegetation, such as *Pandanus andersonii* and *Salacca,* make walking difficult. *Hanguana malayana* flourishes

## Table 2-1
### Tree Species Composition of *Agathis* Forest

| Vernacular Name | Botanical Name | Family | SP-1 (50 × 50) | | SP-2 (50 × 50) | | Total/ha | | % | |
|---|---|---|---|---|---|---|---|---|---|---|
| | | | No. | B.A. | No. | B.A. | No. | B.A. cm² | No. | B.A. |
| Agathis | *Agathis dammara* (Lamb.) L.G. Rich. | Araucariaceae | 83 | 84,431.5 | 30 | 34,611.8 | 226 | 238,086.6 | 37.05 | 62.00 |
| Resak durian | *Cotylelobium burckii* (Heim) Heim | Dipterocarpaceae | 20 | 11,262.1 | 28 | 12,649.7 | 96 | 47,823.6 | 15.74 | 12.45 |
| Nyatoh (1) | | Sapotaceae | 6 | 827.1 | 29 | 8,885.5 | 70 | 19,425.2 | 11.48 | 5.06 |
| Mengilas | *Parastemon urophyllum* A. DC | Rosaceae | | | 12 | 6,600.9 | 24 | 13,201.8 | 3.93 | 3.44 |
| Ubah ribu | *Eugenia bankensis* Hassk. | Myrtaceae | 7 | 820.1 | 4 | 483.0 | 22 | 2,606.2 | 3.61 | 0.68 |
| Nyatoh (2) | | Sapotaceae | 4 | 1,259.6 | 4 | 1,075.1 | 16 | 4,669.4 | 2.62 | 1.22 |
| Merpisang | | | | | 7 | 1,012.9 | 14 | 2,025.8 | 2.30 | 0.53 |
| Mang | *Hopea pentanervia* Sym. | Dipterocarpaceae | 4 | 878.9 | 1 | 143.1 | 10 | 2,044.0 | 1.64 | 0.53 |
| Bintangor (2) | *Calophyllum* sp. | Guttiferae | 2 | 216.2 | 2 | 418.6 | 10 | 1,269.6 | 1.64 | 0.33 |
| Mengkulat | *Ilex hypoglauca* (Miq.) Loes | Anacardiaceae | 4 | 509.8 | 2 | 113.0 | 10 | 1,245.6 | 1.64 | 0.32 |
| Alan | *Shorea albida* Sym. | Dipterocarpaceae | | | 4 | 3,771.5 | 8 | 7,543.0 | 1.31 | 1.96 |
| Dual | *Lophopetalum* sp. | Celastraceae | 2 | 217.3 | 2 | 1,262.9 | 8 | 2,960.4 | 1.31 | 0.77 |
| Rengas | *Melanorrhoea* sp. | Anacardiaceae | | | 3 | 3,776.5 | 6 | 7,753.0 | 0.98 | 2.02 |
| Bintangor (3) | *Calophyllum* sp. | Guttiferae | | | 3 | 3,206.0 | 6 | 6,412.0 | 0.98 | 1.67 |
| Lusi | *Garcinia* sp. | Guttiferae | | | 3 | 1,275.0 | 6 | 2,550.0 | 0.98 | 0.66 |
| Kayu malam | *Diospyros* sp. | Ebenaceae | | | 3 | 852.8 | 6 | 1,705.6 | 0.98 | 0.44 |

*(continued on next page)*

Table 2-1    (continued)

| Vernacular Name | Botanical Name | Family | SP-1 (50 × 50) | | SP-2 (50 × 50) | | Total/ha | | % | |
|---|---|---|---|---|---|---|---|---|---|---|
| | | | No. | B.A. | No. | B.A. | No. | B.A. cm² | No. | B.A. |
| Ubah (1) | Eugenia sp. | Myrtaceae | 3 | 648.8 | | | 6 | 1,297.6 | 0.98 | 0.34 |
| Medang (3) | | | | | 3 | 492.5 | 6 | 985.0 | 0.98 | 0.26 |
| Kedondong (1) | | | 3 | 345.9 | | | 6 | 691.8 | 0.98 | 0.18 |
| Bitis | | Casuarinaceae | | | 2 | 2,072.4 | 4 | 4,144.8 | 0.66 | 1.08 |
| Sempilau | Casuarina nobilis Johnson msc. | | 2 | 1,668.5 | | | 4 | 3,337.0 | 0.66 | 0.87 |
| Bintangor (1) | Calophyllum sp. | Guttiferae | 1 | 1,187.9 | 1 | 171.9 | 4 | 2,719.6 | 0.66 | 0.71 |
| Unknown (1) | | | | | 2 | 324.1 | 4 | 648.2 | 0.66 | 0.17 |
| Medang (1) | | | | | 2 | 314.4 | 4 | 628.8 | 0.66 | 0.16 |
| Ubah (2) | Eugenia sp. | Myrtaceae | 1 | 145.2 | 1 | 137.4 | 4 | 559.8 | 0.66 | 0.15 |
| Medang (2) | | | | | 2 | 253.3 | 4 | 506.6 | 0.66 | 0.13 |
| Keranji | Dialium indum L. | Leguminosae | 1 | 80.1 | 1 | 124.6 | 4 | 409.4 | 0.66 | 0.11 |
| Ru | Casuarina sp. | Casuarinaceae | | | 1 | 1,771.2 | 2 | 3,542.4 | 0.32 | 0.92 |
| Mata ulat | Lophopetalum sp. | Celastraceae | 1 | 356.1 | | | 2 | 712.2 | 0.32 | 0.19 |
| Medang tabak | Dactylocladus stenostachys Oliv. | Crypteroniaceae | | | 1 | 326.7 | 2 | 653.4 | 0.32 | 0.17 |
| Menjalin (2) | Xanthophyllum sp. | Polyglaceae | 1 | 221.6 | | | 2 | 443.2 | 0.32 | 0.12 |
| Lithocarpus | Lithocarpus sp. | Fagaceae | 1 | 156.1 | | | 2 | 312.2 | 0.32 | 0.08 |
| Mempening | Lithocarpus sp. | Fagaceae | | | 1 | 132.7 | 2 | 265.4 | 0.32 | 0.07 |

(continued on next page)

Table 2-1 *(continued)*

| Vernacular name | Botanical name | Family | SP-1 (50 × 50) | | SP-2 (50 × 50) | | Total/ha | | % | |
|---|---|---|---|---|---|---|---|---|---|---|
| | | | No. | B.A. | No. | B.A. | No. | B.A. cm$^2$ | No. | B.A. |
| Unknown (2) | | | | | 1 | 95.0 | 2 | 190.0 | 0.32 | 0.05 |
| Kedondong (2) | | | 1 | 91.6 | | | 2 | 183.2 | 0.32 | 0.05 |
| Geronggang drum | *Cratoxylum* sp. | Hyperiaceae | | | 1 | 88.2 | 2 | 176.4 | 0.32 | 0.05 |
| Menjalin (1) | *Xanthophyllum* sp. | Polyglaceae | 1 | 83.3 | | | 2 | 166.6 | 0.32 | 0.04 |
| Mentang kelait | | | | | 1 | 75.4 | 2 | 150.8 | 0.32 | 0.04 |
| Total | | | 148 | 105,407.7 | 157 | 86,615.4 | 610 | 384,046.2 | 100.00 | 100.00 |

where water has collected, large *Nepenthes* abound, and the peat is soft and deep and contains intertwined root systems, which impede progress even further. In some places, the walkway has begun to rot and nails project, and it is dangerous. In other places, you must walk on fallen trunks lodged on buttresses several meters above the ground, while the large, concealed cavities between buttresses present a further danger. The first approach is thus difficult, but with practice you become accustomed to walking from one buttress to another.

## Physiognomy

The plot was established in a typical *alan batu* forest just 100 m from the railway track. The tallest trees reached 60 m. Many trees had lost the ends of their branches and bore stag-headed crowns, though some had large, intact crowns. This first story projected considerably above the second, which was 20–30 m high and consisted mainly of *Diospyros* and *Xanthophyllum,* many having a DBH of about 30 cm. The third story was sparse and below it was the ground vegetation, with abundant *Pandanus andersonii, Nepenthes,* and, where water had collected, *Hanguana malayana.* Epiphytes were scarce in the upper stories, but mosses, lichens, and epiphytic climbers grew around the buttresses of the larger trees.

Notable on the forest floor were fallen trees and buttress roots, and the networks of roots emerging from the buttresses. Fallen trees were more abundant than in any of the other forest types covered by the survey and included trees of all sizes. When a large tree falls, it naturally takes with it several medium and small trees. Several times during the survey we heard the sound of falling trees. We also saw trees whose fall had been arrested by entangling vines and which appeared to be sinking into the peat. Of the modified roots, the buttresses and associated root system of *alan* are of interest. As Anderson (1961b) has noted, the largest buttresses of *alan batu* are about 4 m high, spreading outward and radiating a network of ramifying feeding roots, and at the same time putting down vertical roots, some of which penetrate deep into the ground. Litter accumulates on the "platform" of feeding roots, forming a false forest floor that appears no different from the real forest floor. If you miss your footing on the buttresses, you can fall through the false floor to the peat, as much as a meter below.

Where a tree has fallen, its buttresses project into the air, forming a high wall below which a pool of water forms that is inhabited by fish. Peat swamp forest, unlike freshwater swamp forest, is not inundated by rivers, so this phenomenon was at first inexplicable. On visiting again in February, however, at the height of the rainy season, the mystery was solved. Water was flowing from the pools to adjacent depressions, providing a route for fish to enter.

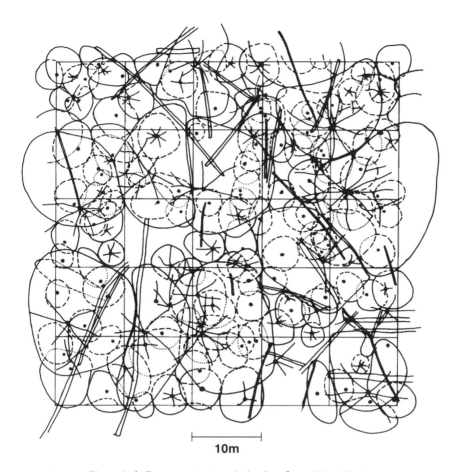

10m

Figure 2-4. Crown projection of *alan batu* forest (50 × 50 m).

TREE SPECIES COMPOSITION

The number of *alan* trees in this plot was not particularly high; *Palaquium* (1) and *Litsea cylindrocarpa* (1) were most abundant. However, these were medium and small trees. The first story was dominated by *alan*, and although few in number, their total basal area accounted for 48 percent of the total. In the second story, *Diospyros, Dyera,* and *Gonystylus bancanus* appeared, and together with *Palaquium* (1) and *Litsea* they formed a fairly open canopy. The second and third stories were intermixed and difficult to differentiate. Below this was the ground vegetation, with *Pandanus andersonii, Salacca,* and *Nepenthes* (figs. 2-4, 2-5).

The tallest trees, at almost 60 m, were the tallest recorded in the present survey. Their diameters were also the largest, as were their buttresses. The

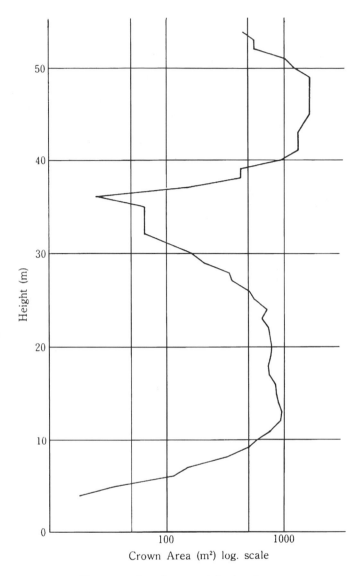

Figure 2-5. Stratification of *alan batu* forest.

number of species was small, 24, indicating a simple structure. The number of trees per ha, 522, was fewer than in *Agathis* forest, but the basal area of 42.6 m² was larger (table 2-2). *Palaquium, Polyalthia, Eugenia,* and other species also found in the peat swamp forest of Sumatra were common, too, indicating that the varying conditions in peat land in different locations limit species considerably, even more than the conditions in *kerangas* forest.

Table 2-2

Tree Species Composition of *Alan Batu* Forest

| VERNACULAR NAME | BOTANICAL NAME | FAMILY | SP-1 (50 × 50) | | SP-2 (50 × 50) | | TOTAL/HA | | % | |
|---|---|---|---|---|---|---|---|---|---|---|
| | | | No. | B.A. | No. | B.A. | No. | B.A. cm² | No. | B.A. |
| Nyatoh (1) | Palaquium leiocarpum Boerl. | Sapotaceae | 65 | 35,097.3 | 66 | 38,118.8 | 262 | 146,432.2 | 50.19 | 34.39 |
| Medang | Litsea resinosa Bl. | Lauraceae | 23 | 3,090.3 | 15 | 2,454.3 | 76 | 11,089.2 | 14.56 | 2.60 |
| Alan | Shorea albida Sym. | Dipterocarpaceae | 10 | 62,504.4 | 7 | 39,985.7 | 34 | 204,980.2 | 6.51 | 48.14 |
| Kayu malam | Diospyros evena Bakh. | Ebenaceae | 11 | 5,727.3 | 6 | 1,448.4 | 34 | 14,351.4 | 6.51 | 3.37 |
| Jelutong | Dyera polyphylla (Miq.) Ashton | Apocynaceae | 6 | 1,581.0 | 5 | 1,518.2 | 22 | 6,198.4 | 4.21 | 1.46 |
| Ludai | Macaranga caladifolia Becc. | Euphorbiaceae | 2 | 413.2 | 4 | 879.0 | 12 | 2,584.6 | 2.30 | 0.61 |
| Medang (2) | Ilex hypoglauca (Miq.) Loes. | Aquifoliaceae | 4 | 881.6 | 1 | 280.4 | 10 | 2,324.0 | 1.92 | 0.55 |
| Merpisang (2) | Polyalthia sp. | Annonaceae | 1 | 224.2 | 4 | 745.1 | 10 | 1,938.6 | 1.92 | 0.46 |
| Merpisang (3) | Polyalthia hypoleuca Hk. f. ex Th. | Annonaceae | 3 | 1,396.0 | 1 | 105.6 | 8 | 3,003.2 | 1.53 | 0.71 |
| Ramin | Gonystylus bancanus (Miq.) Kurz | Thymelaeaceae | 1 | 1,384.7 | 3 | 7,565.5 | 8 | 17,900.4 | 1.53 | 4.20 |
| Medang tabak | Eugenia polyantha Wight | Myrtaceae | 2 | 752.0 | 1 | 486.7 | 6 | 2,477.4 | 1.15 | 0.58 |
| Ubah (1) | Eugenia havilandii Merr. | Myrtaceae | 2 | 544.9 | 1 | 126.6 | 6 | 1,343.0 | 1.15 | 0.32 |
| Menjalin/Mengilas | Austrobuxus nitidus Miq. | Euphorbiaceae | 2 | 215.7 | 1 | 333.1 | 6 | 1,097.6 | 1.15 | 0.25 |
| Keruntum | Combretocarpus rotundatus (Miq.) Danser | Myrtaceae | | | 2 | 3,242.1 | 4 | 6,484.2 | 0.77 | 1.52 |
| Merpisang (1) | Polyalthia sp. | Annonaceae | 1 | 111.2 | 1 | 103.8 | 4 | 430.0 | 0.77 | 0.10 |
| Ubah (2) | Eugenia sarawacensis Merr. | Myrtaceae | 1 | 95.0 | 1 | 84.9 | 4 | 359.8 | 0.77 | 0.08 |

(continued on next page)

Table 2-2    (continued)

| VERNACULAR NAME | BOTANICAL NAME | FAMILY | SP-1 (50 × 50) | | SP-2 (50 × 50) | | TOTAL/HA | | % | |
|---|---|---|---|---|---|---|---|---|---|---|
| | | | No. | B.A. | No. | B.A. | No. | B.A. cm² | No. | B.A. |
| Pendarahan | Gymnacranthera eugenifolia (A. DC) sinclair var. griffithii (Warb.) Sinclair | Myristicaceae | 1 | 314.0 | | | 2 | 628.0 | 0.38 | 0.15 |
| Nyatoh (3) | | Sapotaceae | | | 1 | 277.5 | 2 | 555.0 | 0.38 | 0.13 |
| Ubah (4) | Eugenia sp. | Myrtaceae | | | 1 | 268.7 | 2 | 537.4 | 0.38 | 0.13 |
| Nyatoh (2) | | Sapotaceae | 1 | 143.1 | | | 2 | 286.2 | 0.38 | 0.07 |
| Medang (3) | | Lauraceae | | | 1 | 120.7 | 2 | 241.4 | 0.38 | 0.06 |
| Ubah (3) | Baccaurea bracteata Muell.-Arg. | Euphorbiaceae | 1 | 96.7 | | | 2 | 193.4 | 0.38 | 0.05 |
| Keranji | Endiandra oriacea Merr. | Lauraceae | 1 | 95.0 | | | 2 | 190.0 | 0.38 | 0.04 |
| Mengkulat | Ilex sp. | Aquifoliaceae | | | 1 | 81.7 | 2 | 163.4 | 0.38 | 0.04 |
| Total | | | 138 | 114,667.6 | 123 | 98,226.8 | 522 | 425,789.0 | 100.00 | 100.00 |

Concerning modified roots, the buttresses of *alan* and the reticulate root systems extending from them are characteristic; and the advanced state of butt rot means that many of the trunks must be hollow. *Diospyros* also has small buttresses, while species of *Eugenia* are characterized by tall buttresses. Some also had knee roots.

According to Anderson, this type of forest can attain a height of 70 m, but in the present plot the tallest trees barely reached 60 m.

This plot had the largest number of fallen trees, and Furukawa (1988) judged the peat to be fairly deep. In a large borrow pit near the plot, where sand had been taken for building the highway, the peat could be seen to be almost 3 m deep, below which was a layer of white sand. Roots about 3 cm in diameter descending from the buttresses penetrated deep into the sand.

This forest underwent gregarious flowering from December 1986 into the following January, with fruiting following two to three months later. Red buds opened into small, pale yellow or cream flowers, and the fruits were red, turning pale green before falling. Besides the survey on the ground, observations by helicopter were made at this time and revealed flowering and fruiting by 80–90 percent of trees in the *alan batu* zone, the percentage falling off toward the *alan bunga* zone.

## Alan Bunga Forest

LOCATION

From the *alan batu* forest, the railway proceeded into the interior to a former logging site now under regeneration with low trees about 10 m high. Only one side of the track had been logged, and on the other side the *alan* forest continued, with the trees becoming progressively smaller and closer to a pure stand. In the center of this zone, the *alan bunga* plot was established. The railway ride took about forty minutes, a distance that would take one and a half to two hours on foot and about two hours for the long trolleys carrying logs. About ten minutes farther into the interior is the logging site.

From here to the plot, we cut through a forest with abundant *Pandanus andersonii,* walking on top of buttresses. The pandan was at its densest here, and its spine-edged leaves grew to almost 3 m. This, together with the poor footing on the buttresses, made this the most arduous of locations. The plot was established about ten minutes' distance into the forest.

When the railway was removed, in the spring of 1986, continued survey became difficult.

PHYSIOGNOMY

*Alan* dominates the upper story with trees 70–90 cm in diameter and heights of nearly 50 m. In the tropics, a forest canopy formed only by this species is rare.

Figure 2-6. Crown projection of *alan bunga* forest (50 × 50 m).

Seen from above by helicopter, the characteristically gray-green crown is extremely even, formed of slightly overlapping crowns. Seen from below, it appears as a regular series of crowns with slight spaces between individuals. Most trees have a long space between the lowest branch and the crown and give the impression of bearing a small crown atop a tall, slender trunk (figs. 2-6, 2-7).

Virtually no second story exists below the canopy, with only occasional *Gonystylus bancanus* and *Campnosperma* occupying this space. The third story comes far below, containing scattered small trees of *Lithocarpus, Palaquium* (1), *Litsea,* and others, many with branches broken by falling branches from the upper story. Consequently, this forest can be considered virtually a pure stand of *alan.* On the ground, *Pandanus andersonii* was most abundant, together with *Pandanus helicopus*

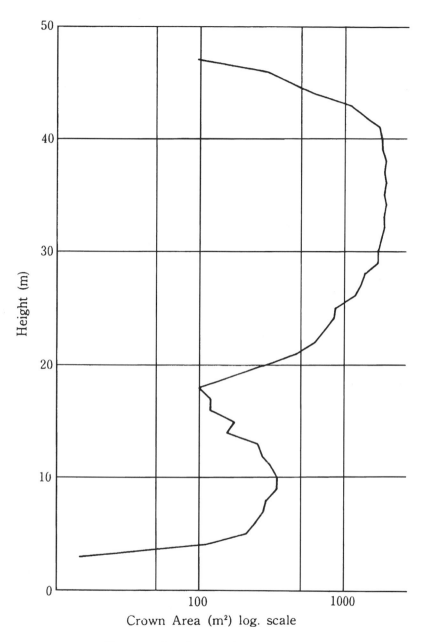

Figure 2-7. Stratification of *alan bunga* forest.

and seedlings of the smaller trees, but no seedlings of *alan*. There were considerable numbers of fallen trees, though fewer than in the *alan batu* forest.

The buttress roots, too, were smaller than those in the *alan batu* forest, making walking easier, but the false forest floor was the same, and without care a person would fall through. The peat was a brighter brown than in the *alan batu* forest, and pools of water were fewer. The forest floor was criss-crossed by the underground portions of *Pandanus andersonii* and *Pandanus helicopus*. Epiphytes were few, but in and around the gaps left by fallen trees were climbing ferns, and on the tree trunks was the lianoid *Timonius flavescens,* which seemed always to be in flower.

The heat and humidity of the *alan bunga* were the same as in the *alan batu* forest, but on occasion, as before rain, a strong wind would blow, making the crowns sway wildly and bringing trees down. Even if trees did not fall, falling branches were a danger, and it was best to leave as quickly as possible.

TREE SPECIES COMPOSITION

*Alan* is overwhelmingly dominant (see table 2-3), accounting for 44 percent of the number of trees and 84 percent of the basal area. These figures are higher even than the 37 percent and 62 percent occupied by *Agathis* in *Agathis* forest and attest to the high degree of dominance by *alan*. Next highest in basal area are *Gonystylus bancanus* and *Combretocarpus rotundatus.* The former is a tall tree that here occupies the story below *alan* and, together with *Calophyllum,* though present in small numbers, attains a considerable stature. Inside the forest, however, many of these trees had died under the pressure of *alan*. Apart from these, most of the remainder were small trees far below the *alan* that appeared barely able to grow. The near absence of a second story meant that the forest floor was fairly bright and provided favorable conditions for *Pandanus andersonii.*

Compared with *alan batu* forest, the total number of 22 species in the *alan bunga* forest is about the same level, while the tallest trees are slightly lower. If *alan batu* is a senescent forest, *alan bunga* can be said to be in its prime: Few trees had decayed roots or hollow trunks, and growth appeared to be vigorous. By contrast, the lower-story trees were all weak, their branches and trunks damaged by falling branches from the upper story, and their growth feeble.

Our observations in these two forest types accord with the progressive seral replacement of *alan bunga* by *alan batu*. Being a tall tree that grows extremely rapidly and is adapted to peat swamp, *alan* first forms the overstory. A large space develops between the overstory and the understory, and the forest lacks a second story, as is seen in the *alan bunga* forest. With time, the upward growth of *alan* slows, and its girth growth also proceeds slowly. This accords with the high density of sapwood of the senescent *alan* trees. As *alan* grows into the large trees

Table 2-3

Tree Species Composition of *Alan Bunga* Forest

| VERNACULAR NAME | BOTANICAL NAME | FAMILY | SP-1 (50 × 50) | | SP-2 (50 × 50) | | TOTAL/HA | | % | |
|---|---|---|---|---|---|---|---|---|---|---|
| | | | No. | B.A. | No. | B.A. | No. | B.A. cm² | No. | B.A. |
| Alan | *Shorea albida* Sym. | Dipterocarpaceae | 34 | 86,923.8 | 34 | 96,192.1 | 136 | 366,231.8 | 44.16 | 84.49 |
| Mempening | *Prunus turfosa* Kalm. | Rosaceae | 10 | 1,651.1 | 6 | 595.9 | 32 | 4,494.0 | 10.32 | 1.04 |
| Ubah (1) | *Eugenia leucoxylon* (Korth.) Miq. var. *phaephyllum* | Myrtaceae | 13 | 1,359.7 | 3 | 292.8 | 32 | 3,305.0 | 10.32 | 0.76 |
| Ramin | *Gonystylus bancanus* (Miq.) Kurz | Thymelaeaceae | 4 | 6,307.9 | 11 | 8,817.7 | 30 | 30,253.2 | 9.68 | 6.98 |
| Nyatoh (1) | *Palaquium cochleariifolium* van Royen | Sapotaceae | 5 | 1,747.6 | 1 | 563.8 | 12 | 4,622.8 | 3.90 | 1.07 |
| Keruntum | *Combretocarpus rotundatus* (Miq.) Danser | Rhizophoraceae | 3 | 5,182.7 | 2 | 1,669.6 | 10 | 13,764.6 | 3.25 | 3.18 |
| Medang (1) | *Xanthophyllum racemosum* | Polygalaceae | 1 | 86.5 | 3 | 527.6 | 8 | 1,228.2 | 2.60 | 0.28 |
| Terentang | *Campnosperma coriacea* (Jack) | Anacardiaceae | 1 | 98.5 | 2 | 992.8 | 6 | 2,182.6 | 1.95 | 0.50 |
| Nyatoh (2) | | Sapotaceae | | | 3 | 900.9 | 6 | 1,801.8 | 1.95 | 0.42 |
| Bintangor | *Calophyllum* sp. | Guttiferae | | | 2 | 538.9 | 4 | 1,077.8 | 1.30 | 0.25 |
| Mengilas | *Parastemon urophyllum* (A. DC) A. Dc./P. spicatum | Rosaceae | 2 | 330.4 | | | 4 | 660.8 | 1.30 | 0.15 |
| Mengkulat | *Ilex hypoglauca* (Miq.) Loes. | Aquifoliaceae | | | 2 | 271.0 | 4 | 542.0 | 1.30 | 0.13 |
| Menjalin | *Xanthophyllum amoena* Chod. | Polygalaceae | | | 2 | 258.5 | 4 | 517.0 | 1.30 | 0.12 |

*(continued on next page)*

Table 2-3   (continued)

| Vernacular name | Botanical name | Family | SP-1 (50 × 50) | | SP-2 (50 × 50) | | Total/ha | | % | |
|---|---|---|---|---|---|---|---|---|---|---|
| | | | No. | B.A. | No. | B.A. | No. | B.A. cm² | No. | B.A. |
| Medang tabak | Dactylocladus stenostachys Oliv. | Crypteroniaceae | 1 | 111.2 | 1 | 128.6 | 4 | 479.6 | 1.30 | 0.11 |
| Mentang lang | | | | | 2 | 188.3 | 4 | 376.6 | 1.30 | 0.09 |
| Jangkang paya (2) | | | | | 1 | 647.0 | 2 | 647.0 | 0.65 | 0.15 |
| Jelutong | Dyera polyphylla (Miq.) Ashton | Apocnaceae | | | 1 | 171.9 | 2 | 343.8 | 0.65 | 0.08 |
| Merpisang | Polyalthia hypoleuca Hook. f. et Th. | Annonaceae | | | 1 | 134.7 | 2 | 269.4 | 0.65 | 0.06 |
| Selunsor | Tristania grandifolia Ridl. | Myrtaceae | | | 1 | 113.0 | 2 | 226.0 | 0.65 | 0.05 |
| Kandis | Garcinia microcarpa Pierre | Guttiferae | | | 1 | 109.3 | 2 | 218.6 | 0.65 | 0.05 |
| Medang (3) | | | | | 1 | 105.6 | 2 | 211.2 | 0.65 | 0.05 |
| Medang (2) | | | | | 1 | 93.3 | 2 | 186.6 | 0.65 | 0.04 |
| Total | | | 74 | 103,799.4 | 81 | 113,313.3 | 310 | 433,640.4 | 100.00 | 100.00 |

that can be seen in the *alan batu* forest, developing large buttresses and hollowing trunks, the slender understory trees gradually grow to occupy the second story, presenting exactly the aspect of medium- and small-diameter trees presently seen in *alan batu* forest. With the filling of the second story, sunlight reaching the forest floor is reduced, and the *Pandanus andersonii* that flourished there declines in vigor and is replaced by other young trees. Such progression appears to be the direction of succession from *alan bunga* to *alan batu*.

Regeneration of *alan* itself was virtually unseen here. In the 50 × 50 m plot, only one sapling and one example of coppice regeneration were seen.

The gregarious fruiting of March 1986 occurred, as noted, mainly in *alan batu* but also, to a lesser though still considerable degree, in *alan bunga*. Only a small proportion of seedlings appears to have become established, however, and in fact the rate seems to have been higher in *alan batu*. It can be hypothesized that, for this even-aged *alan* forest, the senescent *alan* undergoes gregarious flowering and fruiting perhaps only once in several decades, and that a considerable proportion of seedlings take root and grow, forming the kind of *alan bunga* forest that can be seen today.

## *Alan Forest*

LOCATION

From the Hiap Hong factory in Seria, about 13 km down the road that follows the pipeline along the newly constructed railway to the water intake on the Belait River, a wooden walkway enters the forest on the south side. The plot is located about ten minutes' walk into the forest.

The road along the pipeline cuts through *alan* forest, and on both sides are many large *alan* trees of similar size to those of *alan batu*. On the north (seaward) side, in particular, are large senescent trees with stag-headed crowns, and near the intake are slightly lower *alan* with large crowns. On the south side the forest type varies from *alan batu* to secondary forest of the *ulat bulu* type, before again becoming *alan* forest. At one point *kerangas* forest extends from the *Agathis* plot, again giving place to *alan*.

Leaving the jeep, we crossed the roadside ditch, where blackish brown water flowed. Here, the peat reached a depth of several meters. Traversing a hundred meters of secondary forest, we entered the *alan* forest. When we established the plot the wooden walkway was sound, but a year later it had begun to rot, and walking became as difficult as in the *alan batu* forest.

This location was also accessible via a road leading from the road in front of the *Agathis* forest past the entrance to the intake, but had been rendered impassable by construction work on a barrier to prevent saltwater incursion into the intake. By way of Seria, the journey takes about fifty minutes from Sungai Liang.

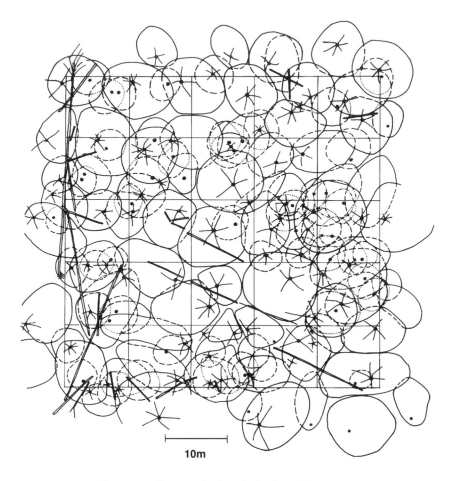

Figure 2-8. Crown projection of *alan* forest (50 × 50 m).

PHYSIOGNOMY

*Alan* forest resembles *alan bunga,* but the interior is somewhat gloomier and the tree crowns somewhat larger, and the forest floor is far drier, with much less *Pandanus andersonii.* The stratification is basically the same as for *alan bunga; alan* occupies the first story, but the second story is less empty, and the medium and small trees appear somewhat larger. There is less *Pandanus andersonii* on the forest floor, and while there are undulations due to buttresses, waterlogged portions are few and walking is easy (figs. 2-8 and 2-9). Epiphytes and climbers are few, as are fallen trees. Boring by Furukawa (1988) revealed thick peat, more than 7 m deep, which appeared slightly drier than that of *alan batu.*

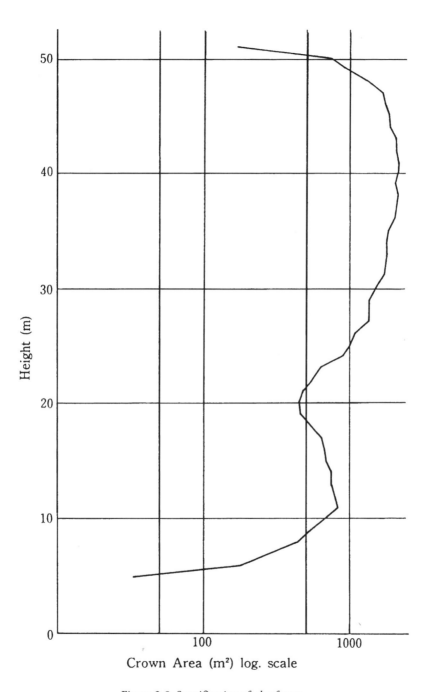

Figure 2-9. Stratification of *alan* forest.

## Table 2-4
### Tree Species Composition of *Alan* Forest

| Vernacular name | Botanical name | Family | SP-1 (50 × 50) | | Total/ha | | % | |
|---|---|---|---|---|---|---|---|---|
| | | | No. | B.A. | No. | B.A. cm² | No. | B.A. |
| Alan | *Shorea albida* Sym. | Dipterocarpaceae | 42 | 105,895.0 | 168 | 423,580.0 | 42.00 | 83.30 |
| Nyatoh | *Palaquium* sp. | Sapotaceae | 29 | 6,345.6 | 116 | 26,382.4 | 29.00 | 5.19 |
| Unknown (1) | | | 7 | 1,101.7 | 28 | 4,406.8 | 7.00 | 0.87 |
| Keruntum | *Combretocarpus rotundatus* (Miq.) Danser | Rhizophoraceae | 6 | 6,964.3 | 24 | 27,857.2 | 6.00 | 5.48 |
| Bitis | | | 3 | 2,749.0 | 12 | 10,996.0 | 3.00 | 2.16 |
| Medang tabak | *Dactylocladus stenostachys* Oliv. | Crypteroniaceae | 3 | 1,215.9 | 12 | 4,863.6 | 3.00 | 0.96 |
| Kandis | *Garcinia* sp. | Guttiferae | 3 | 380.5 | 12 | 1,522.0 | 3.00 | 0.30 |
| Jelutong | *Dyera polyphylla* (Miq.) Ashton | Apocnaceae | 1 | 1,249.7 | 4 | 4,998.8 | 1.00 | 0.99 |
| Ramin | *Gonystylus bancanus* (Miq.) Kurz | Thymelaeaceae | 1 | 326.7 | 4 | 1,306.8 | 1.00 | 0.27 |
| Medang (1) | | Lauraceae | 1 | 201.0 | 4 | 804.0 | 1.00 | 0.16 |
| Mentang lang | | | 1 | 122.7 | 4 | 490.8 | 1.00 | 0.10 |
| Kayu malam | *Diospyros* sp. | Ebenaceae | 1 | 120.7 | 4 | 482.8 | 1.00 | 0.10 |
| Mengkulat | *Ilex hypoglauca* (Miq.) Loes. | Aquifoliaceae | 1 | 96.7 | 4 | 386.6 | 1.00 | 0.08 |
| Ubah | *Eugenia* sp. | Myrtaceae | 1 | 84.9 | 4 | 399.6 | 1.00 | 0.08 |
| Total | | | 100 | 126,854.4 | 400 | 508,477.4 | 100.00 | 100.00 |

TREE SPECIES COMPOSITION

The dominance of *alan* is again high, higher than in *alan bunga* forest, and in fact the highest in the peat swamp forests. This is followed by *Palaquium,* represented by a large number of individuals. *Palaquium* and *Combretocarpus rotundatus* are dominant among the medium and small trees, which also include *Gonystylus bancanus, Dyera lowii,* and *Diospyros.*

Only 14 tree species were found, but this is because of the size of the plot. The actual number is not greatly different from that of *alan bunga.* Farther into the forest, the walkway enters *padang alan* forest, where the peat is even deeper and the number of species even fewer.

The total basal area is high (table 2-4).

## *Padang Alan Forest*

LOCATION

Only 300 m beyond the *Agathis* plot described earlier, at an elevation about 2 m below the *kerangas* forest, you suddenly enter peat swamp forest. Although there is naturally some mixing of species near the boundary, the transition from the rather dry *kerangas* to the wet peat swamp forest is drastic in all respects.

On the forest floor, knee roots and loop roots project from the covering of half-decayed litter, pools of water lie everywhere, and there are many fallen trees, often externally intact but decayed inside. Epiphytic mosses and lichens clothe the trunks, reinforcing for the observer on the ground the dense swamp forest atmosphere. Various palms and epiphytic ferns also appear, and the tall trees change from *Agathis* to *alan* and *Calophyllum.* The atmospheric humidity is close to 100 percent, giving the feeling of being bathed in sweat.

This swamp continues from the *alan* forest seen along the road from the entrance to the liquid natural gas plant. Several forest types can be seen here, which Anderson grouped together as *padang* forest. Locally, the forest where small *alan* grow particularly densely is known as *padang alan,* and I shall use the same term here.

The plot was established about 100 m down into the swamp forest, and about 20 m inside the boundary of the conservation area. From here, a path leads straight to the motorable road, which can be reached in about twenty minutes, a saving of some thirty minutes on the route around the *Agathis.*

PHYSIOGNOMY

This forest lies lower than *alan bunga* and *alan batu* and is characterized by the dense growth of various tree species in what might be called a *kerangas*-like swamp forest. *Alan* declines and other species come to dominate, with a variety appearing in the first story. Below that, the stratification becomes unclear, and

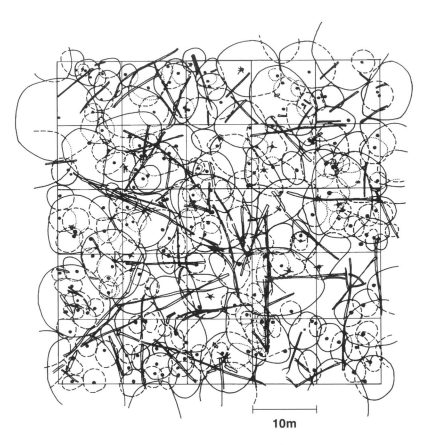

Figure 2-10. Crown projection of *padang alan* forest (50 × 50 m).

the second and third stories form a continuum reaching to the ground vegetation (figs. 2-10 and 2-11).

Also characteristic is the abundance of the epiphytic fern *Asplenium,* which attaches at a height of around 1.5 m. From there down to the forest floor grow many ferns and *Nepenthes,* and epiphytic lichens and mosses are abundant. Walking is easy on the forest floor, but there are many loop roots, and fallen trees are not uncommon. The root layer develops immediately below the surface, and your feet seldom sink into the peat. Furukawa (1988) found peat down to 285 cm.

TREE SPECIES COMPOSITION

The number of individuals increases greatly, indicating how dense the forest is. The basal area, however, is about the same as in *Agathis* forest, and the trees are shorter. Diameters at breast height are at most in the 60-cm class, reflecting the preponderance of medium and small trees. The number of *alan* is less than

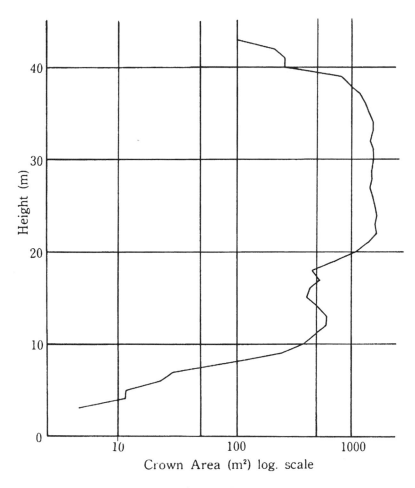

Figure 2-11. Stratification of *padang alan* forest.

half what it is in *alan bunga,* about as abundant as *Gonystylus bancanus, Combreto-carpus rotundatus, Calophyllum,* and others (table 2-5). Of medium and small trees, *Palaquium* is abundant.

Few trees here have modified roots, the most notable being the tall buttresses of *Eugenia,* the flying buttresses of *Casuarina,* and the slender knee and loop roots projecting from the ground. *Alan* has small buttresses about 30 cm high.

## Alan Padang Forest

LOCATION

From the *alan* forest plot, the walkway continues through unbroken *alan* forest, ending after a journey of about one and a half hours. Having passed an

## Table 2-5
### Tree Species Composition of *Padang Alan* Forest

| VERNACULAR NAME | BOTANICAL NAME | FAMILY | SP-1 (50 × 50) | | TOTAL/HA | | % | |
|---|---|---|---|---|---|---|---|---|
| | | | No. | B.A. | No. | B.A. cm² | No. | B.A. |
| Nyatoh (1) | | Sapotaceae | 97 | 17,599.8 | 388 | 70,399.2 | 44.50 | 17.73 |
| Nyatoh (2) | | Sapotaceae | 27 | 12,206.8 | 108 | 48,827.2 | 12.40 | 12.30 |
| Alan | *Shorea albida* Sym. | Dipterocarpaceae | 18 | 20,276.7 | 72 | 81,106.8 | 8.30 | 20.42 |
| Nyatoh (3) | | Sapotaceae | 17 | 2,690.3 | 68 | 10,761.2 | 7.80 | 2.71 |
| Ramin | *Gonystylus bancanus* (Miq.) Kurz | Thymelaeaceae | 16 | 6,857.7 | 64 | 27,430.8 | 7.30 | 6.91 |
| Bintangor (1) | *Calophyllum* sp. | Guttiferae | 12 | 19,823.7 | 48 | 79,294.8 | 5.50 | 19.97 |
| Keruntum | *Combretocarpus rotundatus* (Miq.) Danser | Rhizophoraceae | 10 | 8,454.7 | 40 | 33,818.8 | 4.60 | 8.52 |
| Medang tabak | *Dactylocladus stenostachys* Oliv. | Crypteroniaceae | 7 | 2,026.8 | 28 | 8,107.2 | 3.20 | 2.04 |
| Merpisang (1) | *Polyalthia* sp. | Annonaceae | 4 | 388.8 | 16 | 1,555.2 | 1.80 | 0.39 |
| Rengas | *Melanorrhoea* sp. | Anacardiaceae | 2 | 4,550.0 | 8 | 18,200.0 | 0.90 | 4.58 |
| Kelimpanas paya | *Goniothalamus* sp. | Annonaceae | 2 | 173.5 | 8 | 694.0 | 0.90 | 0.17 |
| Bintangor (2) | *Calophyllum* sp. | Guttiferae | 1 | 2,196.8 | 4 | 8,787.0 | 0.40 | 2.21 |
| Ru | *Casuarina* sp. | Casuarinaceae | 1 | 860.1 | 4 | 3,440.4 | 0.40 | 0.87 |
| Mang | *Hopea pentanervia* Sym. | Dipterocarpaceae | 1 | 730.2 | 4 | 2,920.8 | 0.40 | 0.74 |
| Ubah | *Eugenia* sp. | Myrtaceae | 1 | 226.9 | 4 | 907.6 | 0.40 | 0.23 |
| Selumar | *Jackia ornata* Wall. | | 1 | 113.0 | 4 | 452.0 | 0.40 | 0.11 |
| Kandis | *Garcinia* sp. | Guttiferae | 1 | 103.8 | 4 | 415.2 | 0.40 | 0.10 |
| Total | | | 218 | 99,279.6 | 872 | 397,118.4 | 100.00 | 100.00 |

area damaged by *ulat bulu,* the *alan* becomes slightly smaller, and eventually the forest becomes one of even smaller *alan.* The forest floor is dry and the litter layer is piled deep around the buttresses, creating marked undulations. This area corresponds to that near the center of the Badas peat swamp forest, which can be regarded as the oldest peat belt in the Badas area.

PHYSIOGNOMY

This type presents a completely different aspect from the *alan* forest types described so far. The trees are much lower, around 30 m, and in the stand of medium-sized *alan* trees a number are slanting. Sunlight penetrating the canopy reaches the forest floor, where the peat is xeric and shows little evidence of the growth of humic fungi. The litter layer is thick and has tended to pile up around the lower parts of tree trunks, so that walking has the feel of wading through dry litter rather than peat swamp.

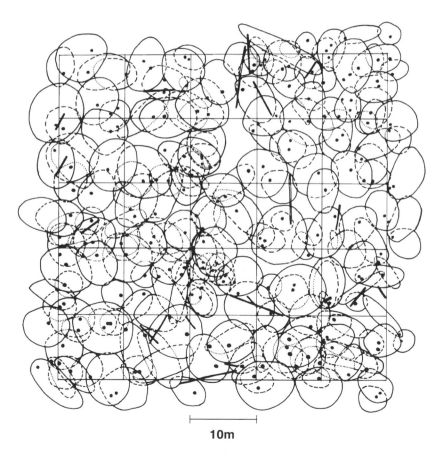

**10m**

Figure 2-12. Crown projection of *alan padang* forest (50 × 50 m).

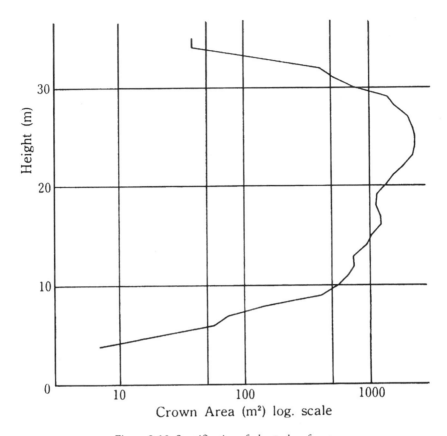

Figure 2-13. Stratification of *alan padang* forest.

Stratification involves three stories: the uppermost, *alan;* the second, other trees; and the third, ground vegetation. The ground vegetation contains few of the species found in the other types of *alan* forest, and includes seedlings of trees other than *alan*. Epiphytic climbers are few, and modified roots are virtually absent (figs. 2-12 and 2-13).

Tree Species Composition

The number of individuals is second to that of *padang alan,* but the basal area is smaller than that of *Agathis*. The tallest tree was 33 m and the largest diameter 46 cm, the smallest found among the peat swamp forest types. The species number was 10, and of the total number of individuals, more than half were *alan,* and 20 percent were *Parastemon*. In addition, *Combretocarpus rotundatus, Garcinia rostrata, Dactylocladus stenostachys,* and other species common to *padang alan* were recorded. The diameter of *alan* ranged from 46 to 14 cm, and other species were similarly of medium or small diameter (table 2-6).

# Table 2-6

## Tree Species Composition of *Alan Padang* Forest

| VERNACULAR NAME | BOTANICAL NAME | FAMILY | SP-1 (50 × 50) No. | B.A. | TOTAL/HA No. | B.A. cm² | % No. | % B.A. |
|---|---|---|---|---|---|---|---|---|
| Alan | *Shorea albida* Sym. | Dipterocarpaceae | 99 | 60,020.2 | 396 | 240,080.8 | 53.23 | 69.65 |
| Mengilas | *Austrobuxus nitidus* Miq. | Euphorbiaceae | 55 | 19,755.7 | 220 | 79,022.8 | 29.57 | 22.93 |
| Keruntum | *Combretocarpus rotundatus* (Miq.) Danser | Rhizophoraceae | 10 | 3,757.0 | 40 | 15,028.0 | 5.38 | 4.36 |
| Kandis (1) | *Garcinia* sp. | Guttiferae | 9 | 816.2 | 36 | 3,264.8 | 4.84 | 0.95 |
| Medang tabak | *Dactylocladus stenostachys* Oliv. | Crypteroniaceae | 6 | 969.7 | 24 | 3,878.8 | 3.23 | 1.13 |
| Nyatoh (1) | *Palaquium* sp. | Sapotaceae | 2 | 246.6 | 8 | 986.4 | 1.08 | 0.29 |
| Medang (1) | *Litsea* sp. | Lauraceae | 2 | 178.1 | 8 | 712.4 | 1.08 | 0.21 |
| Merpisang (1) | *Polyalthia* sp. | Annonaceae | 1 | 248.7 | 4 | 994.8 | 0.54 | 0.29 |
| Ubah (2) | *Eugenia* sp. | Myrtaceae | 1 | 93.3 | 4 | 373.2 | 0.54 | 0.11 |
| Ubah (1) | *Eugenia* sp. | Myrtaceae | 1 | 88.2 | 4 | 352.8 | 0.54 | 0.10 |
| Total | | | 186 | 86,173.4 | 744 | 344,694.8 | 100.00 | 100.00 |

The abundance of leaning trees is a phenomenon often seen in dense, even-aged plantation forests, and the *alan padang* forest showed a similar pattern. Crowns were small, and all species presented a similar appearance, which made immediate identification difficult.

## *Ulat Bulu* Forest

LOCATION

About forty minutes' walk from the *alan* forest on the route to the *alan padang* forest is the remains of defoliation damage by *ulat bulu.* As mentioned earlier, it is a pleasant walk on the wooden walkway under the forest canopy of an outstanding *alan* forest, until, on reaching the *ulat bulu,* you are bathed in direct sunlight and enter secondary forest. On either side of the walkway, the forest has been cleared for several meters and the ground is littered with fallen trunks, branches, and

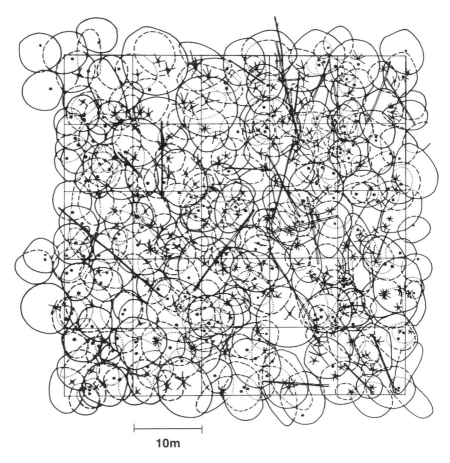

**10m**

Figure 2-14. Crown projection of *ulat bulu* forest (50 × 50 m).

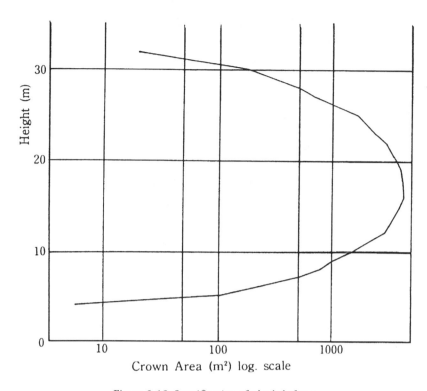

Figure 2-15. Stratification of *ulat bulu* forest.

leaves, no different from the usual situation in a secondary forest. Similar areas of damage in Brunei can be found in the Badas and Sungai Damit areas. Seen from a helicopter, they appear as large, clearly defined circular or oval craters.

PHYSIOGNOMY

The interior of the forest is gloomy, darker than any of the other types of *alan* forest. At eye level only the trunks of standing trees can be seen, and the trees are all of uniform height. This can probably be regarded as twenty-five-year-old secondary forest. It is clear at a glance that, following defoliation by moth caterpillars known as *ulat bulu,* a fairly uniform, even-aged community developed.

Trees close to 30 m high constitute the first story, but the vast majority appear to be around 20 m, forming a pure forest that can be regarded as having a single canopy. On the forest floor, no ground vegetation grows. It is scattered with dead trees and branches from above, the peat is dry, and walking is easy. Epiphytes and climbers are not seen. The forest is presently growing vigorously and shedding branches and leaves, and is easily recognized as *ulat balu* forest (figs. 2-14 and 2-15).

## Table 2-7
### Tree Species Composition of *Ulat Bulu* Forest

| Vernacular name | Botanical name | Family | SP-1 (50 × 50) | | Total/ha | | % | |
|---|---|---|---|---|---|---|---|---|
| | | | No. | B.A. | No. | B.A. cm² | No. | B.A. |
| Nyatoh (1) | | Sapotaceae | 74 | 16,400.5 | 296 | 65,602.0 | 25.34 | 20.24 |
| Medang (1) | | Lauraceae | 65 | 20,908.1 | 260 | 83,632.4 | 22.26 | 25.81 |
| Mempening | *Lithocarpus sundaicus* (Bl.) Rehder | Fagaceae | 55 | 11,289.5 | 220 | 45,158.0 | 18.84 | 13.93 |
| Keyu malam | *Diospyros* sp. | Ebenaceae | 32 | 13,036.2 | 128 | 52,144.8 | 10.96 | 16.09 |
| Bitis | | | 10 | 2,171.7 | 40 | 8,686.8 | 3.42 | 2.68 |
| Ubah air | *Eugenia* sp. | Myrtaceae | 8 | 1,084.7 | 32 | 4,338.8 | 2.74 | 1.34 |
| Keruntum | *Combretocarpus rotundatus* (Miq.) Danser | Rhizophoraceae | 7 | 9,296.3 | 28 | 37,185.20 | 2.40 | 11.47 |
| Mentang lang | | | 7 | 848.6 | 28 | 3,394.4 | 2.40 | 1.05 |
| Ubah (2) | *Eugenia* sp. | Myrtaceae | 4 | 728.3 | 16 | 2,913.2 | 1.37 | 0.90 |
| Medang (2) | | | 4 | 504.2 | 16 | 2,016.8 | 1.37 | 0.62 |
| Mengilas | *Parastemon urophyllum* A. DC. | Rosaceae | 3 | 917.2 | 12 | 3,668.8 | 1.03 | 1.13 |
| Ubah (1) | *Eugenia* sp. | Myrtaceae | 3 | 479.0 | 12 | 1,916.0 | 1.03 | 0.59 |
| Medang tabak | *Dactylocladus stenostachys* Oliv. | Crypteroniaceae | 3 | 312.8 | 12 | 1,251.2 | 1.03 | 0.39 |
| Merpisang (1) | *Polyalthia* sp. | Annonaceae | 3 | 294.3 | 12 | 1,177.2 | 1.03 | 0.36 |
| Jangkang paya | | | 2 | 966.0 | 8 | 3,864.0 | 0.68 | 1.19 |

*(continued on next page)*

Table 2-7  (*continued*)

| VERNACULAR NAME | BOTANICAL NAME | FAMILY | SP-1 (50 × 50) | | TOTAL/HA | | % | |
|---|---|---|---|---|---|---|---|---|
| | | | No. | B.A. | No. | B.A. cm$^2$ | No. | B.A. |
| Terentang | *Campnosperma coriaceae* (Jack) Hall. f. ex v. Steen | Anacardiaceae | 2 | 436.5 | 8 | 1,746.0 | 0.68 | 0.54 |
| Medang (4) | | | 2 | 188.9 | 8 | 755.6 | 0.68 | 0.23 |
| Geronggang | *Cratoxylum arborescens* (Vahl) Bl. | Hypericaceae | 1 | 292.4 | 4 | 1,169.6 | 0.34 | 0.36 |
| Unknown (1) | | | 1 | 216.3 | 4 | 865.2 | 0.34 | 0.27 |
| Ramin | *Gonystylus bancanus* (Miq.) Kurz | Thymelaeaceae | 1 | 145.2 | 4 | 580.8 | 0.34 | 0.18 |
| Somah | *Ploiarium alternifolium* (Vahl) Melch. | Theaceae | 1 | 126.6 | 4 | 506.4 | 0.34 | 0.16 |
| Medang (3) | | | 1 | 109.3 | 4 | 437.2 | 0.34 | 0.13 |
| Menjalin | *Xanthophyllum* sp. | Polygalaceae | 1 | 96.7 | 4 | 386.8 | 0.34 | 0.12 |
| Ubah (3) | *Eugenia* sp. | Myrtaceae | 1 | 89.8 | 4 | 359.2 | 0.34 | 0.11 |
| Alan | *Shorea albida* Sym. | Dipterocarpaceae | 1 | 81.7 | 4 | 326.8 | 0.34 | 0.10 |
| Total | | | 292 | 81,020.8 | 1,168 | 324,083.2 | 100.00 | 100.00 |

TREE SPECIES COMPOSITION

For a forest regenerating after insect damage, the number of species is enormously large. The basal area is disproportionately small, about the same value as in *alan padang*, reflecting the dense growth of slender trees.

Species number is high at 26, among which *Palaquium, Litsea,* and *Lithocarpus* are predominant. *Diospyros, Eugenia, Parastemon, Combretocarpus rotundatus,* and others also appear. Only one individual of *alan* was found, but in view of its state of regeneration in *alan bunga* and other forest types, this figure is probably reasonable (table 2-7).

## *Mixed Peat Swamp Forest (1)*

LOCATION

The trip from Sungai Liang through Seria to Kuala Belait takes about thirty minutes, and from there we traveled by speedboat up the Belait River for about fifteen minutes, past the second big turn, to a small tributary on the right-hand side. A short distance up the tributary, we left the boat. From the estuary, the riverbanks had been lined with *nipa,* and where we disembarked it grew to a height of about 10 m. At low tide, we landed on the mud below the *nipa,* while at high tide we reached the bank by swimming through the *nipa* leaves. Just beyond a belt of spiny *Salacca* lay the riparian forest, through which we passed. After about thirty minutes, or 300 m from the bank, we established our plot. The ground up to this point was subject to flooding and perennially waterlogged, and with knee roots, loop roots, and other obstacles, walking was extremely difficult. Within the plot we were always ankle-deep in peat.

PHYSIOGNOMY

The plot was uninfluenced by the river and appeared drier than the riverbank. Among mixed peat swamp forest, the forest type here lacked tall trees with large crowns. The canopy was relatively low, and individual heights were mainly of the medium and small classes. The forest thus lacked remarkable features, although there were many types of modified roots, and it was possible to observe stilt and knee roots, flying buttresses, buttress roots, and other interesting forms peculiar to swamp forest (figs. 2-16 and 2-17).

Stratification presented a continuum from the first to fourth stories, and ground vegetation was also abundant. Fallen trees were frequent, and their trunks and fallen branches were scattered across the forest floor, forming slight elevations.

Climbers were observed in the gaps. Epiphytes and climbers were generally few, but many epiphytes were found in the forks of the larger trees.

This type of peat swamp forest appeared on the shallowest peat. In Brunei, it was found only at this location on the Belait River.

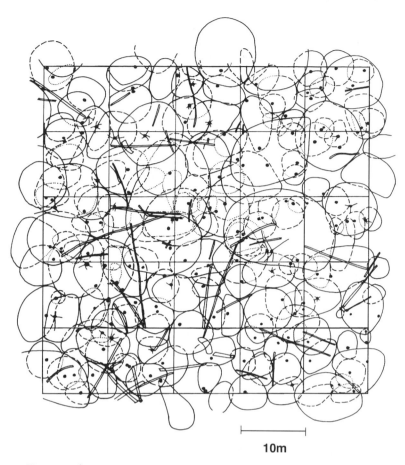

**10m**

Figure 2-16. Crown projection of mixed peat swamp forest (1) (50 × 50 m).

TREE SPECIES COMPOSITION

The forest was low, the tallest trees reaching only 36 m, and the largest diameter was of medium class at 73 cm. Trees of medium diameter were abundant, and the total number of species was 41. The number of individuals was comparable to that in *padang alan,* but the total basal area was slightly smaller. This can be regarded as a fairly dense forest, with many medium and small trees.

*Copaifera palustris,* a typical species of mixed swamp forest in Brunei and Sarawak, was present in large numbers and with a large basal area. Various types of *Shorea* were also present but their numbers were few. The tall buttresses of *Eugenia,* and the stilts of *Xanthophyllum amoena, nantungan* (unidentified), and others were notable. Fifty percent of the trees had buttresses, stilts, or other modified roots (table 2-8).

Boring by Furukawa (1988) showed the peat is shallow, and at 150 cm man-

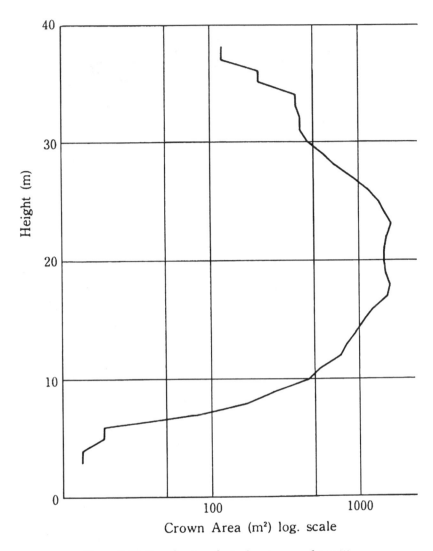

Figure 2-17. Stratification of mixed peat swamp forest (1).

grove species appeared, indicating that this is the most recent type of peat swamp forest in Brunei.

## Mixed Peat Swamp Forest (2)

LOCATION

This plot is also located in the Rassau district, about fifteen minutes' ride by boat up the Belait River, and about 500 m inland from the Dayak houses on the left bank.

## Table 2-8

Tree Species Composition of Mixed Peat Swamp Forest (1)

| VERNACULAR NAME | BOTANICAL NAME | FAMILY | SP-1 (50 × 50) | | TOTAL/HA | | % | |
|---|---|---|---|---|---|---|---|---|
| | | | No. | B.A. | No. | B.A. cm$^2$ | No. | B.A. |
| Sepetir | Sindora leiocarpa Becker ex De Wit | Leguminosae | 43 | 33,702.9 | 172 | 134,811.6 | 20.77 | 36.91 |
| Nantungan | | | 13 | 2,676.2 | 52 | 10,704.8 | 6.28 | 2.93 |
| Tampoi paya | Blumeodendron tokbrai (Kurz) J.J.S. | Euphorbiaceae | 13 | 1,848.6 | 52 | 7,394.4 | 6.28 | 2.02 |
| Amat | Tetramerista glabra Miq. | Tetrameristaceae | 12 | 7,058.4 | 48 | 28,233.6 | 5.80 | 7.73 |
| Mengkulat | Ilex hypoglauca (Miq.) | Aquifoliaceae | 12 | 3,510.7 | 48 | 14,042.8 | 5.80 | 3.84 |
| Pendarahan (1) | Gymnacranthera sp. | Myristicaceae | 12 | 2,844.9 | 48 | 11,379.6 | 5.80 | 3.12 |
| Menjalin | Xanthophyllum amoena Chod. | Polygalaceae | 11 | 2,216.8 | 44 | 8,866.2 | 5.31 | 2.43 |
| Pendarahan (2) | | | 10 | 2,829.9 | 40 | 11,319.6 | 4.83 | 3.10 |
| Ubah (4) | Eugenia sp. | Myrtaceae | 9 | 5,270.5 | 36 | 21,082.0 | 4.34 | 5.77 |
| Resak air | Vatica umbonata (Hook. f.) Burck. | Dipterocarpaceae | 7 | 1,173.0 | 28 | 4,692.0 | 3.38 | 1.28 |
| Lusi | Garcinia sp. | Guttiferae | 5 | 2,323.3 | 20 | 9,293.2 | 2.42 | 2.54 |
| Merpisang (2) | Polyalthia sp. | Annonaceae | 5 | 1,498.6 | 20 | 5,944.4 | 2.42 | 1.64 |
| Medang (1) | | | 5 | 679.5 | 20 | 2,718.0 | 2.42 | 0.74 |
| Kedondong (2) | | | 4 | 2,763.9 | 16 | 11,055.6 | 1.93 | 3.03 |
| Kayu malam | Diospyros sp. | Ebenaceae | 4 | 2,608.4 | 16 | 10,433.6 | 1.93 | 2.86 |
| Kedondong (1) | | | 3 | 2,507.8 | 12 | 10,031.2 | 1.45 | 2.75 |
| Ubah (3) | | | 3 | 1,395.6 | 12 | 5,582.4 | 1.45 | 1.53 |

(continued on next page)

Table 2-8   (*continued*)

| VERNACULAR NAME | BOTANICAL NAME | FAMILY | SP-1 (50 × 50) | | TOTAL/HA | | % | |
|---|---|---|---|---|---|---|---|---|
| | | | No. | B.A. | No. | B.A. cm² | No. | B.A. |
| Merpisang (3) | | | 3 | 856.8 | 12 | 3,427.2 | 1.45 | 0.94 |
| Mengilas | *Parastemon urophylllum* A. DC. | Rosaceae | 2 | 2,960.1 | 8 | 11,840.4 | 1.00 | 3.24 |
| Meranti (1) | *Shorea* sp. | Dipterocarpaceae | 2 | 1,669.8 | 8 | 6,679.2 | 1.00 | 1.83 |
| Merpisang (4) | *Polyalthia* sp. | Annonaceae | 2 | 1,085.2 | 8 | 4,322.0 | 1.00 | 1.18 |
| Simpor | *Dillenia pulchella* (Jack) Gilg. | Delleniaceae | 2 | 1,076.4 | 8 | 4,305.6 | 1.00 | 1.18 |
| Ubah (2) | *Eugenia* sp. | Myrtaceae | 2 | 893.3 | 8 | 3,573.3 | 1.00 | 0.98 |
| Jelutong paya | *Dyera polyphylla* (Miq.) Ashton | Apocynaceae | 2 | 717.1 | 8 | 2,868.4 | 1.00 | 0.79 |
| Merpisang (1) | *Polyalthia* sp. | Annonaceae | 2 | 672.8 | 8 | 2,691.2 | 1.00 | 0.74 |
| Rambutan hutan | *Nephelium* sp. | Sapindaceae | 2 | 623.2 | 8 | 2,492.8 | 1.00 | 0.68 |
| Jangtongan | | | 2 | 590.5 | 8 | 2,362.0 | 1.00 | 0.65 |
| Merpisang (5) | | | 2 | 193.4 | 8 | 773.6 | 1.00 | 0.21 |
| Kedondong (3) | | | 1 | 972.6 | 4 | 3,890.4 | 0.48 | 1.07 |
| Selumar | *Jackia ornata* Wall. | Rubiaceae | 1 | 422.5 | 4 | 1,690.0 | 0.48 | 0.46 |
| Ubah (5) | *Eugenia* sp. | Myrtaceae | 1 | 376.5 | 4 | 1,506.1 | 0.48 | 0.41 |
| Kalim paya | | | 1 | 286.4 | 4 | 1,145.6 | 0.48 | 0.31 |
| Merbau | *Intsia palembanica* Miq. | Leguminosae | 1 | 174.3 | 4 | 697.2 | 0.48 | 0.19 |
| Ubah (1) | *Eugenia* sp. | Myrtaceae | 1 | 134.7 | 4 | 538.8 | 0.48 | 0.15 |
| Dual | *Lophopetalum multinervium* Ridl. | Celastraceae | 1 | 128.6 | 4 | 514.4 | 0.48 | 0.14 |

(*continued on next page*)

Table 2-8  (continued)

| VERNACULAR NAME | BOTANICAL NAME | FAMILY | SP-1 (50 × 50) | | TOTAL/HA | | | % | |
|---|---|---|---|---|---|---|---|---|---|
| | | | No. | B.A. | No. | B.A. cm² | | No. | B.A. |
| Keranji | *Dialium laurinum* Baker | Leguminosae | 1 | 114.9 | 4 | 459.6 | | 0.48 | 0.13 |
| Mempait | | | 1 | 105.6 | 4 | 422.4 | | 0.48 | 0.12 |
| Asam piang | *Mangifera* sp. | Anacardiaceae | 1 | 100.2 | 4 | 400.8 | | 0.48 | 0.11 |
| Jangkang paya | | | 1 | 86.5 | 4 | 346.0 | | 0.48 | 0.09 |
| Mantangkulait | | | 1 | 83.3 | 4 | 333.2 | | 0.48 | 0.09 |
| Medang (2) | | | 1 | 80.1 | 4 | 320.4 | | 0.48 | 0.09 |
| Total | | | 207 | 91,313.8 | 828 | 365,185.6 | | 100.00 | 100.00 |

An old walkway leads into the forest, extending almost to the national border, but most of it is decayed and dangerous. The swamp forest begins immediately beyond the riparian *Pandanus helicopus* and the belt of *Salacca*. Medium-diameter trees are at first abundant, but larger trees soon appear, and the plot was established where the large trees of *Dyera lowii* showed signs of tapping for *jelutong* and labels remaining from Anderson's survey. The forest floor is peat, with many pools of water, and we often sank up to our knees and caught our feet in the tangle of roots. The forest is dark, with abundant humic litter. As in the mixed peat swamp forest (1), there are many types of buttresses and prop roots.

We reached the plot after walking twenty minutes on the walkway and twenty minutes on the forest floor. At the height of the rainy season, black water streamed through the forest toward the river. Ten minutes farther into the forest, *alan* appeared.

## PHYSIOGNOMY

Anderson notes that this type of forest resembles mixed dipterocarp forest. Indeed, the forest contains tall trees and is less swampy than *alan bunga* and *alan batu* forests. Species composition is also varied. The biggest trees are *Dryobalanops rappa*, which have large buttresses and reach heights of 50 m, being comparable with the *Dryobalanops beccarii* of the mixed dipterocarp forest (figs. 2-18 and 2-19).

Stratification is also similar to mixed dipterocarp forest, with a somewhat emergent first story, and second to fourth stories forming a continuum down to the ground vegetation. On the forest floor, old fallen trees form slight elevations, providing slight relief on the level ground, across which water flows at the height of the rainy season. There are again many pools of water, and in these places many knee roots project from the ground; buttresses and stilt roots are found on the tree trunks. Climbers are abundant where trees have fallen. Epiphytes are scarce.

## TREE SPECIES COMPOSITION

The density of 645 stems per ha and the basal area of 33.9 m² are comparatively low, but the species number, 58, is the highest for peat swamp forest. Tree heights approaching 50 m with maximum diameters of 140 cm are of the mixed dipterocarp forest class.

The most dominant tree is *Copaifera palustris*, which accounts for 12–13 percent of the total, far from the dominance of more than 50 percent by a single species in *alan* forest. *Copaifera palustris* is followed by *nantungan, Polyalthia, Eugenia, Diospyros, Litsea, Tetramerista*, and others; and large trees of *Dryobalanops rappa* and *Gonystylus bancanus*, though few, are also seen (table 2-9). *Shorea* appear, too, the situation being similar to that in the mixed peat swamp forest (1) plot.

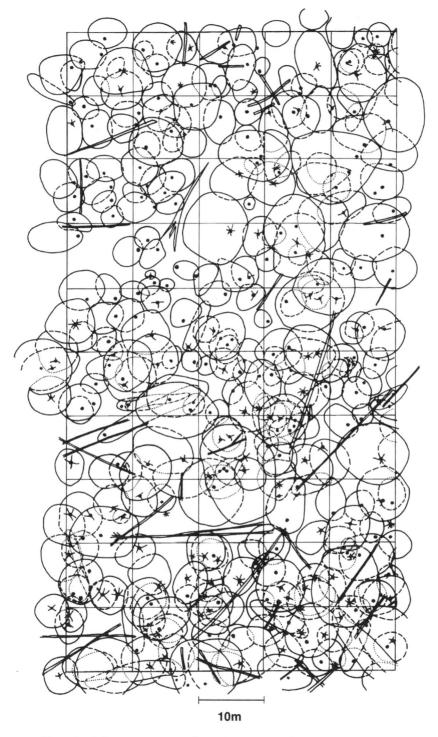

**10m**

Figure 2-18. Crown projection of mixed peat swamp forest (2) (50 × 100 m).

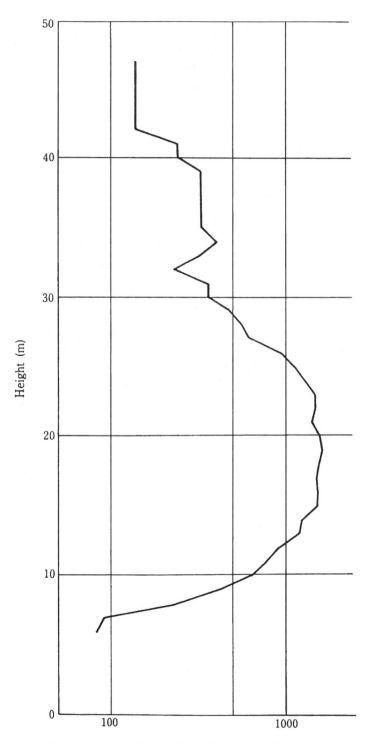

Figure 2-19. Stratification of mixed peat swamp forest (2).

# Table 2-9
## Tree Species Composition of Mixed Peat Swamp Forest (2)

| VERNACULAR NAME | BOTANICAL NAME | FAMILY | SP-1 (50 × 50) | | SP-2 (50 × 50) | | SP-3 (50 × 50) | | SP-4 (50 × 50) | | TOTAL/HA | | % | |
|---|---|---|---|---|---|---|---|---|---|---|---|---|---|---|
| | | | No. | B.A. | No. | B.A. | No. | B.A. | No. | B.A. | No. | B.A. cm$^2$ | No. | B.A. |
| Sepetir | *Sindora leiocarpa* Becker ex de Wit | Leg. | 25 | 18,647.5 | 24 | 8,471.8 | 18 | 7,883.3 | 18 | 8,049.2 | 85 | 43,051.8 | 13.18 | 12.72 |
| Merpisang (1) | | | 21 | 8,935.8 | 19 | 6,696.8 | 10 | 4,251.8 | 14 | 6,469.6 | 64 | 26,354.0 | 9.92 | 7.78 |
| Nantungan | | | 18 | 5,645.8 | 5 | 1,131.3 | 11 | 4,303.6 | 26 | 6,285.3 | 60 | 17,366.0 | 9.30 | 5.13 |
| Jelutong | *Dyera polyphylla* (Miq.) Ashton | Apocy. | 16 | 5,817.8 | 20 | 8,413.0 | 10 | 1,714.8 | 11 | 5,392.4 | 57 | 21,338.0 | 8.84 | 6.30 |
| Ubah (4) | *Eugenia* sp. | Myrt. | 4 | 2,406.8 | 5 | 857.9 | 12 | 7,813.8 | 16 | 8,396.1 | 37 | 19,474.6 | 5.74 | 5.75 |
| Kayu malam | *Diospyros* sp. | Ebe. | 11 | 2,884.9 | 10 | 2,184.9 | 7 | 2,492.8 | 6 | 1,408.6 | 34 | 8,971.2 | 5.27 | 2.65 |
| Medang (3) | | | 8 | 1,396.6 | 8 | 1,802.5 | 7 | 1,124.3 | 5 | 1,220.7 | 28 | 5,544.1 | 4.34 | 1.64 |
| Amat | *Terameristea glabra* Miq. | Tetr. | 4 | 7,089.3 | 13 | 8,749.7 | 2 | 303.9 | 7 | 2,839.9 | 26 | 18,982.8 | 4.03 | 5.61 |
| Mengkulat | *Ilex hypoglauca* (Miq.) Loes. | Aqui. | 6 | 1,464.2 | 5 | 1,744.2 | 5 | 1,003.0 | 5 | 1,152.2 | 21 | 5,363.6 | 3.26 | 1.58 |
| Ubah (5) | *Eugenia* sp. | Myrt. | 9 | 1,415.2 | 1 | 426.2 | 7 | 3,422.4 | 2 | 2,099.2 | 19 | 7,363.0 | 2.95 | 2.17 |
| Menjalin | *Xanthophyllum amoena* Chod. | Poly. | 5 | 651.5 | 7 | 1,442.1 | 3 | 961.6 | 4 | 575.2 | 19 | 3,630.4 | 2.95 | 1.07 |
| Medang (5) | | | | | 6 | 1,843.2 | 3 | 1,189.3 | 6 | 4,932.6 | 15 | 7,965.1 | 2.33 | 2.35 |
| Tampoi paya | *Blumeodendron tokbrai* (Kurz) J. J. S. | Eup. | | | 4 | 781.8 | 5 | 899.1 | 4 | 421.3 | 13 | 2,102.2 | 2.02 | 0.62 |

(continued on next page)

Table 2-9  (*continued*)

| Vernacular Name | Botanical Name | Family | SP-1 (50 × 50) | | SP-2 (50 × 50) | | SP-3 (50 × 50) | | SP-4 (50 × 50) | | Total/ha | | % | |
|---|---|---|---|---|---|---|---|---|---|---|---|---|---|---|
| | | | No. | B.A. | No. | B.A. | No. | B.A. | No. | B.A. | No. | B.A. cm² | No. | B.A. |
| Nyatoh (2) | | Sapo. | 3 | 15,800.0 | 3 | 2,252.7 | 1 | 379.9 | 2 | 209.2 | 9 | 18,641.8 | 1.40 | 5.51 |
| Medang (4) | | | 2 | 183.2 | 3 | 1,167.2 | 2 | 195.3 | 2 | 1,058.5 | 9 | 2,604.2 | 1.40 | 0.77 |
| Merpisang (2) | *Polyalthia* sp. | Ann. | | | 2 | 170.1 | 7 | 1,461.7 | | | 9 | 1,631.8 | 1.40 | 0.48 |
| Kedondong (1) | | | 1 | 1,962.5 | 2 | 543.0 | 5 | 4,909.1 | | | 8 | 7,414.6 | 1.24 | 2.19 |
| Lusi | *Garcinia* sp. | Guti. | 1 | 376.5 | 2 | 1,580.7 | 2 | 1,222.4 | 3 | 2,254.7 | 8 | 5,434.3 | 1.24 | 1.61 |
| Kapor paya | *Dryobalanops rappa* Becc. | Dipt. | 1 | 15,386.0 | 1 | 1,962.5 | 1 | 105.6 | 4 | 10,101.4 | 7 | 27,555.5 | 1.09 | 8.14 |
| Kedondong (2) | | | 2 | 2,245.3 | 3 | 7,624.4 | 1 | 426.2 | 1 | 116.8 | 7 | 10,412.7 | 1.09 | 3.08 |
| Nyatoh (1) | | Sapo. | 3 | 266.6 | 1 | 2,659.0 | 3 | 6,532.3 | | | 7 | 9,457.9 | 1.09 | 2.79 |
| Tampoi antu | *Baccaurea bracteata* Muell.-Arg | Euph. | 6 | 942.2 | | | 1 | 265.8 | | | 7 | 1,208.0 | 1.09 | 0.36 |
| Kedondong (3) | | | | | 2 | 2,446.4 | | | 4 | 3,069.0 | 6 | 5,515.4 | 0.93 | 1.63 |
| Jangkang paya | | | 2 | 1,819.1 | 2 | 153.9 | 1 | 1,704.7 | 2 | 1,500.6 | 6 | 5,178.3 | 0.93 | 1.53 |
| Ramin | *Gonystylus bacanus* (Miq.) Kurz | Thyme. | | | 1 | 7,261.7 | 1 | 1,682.8 | 2 | 6,319.1 | 5 | 15,263.6 | 0.78 | 4.51 |
| Pendarahan (1) | *Gymnacrathera* sp. | Myris. | 1 | 283.4 | | | 4 | 2,853.2 | | | 5 | 3,136.6 | 0.78 | 0.93 |
| Medang (1) | | | 1 | 98.5 | 1 | 274.5 | 2 | 1,288.4 | | | 4 | 1,661.4 | 0.62 | 0.49 |
| Ubah (3) | *Eugenia* sp. | Myrt. | 2 | 1,051.2 | 2 | 379.3 | | | | | 4 | 1,430.5 | 0.62 | 0.42 |
| Meranti (2) | *Shorea* sp. | Dipt. | | | | | 2 | 264.0 | 2 | 935.3 | 4 | 1,199.3 | 0.62 | 0.35 |

(continued on next page)

Table 2-9  *(continued)*

| VERNACULAR NAME | BOTANICAL NAME | FAMILY | SP-1 (50 × 50) | | SP-2 (50 × 50) | | SP-3 (50 × 50) | | SP-4 (50 × 50) | | TOTAL/HA | | % | |
|---|---|---|---|---|---|---|---|---|---|---|---|---|---|---|
| | | | No. | B.A. | No. | B.A. | No. | B.A. | No. | B.A. | No. | B.A. cm² | No. | B.A. |
| Resak Air | *Vatica umbonata* (Hook. f.) Burck. | Dipt. | 2 | 534.6 | | | 1 | 181.4 | 1 | 128.6 | 4 | 844.6 | 0.62 | 0.25 |
| Mentang keleit | | | 2 | 353.6 | | | 2 | 265.8 | | | 4 | 619.4 | 0.62 | 0.18 |
| Rambutan hutan | *Nephelium* sp. | Sapi. | 1 | 120.7 | | | 2 | 295.5 | 1 | 128.6 | 4 | 544.8 | 0.62 | 0.16 |
| Mengilas | *Parastemon urophyllum* A. DC. | Rosa. | | | | | 1 | 3,235.5 | 2 | 4,865.5 | 3 | 8,101.0 | 0.47 | 2.39 |
| Ubah (2) | *Eugenia* sp. | Myr. | 1 | 113.0 | 2 | 2,109.1 | | | | | 3 | 2,222.1 | 0.47 | 0.66 |
| Medang (7) | | | | | | | 2 | 1,992.4 | 1 | 179.0 | 3 | 2,171.4 | 0.47 | 0.64 |
| Ubah (1) | *Eugenia* sp. | Myr. | 1 | 151.7 | | | 2 | 1,430.0 | | | 3 | 1,581.7 | 0.47 | 0.47 |
| Medang (2) | | | 1 | 866.2 | 1 | 323.5 | | | 1 | 301.6 | 3 | 1,511.3 | 0.47 | 0.45 |
| Pendarahan (2) | *Gymnancranthera* sp. | Myri. | 1 | 234.9 | 2 | 1,242.4 | | | | | 3 | 1,477.3 | 0.47 | 0.44 |
| Perawa | | | | | 2 | 385.6 | 1 | 563.8 | | | 3 | 949.4 | 0.47 | 0.28 |
| Pendarahan (3) | *Gymnancranthera* sp. | Myri. | 2 | 551.6 | 1 | 251.5 | | | | | 3 | 803.1 | 0.47 | 0.24 |
| Sabong ribut | *Ctenolophon parvifolius* Oliv. | Lina. | 1 | 808.9 | | | | | 1 | 1,589.6 | 2 | 2,398.5 | 0.31 | 0.71 |
| Dual | *Lophopetalum multinervium* Riu. | Celas. | 2 | 1,286.0 | | | | | | | 2 | 1,286.0 | 0.31 | 0.38 |
| Asam piang | *Mangifera* sp. | Anac. | 2 | 1,092.2 | | | | | | | 2 | 1,092.2 | 0.31 | 0.32 |

*(continued on next page)*

Table 2-9   (continued)

| Vernacular name | Botanical name | Family | SP-1 (50 × 50) | | SP-2 (50 × 50) | | SP-3 (50 × 50) | | SP-4 (50 × 50) | | Total/ha | | % | |
|---|---|---|---|---|---|---|---|---|---|---|---|---|---|---|
| | | | No. | B.A. | No. | B.A. | No. | B.A. | No. | B.A. | No. | B.A. cm$^2$ | No. | B.A. |
| Simpor | Dillenia pulchella (Jack) Gilg. | Dill. | | | 1 | 918.2 | | | 1 | 109.3 | 2 | 1,027.5 | 0.31 | 0.30 |
| Meranti (1) | Shorea sp. | Dipt. | 1 | 555.4 | 1 | 455.9 | | | | | 2 | 1,011.3 | 0.31 | 0.30 |
| Ubah (6) | Eugenia sp. | Myr. | 1 | 179.0 | | | 1 | 307.8 | | | 2 | 486.8 | 0.31 | 0.14 |
| Kandis (1) | Garcinia sp. | Gut. | 2 | 341.7 | | | | | | | 2 | 341.7 | 0.31 | 0.10 |
| Kelimpanas paya | | | | | 1 | 100.2 | | | 1 | 81.7 | 2 | 181.9 | 0.31 | 0.05 |
| Medang (8) | | | | | | | | | 1 | 2,180.2 | 1 | 2,180.2 | 0.16 | 0.64 |
| Somah | Ploiarium alternifolium (Vahl) Melch. | Thea. | | | | | | | 1 | 774.0 | 1 | 774.0 | 0.16 | 0.23 |
| Selunsor | Tristania sp. | | | | 1 | 568.0 | | | | | 1 | 568.0 | 0.16 | 0.17 |
| Unknown | | | | | | | | | 1 | 277.5 | 1 | 277.5 | 0.16 | 0.08 |
| Ubah (7) | Eugenia sp. | Myr. | | | | | 1 | 244.2 | | | 1 | 244.2 | 0.16 | 0.07 |
| Meranti paya (3) | Shorea sp. | Dipt. | | | | | 1 | 208.6 | | | 1 | 208.6 | 0.16 | 0.06 |
| Geronggang | Cratoxylum arborescens (Vahl) Bl. | Hype. | | | | | | | 1 | 151.7 | 1 | 151.7 | 0.16 | 0.04 |
| Medang (6) | | | | | 1 | 107.5 | | | | | 1 | 107.5 | 0.16 | 0.03 |
| Ara | Ficus sp. | Mor. | | | | | | | 1 | 81.7 | 1 | 81.7 | 0.16 | 0.02 |
| Bintangor | Calophyllum sp. | | | | | | | | 1 | 81.7 | 1 | 81.7 | 0.16 | 0.02 |
| Total | | | 172 | 103,979.4 | 165 | 79,482.7 | 147 | 69,360.1 | 161 | 85,737.6 | 645 | 338,559.8 | 100.00 | 100.00 |

A major feature of this forest is again modified roots, with stilt roots and knee roots particularly notable.

## Mixed Dipterocarp Forest

LOCATION

The Labi road running south from Sungai Liang is lined first by *Melaleuca leucadendron,* then by *Acacia auriculiformis,* before rising slightly into a plantation of Caribbean pine. Here we turned left onto the logging road to K7 (compartment 7), following the mountain ridge that separates the forest reserve from secondary forest, and left the car in front of a fallen tree. The drive to this point took twenty minutes from Sungai Liang, but subsequently, with the deterioration of the logging road, we had to walk for about forty minutes from the turnoff

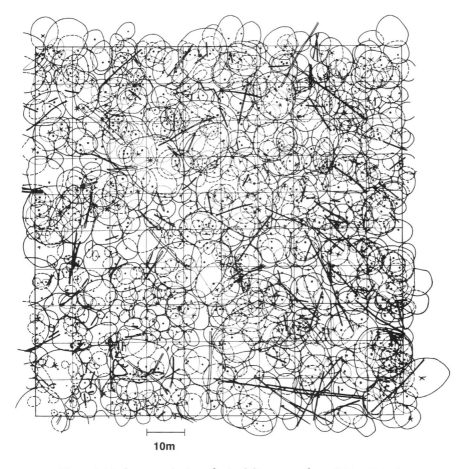

**10m**

Figure 2-20. Crown projection of mixed dipterocarp forest (100 × 100 m).

to reach this point. A short climb led to a red signboard announcing Virgin Jungle Reserve on the right-hand side, and from there a path led down to the plot, which we reached in about three minutes. The forest outside the reserve had been logged and was regenerating in places, but the former log-collection point remained bare. Overall, the forest was in good condition, but the road was in poor repair, and the absence of roadside ditches had allowed rain to destroy the surface

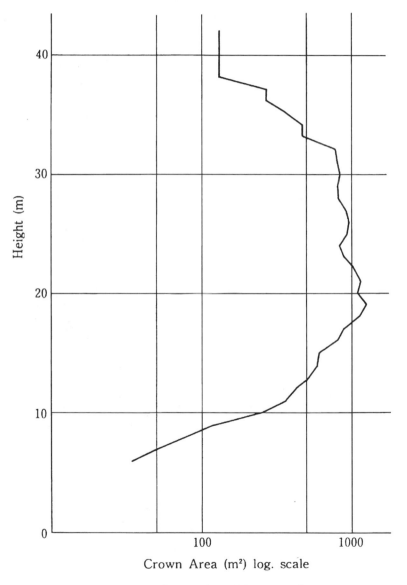

Figure 2-21. Stratification of mixed dipterocarp forest.

in many places. If you could approach by car, the plot would be the closest to Sungai Liang.

PHYSIOGNOMY

This plot is also one that Ashton (1964a) surveyed, and as totally undisturbed Virgin Jungle Reserve, it contains typical mixed dipterocarp forest. Its structure is continuous from the emergent story of 50-m class trees to the second story and ground vegetation, making the various strata difficult to distinguish. Many trees have straight trunks, and many of the large trees have buttresses. Climbers are relatively abundant, and epiphytes can be seen near the forks of the larger trees. Various palms are abundant on the ground, and seedlings of the overstory trees congregate here and there. The forest floor is comparatively bright and the litter layer comparatively thick. Fallen trees are as abundant as in mixed swamp forest, and several trees fell during the two years of the survey. Strong winds blow in from the nearby ocean, and particularly before rain, the tree crowns sway frighteningly (figs. 2-20 and 2-21).

TREE SPECIES COMPOSITION

The density of 736 stems per ha is on a par with *alan padang* forest, but the basal area is larger, the second highest of any plot. The maximum height was 50 m and the greatest diameter 114 cm, on the small side for a mixed dipterocarp forest. However, the species number, at 174, was typically high, showing the highest value of any forest type. Dipterocarpaceae occupied a high proportion of the species composition, being represented in every layer from the overstory to the small-diameter trees.

The characteristic of this forest type, as is often noted, is the lack of a notable dominant. As shown in table 2-10, no species accounts for more than 6 percent of all individuals; and in basal area, apart from the 18 percent of *Dryobalanops beccarii,* most occupy a small proportion, a few percent or less. The soil is yellow podzolic soil, which is far from good, but it supports a much greater range of species than peat.

## *Kapor Paya Forest*

LOCATION

Fifteen minutes from Sungai Liang in the direction of Seria is the Anduki Forest Reserve. This faces onto the busiest national highway, along which is a belt of forest that has burned and regenerated on several occasions as the result of frequent forest fires caused by discarded cigarette ends and other factors. The road was built on banked-up earth along the line where peat begins to mix with coastal sand, and in the vicinity are *Acrostichum aureum, nipa,* Rhizophoraceae, and other coastal vegetation. Here, the forest appears to be in a secondary succes-

Table 2-10

Tree Species Composition of Mixed Dipterocarp Forest

| VERNACULAR NAME | BOTANICAL NAME | SP-1 (50 × 50) | | SP-2 (50 × 50) | | SP-3 (50 × 50) | | SP-4 (50 × 50) | | TOTAL/HA | | % | |
|---|---|---|---|---|---|---|---|---|---|---|---|---|---|
| | | No. | B.A. | No. | B.A. | No. | B.A. | No. | B.A. | No. | B.A. cm² | No. | B.A. |
| Sabal | *Dacryodes expansa* (Ridl.) H. J. Lam | 4 | 2,544.8 | 4 | 991.5 | 24 | 10,297.6 | 9 | 4,025.1 | 41 | 17,859.0 | 5.57 | 3.61 |
| Asam panas (1) | *Mangifera* sp. | 9 | 2,511.4 | 8 | 1,504.6 | 10 | 2,678.8 | 14 | 4,247.1 | 41 | 10,941.9 | 5.57 | 2.21 |
| Kayu tulang | *Allantospermum borneensis* Forman | 28 | 14,238.5 | | | | | 4 | 1,239.4 | 32 | 15,477.9 | 4.35 | 3.13 |
| Kedondong (4) | | 3 | 295.3 | 13 | 5,469.8 | 11 | 3,026.4 | 4 | 570.9 | 31 | 9,362.6 | 4.21 | 1.90 |
| Kapor bukit | *Dryobalanops beccarii* Dyer | 7 | 1,349.3 | 5 | 32,460.6 | 7 | 33,003.2 | 8 | 22,415.9 | 27 | 89,229.0 | 3.67 | 18.07 |
| Ubah (9) | *Eugenia* sp. | | | 12 | 7,165.4 | 5 | 6,192.2 | 6 | 6,567.6 | 23 | 19,925.2 | 3.13 | 4.03 |
| Bintangor (2) | *Calophyllum* sp. | 5 | 5,363.4 | 3 | 380.5 | 5 | 6,506.8 | 5 | 1,739.7 | 18 | 13,990.4 | 2.45 | 2.83 |
| Limau sebayan | *Gonocaryum* sp. | 5 | 795.5 | 6 | 889.6 | 5 | 911.3 | 2 | 225.9 | 18 | 2,822.3 | 2.45 | 0.57 |
| Kedondong (6) | | | | 3 | 3,419.9 | 9 | 3,819.9 | 4 | 609.1 | 16 | 7,848.9 | 2.17 | 1.59 |
| Belian landak | *Strombosia rotundifolia* King | 2 | 924.0 | 7 | 1,708.0 | 5 | 1,432.2 | 1 | 203.5 | 15 | 4,267.7 | 2.03 | 0.86 |
| Tinjau belukar | | 5 | 826.3 | 4 | 449.4 | 3 | 403.4 | 3 | 490.7 | 15 | 2,169.8 | 2.03 | 0.44 |
| Meranti sudu | *Shorea quadrinervis* V. sl. | 2 | 3,852.3 | 3 | 3,182.6 | 3 | 1,128.1 | 4 | 551.9 | 12 | 8,714.9 | 1.63 | 1.76 |
| Medang sisik (1) | | 3 | 2,859.5 | | | 3 | 1,090.0 | 4 | 1,027.7 | 10 | 4,977.2 | 1.36 | 1.01 |
| Medang (1) | | | | 8 | 3,764.3 | 2 | 239.8 | | | 10 | 4,004.1 | 1.36 | 0.81 |
| Medang (2) | | 3 | 756.4 | 2 | 920.5 | 2 | 263.5 | 3 | 926.9 | 10 | 2,867.3 | 1.36 | 0.58 |
| Seraya | *Shorea curtisii* Dyer | 2 | 1,741.2 | 3 | 11,837.9 | 3 | 5,899.9 | 1 | 292.4 | 9 | 19,771.4 | 1.22 | 4.00 |

(continued on next page)

Table 2-10   (continued)

| VERNACULAR NAME | BOTANICAL NAME | SP-1 (50 × 50) No. | B.A. | SP-2 (50 × 50) No. | B.A. | SP-3 (50 × 50) No. | B.A. | SP-4 (50 × 50) No. | B.A. | TOTAL/HA No. | B.A. cm² | % No. | B.A. |
|---|---|---|---|---|---|---|---|---|---|---|---|---|---|
| Kembang semankok (2) | Scaphium sp. | 1 | 98.5 | | | 7 | 4,506.3 | 1 | 1,682.8 | 9 | 6,287.6 | 1.22 | 1.27 |
| Bintawak | Artocarpus anisophyllus Miq. | 1 | 642.1 | 3 | 808.6 | 3 | 1,030.2 | 1 | 1,352.0 | 8 | 3,832.9 | 1.09 | 0.78 |
| Sireh-sireh (1) | | | | 4 | 1,363.8 | 2 | 1,049.0 | 2 | 1,209.7 | 8 | 3,622.5 | 1.09 | 0.73 |
| Kedondong (3) | | 4 | 1,283.1 | | | | | 4 | 972.9 | 8 | 2,256.0 | 1.09 | 0.46 |
| Tampoi antu (1) | Baccaurea bracteata Muell.-Arg | 3 | 573.0 | 2 | 192.1 | 1 | 122.7 | 2 | 561.1 | 8 | 1,448.9 | 1.09 | 0.29 |
| Kedondong (2) | | 1 | 89.8 | 1 | 98.5 | 4 | 983.9 | 2 | 195.3 | 8 | 1,367.5 | 1.09 | 0.28 |
| Damar hitam timbul | Shorea laxa V. sl. | 1 | 6,065.2 | | | | | 6 | 7,334.8 | 7 | 13,400.0 | 0.95 | 2.71 |
| Kayu malam (2) | Diospyros sp. | 1 | 132.7 | 2 | 1,440.2 | 1 | 188.6 | 2 | 2,364.4 | 6 | 4,125.9 | 0.83 | 0.83 |
| Ubah (12) | Eugenia sp. | | | 2 | 2,946.5 | 3 | 616.0 | 1 | 356.1 | 6 | 3,918.6 | 0.82 | 0.79 |
| Medang (8) | | | | 3 | 1,971.2 | 2 | 485.4 | 1 | 1,051.6 | 6 | 3,508.2 | 0.82 | 0.71 |
| Tarap | Artocarpus odoratissimus (Blanco) Merr. | 1 | 218.9 | 1 | 257.2 | 2 | 616.0 | 2 | 647.8 | 6 | 1,739.9 | 0.82 | 0.35 |
| Merpisang (2) | | 3 | 651.4 | | | 3 | 398.7 | | | 6 | 1,050.1 | 0.82 | 0.21 |
| Ubah (3) | | 6 | 1,049.9 | | | | | | | 6 | 1,049.9 | 0.82 | 0.21 |
| Keruing buah bulat | Dipterocarpus globosus Vesque | 2 | 11,626.5 | | | | | 3 | 17,394.0 | 5 | 29,020.5 | 0.68 | 5.88 |

(continued on next page)

Table 2-10   *(continued)*

| Vernacular Name | Botanical Name | SP-1 (50 × 50) | | SP-2 (50 × 50) | | SP-3 (50 × 50) | | SP-4 (50 × 50) | | Total/ha | | % | |
|---|---|---|---|---|---|---|---|---|---|---|---|---|---|
| | | No. | B.A. | No. | B.A. | No. | B.A. | No. | B.A. | No. | B.A. cm² | No. | B.A. |
| Nyatoh (2) | | 1 | 923.5 | | | 3 | 4,592.0 | 1 | 107.5 | 5 | 5,623.0 | 0.68 | 1.14 |
| Pendarahan (8) | | | | 2 | 1,277.0 | 2 | 2,317.5 | 1 | 896.8 | 5 | 4,491.3 | 0.68 | 0.91 |
| Kembang semankok jangtong (1) | *Scaphium* sp. | 1 | 151.7 | 3 | 3,752.9 | | | 1 | 151.7 | 5 | 4,056.3 | 0.68 | 0.82 |
| Pendarahan (2) | | 4 | 1,602.7 | | | | | 1 | 165.0 | 5 | 1,767.7 | 0.68 | 0.36 |
| Mengkulat | *Ilex hypoglauca* (Miq.) Loes. | | | | | 2 | 844.3 | 3 | 857.1 | 5 | 1,701.4 | 0.68 | 0.34 |
| Pallih putat | | 1 | 122.7 | 1 | 687.8 | | | 3 | 675.6 | 5 | 1,486.1 | 0.68 | 0.30 |
| Medang (3) | | 1 | 352.8 | | | | | 4 | 968.0 | 5 | 1,320.8 | 0.68 | 0.27 |
| Kayu malam (1) | *Diosypros* sp. | 1 | 91.6 | 2 | 483.0 | 1 | 186.2 | 1 | 216.3 | 5 | 977.1 | 0.68 | 0.20 |
| Merpisang (1) | | 3 | 734.2 | 2 | 185.3 | | | | | 5 | 919.5 | 0.68 | 0.19 |
| Resak (1) | *Vatica* sp. | 2 | 199.5 | | | | | 3 | 498.5 | 5 | 698.0 | 0.68 | 0.14 |
| Pendarahan (3) | | 2 | 175.0 | 2 | 290.3 | | | 1 | 151.7 | 5 | 617.0 | 0.68 | 0.12 |
| Dual | *Lophopetalum javanica* (Zoll) Turcz. | 3 | 372.2 | 1 | 114.9 | | | 1 | 120.7 | 5 | 607.8 | 0.68 | 0.12 |
| Kandis | *Garcinia* sp. | 2 | 248.7 | 1 | 91.6 | 1 | 147.3 | 1 | 84.9 | 5 | 572.5 | 0.68 | 0.12 |
| Ubah (25) | *Eugenia* sp. | | | | | 1 | 81.7 | 4 | 373.2 | 5 | 454.9 | 0.68 | 0.09 |
| Keruing mempeles | *Dipterocarpus crinitus* Dyer | | | 1 | 174.3 | 3 | 4,489.3 | | | 4 | 4,663.6 | 0.54 | 0.94 |

*(continued on next page)*

Table 2-10  (continued)

| Vernacular Name | Botanical Name | SP-1 (50 × 50) | | SP-2 (50 × 50) | | SP-3 (50 × 50) | | SP-4 (50 × 50) | | Total/ha | | % | |
|---|---|---|---|---|---|---|---|---|---|---|---|---|---|
| | | No. | B.A. | No. | B.A. | No. | B.A. | No. | B.A. | No. | B.A. cm² | No. | B.A. |
| Rengas (2) | | | | 2 | 359.2 | 1 | 1,793.6 | 1 | 1,884.8 | 4 | 4,037.6 | 0.54 | 0.82 |
| Bintangor (1) | *Calophyllum* sp. | 1 | 1,575.5 | 1 | 83.3 | | | 2 | 1,763.2 | 4 | 3,422.0 | 0.54 | 0.69 |
| Keranji | *Dialium cochinchinense* Pierre | | | 2 | 2,708.8 | | | 2 | 470.1 | 4 | 3,178.9 | 0.54 | 0.64 |
| Meranti sarang punai | *Shorea parvifolia* Dyer | | | 1 | 203.5 | 3 | 2,912.4 | | | 4 | 3,115.9 | 0.54 | 0.63 |
| Putat | *Barringtonia* sp. | 1 | 226.9 | | | 2 | 1,247.0 | 1 | 1,632.3 | 4 | 3,106.2 | 0.54 | 0.63 |
| Ubah (22) | *Eugenia* sp. | | | | | 1 | 1,218.6 | 3 | 892.6 | 4 | 2,111.2 | 0.54 | 0.43 |
| Ubah (2) | *Eugenia* sp. | 2 | 389.8 | | | 1 | 96.7 | 1 | 860.1 | 4 | 1,346.6 | 0.54 | 0.27 |
| Kedondong (7) | | | | | | 3 | 1,145.8 | 1 | 153.9 | 4 | 1,299.7 | 0.54 | 0.26 |
| Pendarahan (9) | | | | | | 1 | 91.6 | 3 | 1,068.0 | 4 | 1,159.6 | 0.54 | 0.23 |
| Ubah (10) | *Eugenia* sp. | | | 3 | 823.1 | 1 | 268.7 | | | 4 | 1,091.8 | 0.54 | 0.22 |
| Ubah (20) | *Eugenia* sp. | | | | | 1 | 642.1 | 3 | 301.0 | 4 | 943.1 | 0.54 | 0.19 |
| Ubah (5) | *Eugenia* sp. | 1 | 615.4 | 1 | 109.3 | | | 2 | 114.9 | 4 | 839.6 | 0.54 | 0.17 |
| Pendarahan (7) | | | | 1 | 111.2 | 2 | 304.0 | 1 | 376.5 | 4 | 791.7 | 0.54 | 0.16 |
| Bantas | *Blumeodendron* sp. | 1 | 86.5 | | | 1 | 93.3 | 2 | 170.1 | 4 | 349.9 | 0.54 | 0.07 |
| Ubah (16) | *Eugenia* sp. | | | | | 1 | 196.0 | 2 | 2,299.9 | 3 | 2,495.9 | 0.41 | 0.51 |
| Somah | *Ploiarium alternifolium* (Vahl) Melch. | 3 | 2,276.9 | | | | | | | 3 | 2,276.9 | 0.41 | 0.46 |
| Rengas (1) | | | | 1 | 1,915.7 | 2 | 186.5 | | | 3 | 2,102.2 | 0.41 | 0.43 |

(continued on next page)

Table 2-10   *(continued)*

| Vernacular name | Botanical name | SP-1 (50 × 50) No. | B.A. | SP-2 (50 × 50) No. | B.A. | SP-3 (50 × 50) No. | B.A. | SP-4 (50 × 50) No. | B.A. | Total/ha No. | B.A. cm² | % No. | B.A. |
|---|---|---|---|---|---|---|---|---|---|---|---|---|---|
| Merbatu | *Xanthophyllum* sp. | 2 | 1,921.8 | | | 1 | 174.3 | | | 3 | 2,096.1 | 0.41 | 0.42 |
| Resak (2) | *Vatica* sp. | 2 | 292.0 | | | | | 1 | 1,632.3 | 3 | 1,924.3 | 0.41 | 0.39 |
| Ubah (11) | *Eugenia* sp. | | | 1 | 759.3 | 2 | 996.0 | | | 3 | 1,755.3 | 0.41 | 0.36 |
| Medang serukam | *Chaetocarpus castanocarpus* (Roxb.) Thw. | 2 | 1,305.5 | | | 1 | 307.8 | | | 3 | 1,613.3 | 0.41 | 0.33 |
| Pendarahan (1) | | 2 | 735.3 | | | 1 | 655.6 | | | 3 | 1,390.9 | 0.41 | 0.28 |
| Minyak borok | *Xanthophyllum* sp. | 1 | 580.8 | 2 | 738.4 | | | | | 3 | 1,319.2 | 0.41 | 0.27 |
| Medang (11) | | | | 1 | 589.3 | 2 | 589.4 | | | 3 | 1,178.7 | 0.41 | 0.24 |
| Ubah (7) | | 2 | 380.1 | 1 | 314.0 | | | | | 3 | 694.1 | 0.41 | 0.14 |
| Medang (5) | | 1 | 326.7 | | | 2 | 311.7 | | | 3 | 638.4 | 0.41 | 0.13 |
| Nyatoh (7) | | 3 | 617.1 | | | | | | | 3 | 617.1 | 0.41 | 0.12 |
| Merpisang (4) | | | | | | 1 | 105.6 | 2 | 434.3 | 3 | 539.9 | 0.41 | 0.11 |
| Entabuloh | *Gironniera nervosa* Planch. | 1 | 158.3 | | | 2 | 322.4 | | | 3 | 480.7 | 0.41 | 0.10 |
| Ubah (17) | | | | 2 | 313.2 | 1 | 156.1 | | | 3 | 469.3 | 0.41 | 0.10 |
| Saga saga | *Ormosia sumatrana* (Miq.) Prain | 2 | 244.8 | 1 | 153.9 | | | | | 3 | 398.7 | 0.41 | 0.08 |
| Meranti kawang tikus | *Shorea acuta* Ashton | 1 | 122.7 | 1 | 130.6 | | | 1 | 126.6 | 3 | 379.9 | 0.41 | 0.08 |
| Pendarahan (5) | | 1 | 107.5 | | | 1 | 95.0 | 1 | 120.7 | 3 | 323.2 | 0.41 | 0.07 |

*(continued on next page)*

Table 2-10   (continued)

| Vernacular name | Botanical name | SP-1 (50 × 50) No. | B.A. | SP-2 (50 × 50) No. | B.A. | SP-3 (50 × 50) No. | B.A. | SP-4 (50 × 50) No. | B.A. | Total/ha No. | B.A. cm² | % No. | B.A. |
|---|---|---|---|---|---|---|---|---|---|---|---|---|---|
| Ubah (21) | | | | | | 3 | 276.2 | | | 3 | 276.2 | 0.41 | 0.06 |
| Unknown 313/369 | | | | 2 | 7,785.4 | | | | | 2 | 7,785.4 | 0.27 | 1.58 |
| Kelidang babi | *Artocarpus anisophyllus* Miq. | 1 | 4,462.9 | | | | | 1 | 1,157.5 | 2 | 5,620.4 | 0.27 | 1.14 |
| Bintangor (3) | *Calophyllum* sp. | | | 1 | 3,629.8 | | | 1 | 1,846.5 | 2 | 5,476.3 | 0.27 | 1.11 |
| Sabong ribut | *Ctenolophon parvifolius* Oliv. | | | 1 | 3,846.5 | 1 | 400.9 | | | 2 | 4,247.4 | 0.27 | 0.86 |
| Kempas | *Koompassia malaccensis* Maing. ex Benth. | | | | | 2 | 3,943.2 | | | 2 | 3,943.2 | 0.27 | 0.80 |
| Bitis | | 1 | 1,561.5 | | | 1 | 1,358.5 | | | 2 | 2,920.0 | 0.27 | 0.59 |
| Meranti laut puteh | *Shorea rubella* Ashton | | | | | | | 2 | 1,830.7 | 2 | 1,830.7 | 0.27 | 0.37 |
| Ubah (24) | | | | | | 2 | 1,450.4 | | | 2 | 1,450.4 | 0.27 | 0.29 |
| Medang (15) | | | | | | | | 2 | 1,034.0 | 2 | 1,034.0 | 0.27 | 0.21 |
| Ubah (6) | *Eugenia* sp. | 1 | 176.6 | | | 1 | 854.9 | | | 2 | 1,031.5 | 0.27 | 0.21 |
| Medang (4) | | 2 | 939.0 | | | | | | | 2 | 939.0 | 0.27 | 0.19 |
| Rambutan hutan | *Nephelium* sp. | | | 1 | 452.2 | 1 | 339.6 | | | 2 | 791.8 | 0.27 | 0.16 |
| Mer. sarang punai bukit | *Shorea ovata* Dyer ex Brandis | | | 1 | 551.3 | | | 1 | 179.0 | 2 | 730.3 | 0.27 | 0.15 |

(continued on next page)

Table 2-10   (continued)

| VERNACULAR NAME | BOTANICAL NAME | SP-1 (50 × 50) | | SP-2 (50 × 50) | | SP-3 (50 × 50) | | SP-4 (50 × 50) | | TOTAL/HA | | % | |
|---|---|---|---|---|---|---|---|---|---|---|---|---|---|
| | | No. | B.A. | No. | B.A. | No. | B.A. | No. | B.A. | No. | B.A. cm² | No. | B.A. |
| Medang (12) | | | | | | 1 | 109.3 | 1 | 534.7 | 2 | 644.0 | 0.27 | 0.13 |
| Kedondong (1) | | 1 | 415.3 | 2 | 7,785.4 | 1 | 124.6 | | | 2 | 539.9 | 0.27 | 0.11 |
| Asam damaran | Mangifera quadrifida Jack | | | | | 1 | 366.2 | 1 | 160.5 | 2 | 526.7 | 0.27 | 0.11 |
| Pendarahan (10) | | | | | | | | 2 | 503.3 | 2 | 503.3 | 0.27 | 0.10 |
| Nipis kulit | Memecylon borneense Merr. | 1 | 83.3 | | | | | 1 | 323.5 | 2 | 406.8 | 0.27 | 0.08 |
| Merawan daun tebal | Hopea treubii Heim | | | | | | | 2 | 403.2 | 2 | 403.2 | 0.27 | 0.08 |
| Tempinis | Artocarpus kemando Miq. | | | | | | | 2 | 325.4 | 2 | 325.4 | 0.27 | 0.07 |
| Pendarahan (4) | | 2 | 321.1 | | | | | | | 2 | 321.1 | 0.27 | 0.06 |
| Sawar bubu | Canthium confertum Kunth. | | | 2 | 305.3 | | | | | 2 | 305.3 | 0.27 | 0.06 |
| Sireh-sireh (2) | | | | | | | | 2 | 259.1 | 2 | 259.1 | 0.27 | 0.05 |
| Tampoi antu (2) | Baccaurea sp. | | | 1 | 80.1 | 1 | 169.6 | | | 2 | 249.7 | 0.27 | 0.05 |
| Meranti paya | Shorea platycarpa Heim | | | | | 2 | 226.1 | | | 2 | 226.1 | 0.27 | 0.05 |
| Ubah (14) | Eugenia sp. | | | 2 | 197.3 | | | | | 2 | 197.3 | 0.27 | 0.04 |
| Ramin | Gonystylus velutinus Airy Shaw | | | | | 1 | 111.2 | 1 | 78.5 | 2 | 189.7 | 0.27 | 0.04 |
| Pupis | | | | | | 2 | 186.7 | | | 2 | 186.7 | 0.27 | 0.04 |
| Ubah (1) | Eugenia sp. | 2 | 186.6 | | | | | | | 2 | 186.6 | 0.27 | 0.04 |

(continued on next page)

Table 2-10   (continued)

| Vernacular Name | Botanical Name | SP-1 (50 × 50) No. | SP-1 B.A. | SP-2 (50 × 50) No. | SP-2 B.A. | SP-3 (50 × 50) No. | SP-3 B.A. | SP-4 (50 × 50) No. | SP-4 B.A. | Total/Ha No. | Total/Ha B.A. cm² | % No. | % B.A. |
|---|---|---|---|---|---|---|---|---|---|---|---|---|---|
| Kapor paringgi | Dryobalanops aromatica Gaertn. f. | | | | | | | 1 | 10,381.6 | 1 | 10,381.6 | 0.14 | 2.10 |
| Damar laut merah | Shorea kunstleri King | 1 | 6,079.0 | | | | | | | 1 | 6,079.0 | 0.14 | 1.23 |
| Kerung kerubong | Dipterocarpus geniculatus Vesque | 1 | 4,800.5 | | | | | | | 1 | 4,800.5 | 0.14 | 0.97 |
| Resak hitam | Cotylelobium melanoxylon (Hook. f.) Pierre | 1 | 4,510.3 | | | | | | | 1 | 4,510.3 | 0.14 | 0.91 |
| Bintangor (5) | Calophyllum sp. | | | | | | | 1 | 3,316.6 | 1 | 3,316.6 | 0.14 | 0.67 |
| Meranti paya bersisik | Shorea scaberrima Burck | | | | | | | 1 | 3,225.4 | 1 | 3,225.4 | 0.14 | 0.65 |
| Unknown 695 | | | | | | | | 1 | 2,374.6 | 1 | 2,374.6 | 0.14 | 0.48 |
| Unknown 771 | | | | | | | | 1 | 2,374.6 | 1 | 2,374.6 | 0.14 | 0.48 |
| Nyatoh (1) | | 1 | 1,358.5 | | | | | | | 1 | 1,358.5 | 0.14 | 0.27 |
| Meranti langgai | Shorea beccariana Burck | 1 | 1,915.7 | | | | | | | 1 | 1,915.7 | 0.14 | 0.39 |
| Meranti lop | Shorea scabrida Sym. | | | | | 1 | 1,431.3 | | | 1 | 1,431.3 | 0.14 | 0.29 |
| Ubah (19) | Eugenia sp. | | | | | 1 | 1,313.2 | | | 1 | 1,313.2 | 0.14 | 0.27 |
| Unknown 703 | | | | | | | | 1 | 1,212.4 | 1 | 1,212.4 | 0.14 | 0.25 |
| Mengilas | Parastemon urophyllum A. DC. | | | 1 | 989.3 | | | | | 1 | 989.3 | 0.14 | 0.20 |

(continued on next page)

Table 2-10  (continued)

| Vernacular name | Botanical name | SP-1 (50 × 50) | | SP-2 (50 × 50) | | SP-3 (50 × 50) | | SP-4 (50 × 50) | | Total/ha | | % | |
|---|---|---|---|---|---|---|---|---|---|---|---|---|---|
| | | No. | B.A. | No. | B.A. | No. | B.A. | No. | B.A. | No. | B.A. cm² | No. | B.A. |
| Nyatoh (9) | | | | | | | | 1 | 939.8 | 1 | 939.8 | 0.14 | 0.19 |
| Unknown 634 | | | | | | | | 1 | 912.8 | 1 | 912.8 | 0.14 | 0.18 |
| Benchaloi | Anisoptera grossivenia V. Sl. | 1 | 886.2 | | | | | | | 1 | 886.2 | 0.14 | 0.18 |
| Berangan | Castanopsis sp. | 1 | 881.0 | | | | | | | 1 | 881.0 | 0.14 | 0.18 |
| Kedondong (5) | | 1 | 764.2 | | | | | | | 1 | 764.2 | 0.14 | 0.15 |
| Ubah (23) | Eugenia sp. | | | | | 1 | 678.5 | | | 1 | 678.5 | 0.14 | 0.14 |
| Medang (13) | | | | | | 1 | 619.8 | | | 1 | 619.8 | 0.14 | 0.13 |
| Ubah (4) | Eugenia sp. | 1 | 585.0 | | | | | | | 1 | 585.0 | 0.14 | 0.12 |
| Ranchah-ranchah | Mangifera havilandii Ridl. | 1 | 568.0 | | | | | | | 1 | 568.0 | 0.14 | 0.12 |
| Unknown 623 | | | | | | | | 1 | 482.8 | 1 | 482.8 | 0.14 | 0.10 |
| Nyatoh (5) | | 1 | 432.4 | | | | | | | 1 | 432.4 | 0.14 | 0.09 |
| Nyatoh (4) | | 1 | 408.1 | | | | | | | 1 | 408.1 | 0.14 | 0.08 |
| Lakong | Pseudonephelium fumatum (Bl.) Radlk. | | | | | 1 | 317.1 | | | 1 | 317.1 | 0.14 | 0.06 |
| Kedondong (8) | | | | | | | | 1 | 298.5 | 1 | 298.5 | 0.14 | 0.06 |
| Nyatoh (6) | | 1 | 295.4 | | | | | | | 1 | 295.4 | 0.14 | 0.06 |
| Merpisang (3) | | 1 | 286.4 | | | | | | | 1 | 286.4 | 0.14 | 0.06 |
| Nyatoh (8) | | | | 1 | 286.4 | | | | | 1 | 286.4 | 0.14 | 0.06 |

(continued on next page)

Table 2-10   (continued)

| VERNACULAR NAME | BOTANICAL NAME | SP-1 (50 × 50) | | SP-2 (50 × 50) | | SP-3 (50 × 50) | | SP-4 (50 × 50) | | TOTAL/HA | | % | |
|---|---|---|---|---|---|---|---|---|---|---|---|---|---|
| | | No. | B.A. | No. | B.A. | No. | B.A. | No. | B.A. | No. | B.A. cm² | No. | B.A. |
| Ubah (15) | *Eugenia* sp. | | | 1 | 283.4 | | | | | 1 | 283.4 | 0.14 | 0.06 |
| Ludai | | | | 1 | 277.5 | | | | | 1 | 277.5 | 0.14 | 0.06 |
| Pendarahan (6) | | 1 | 277.5 | | | | | | | 1 | 277.5 | 0.14 | 0.06 |
| Ubah (8) | *Eugenia* sp. | 1 | 277.5 | | | | | | | 1 | 277.5 | 0.14 | 0.06 |
| Ubah (26) | *Eugenia* sp. | | | | | | | 1 | 271.6 | 1 | 271.6 | 0.14 | 0.05 |
| Ubah (13) | *Eugenia* sp. | | | 1 | 232.2 | | | | | 1 | 232.2 | 0.14 | 0.05 |
| Mempening | *Litbocarpus* sp. | | | | | 1 | 229.5 | | | 1 | 229.5 | 0.14 | 0.05 |
| Keruing keset | *Dipterocarpus gracilis* Bl. | | | | | 1 | 224.2 | | | 1 | 224.2 | 0.14 | 0.05 |
| Durian borong | *Durio* sp. | 1 | 203.5 | | | | | | | 1 | 203.5 | 0.14 | 0.04 |
| Bintangor (4) | *Calophyllum* sp. | | | 1 | 165.0 | | | | | 1 | 165.0 | 0.14 | 0.03 |
| Petaling | *Ochanostachys amentacea* Mast. | 1 | 162.8 | | | | | | | 1 | 162.8 | 0.14 | 0.03 |
| Tampoi (1) | *Baccaurea* sp. | | | 1 | 158.3 | | | | | 1 | 158.3 | 0.14 | 0.03 |
| Resak hijau | *Vatica micrantha* V. Sl. | 1 | 156.1 | | | | | | | 1 | 156.1 | 0.14 | 0.03 |
| Tampoi (2) | *Baccaurea* sp. | 1 | 153.9 | | | | | | | 1 | 153.9 | 0.14 | 0.03 |
| Medang (7) | | | | 1 | 153.9 | | | | | 1 | 153.9 | 0.14 | 0.03 |
| Ubah | *Eugenia* sp. | | | 1 | 138.9 | | | | | 1 | 138.9 | 0.14 | 0.03 |
| Lanang | | | | 1 | 136.8 | | | | | 1 | 136.8 | 0.14 | 0.03 |

(continued on next page)

Table 2-10   (*continued*)

| VERNACULAR NAME | BOTANICAL NAME | SP-1 (50 × 50) | | SP-2 (50 × 50) | | SP-3 (50 × 50) | | SP-4 (50 × 50) | | TOTAL/HA | | % | |
|---|---|---|---|---|---|---|---|---|---|---|---|---|---|
| | | No. | B.A. | No. | B.A. | No. | B.A. | No. | B.A. | No. | B.A. cm$^2$ | No. | B.A. |
| Medang (9) | | | | 1 | 132.7 | | | | | 1 | 132.7 | 0.14 | 0.03 |
| Peragan-peragan | | | | | | | | 1 | 128.6 | 1 | 128.6 | 0.14 | 0.03 |
| Medang (6) | | 1 | 118.8 | | | | | | | 1 | 118.8 | 0.14 | 0.02 |
| Medang (21) | | | | | | | | 1 | 116.8 | 1 | 116.8 | 0.14 | 0.02 |
| Kelimpanas | | | | 1 | 114.9 | | | | | 1 | 114.9 | 0.14 | 0.02 |
| Tempagas | *Memecylon borneense* Merr. | | | 1 | 114.9 | | | | | 1 | 114.9 | 0.14 | 0.02 |
| Ubah (18) | *Eugenia* sp. | | | | | | | 1 | 113.0 | 1 | 113.0 | 0.14 | 0.02 |
| Unknown 673 | | | | | | | | 1 | 107.5 | 1 | 107.5 | 0.14 | 0.02 |
| Asam (1) | *Mangifera* sp. | 1 | 103.8 | | | | | | | 1 | 103.8 | 0.14 | 0.02 |
| Nyatoh (3) | | 1 | 103.8 | | | | | | | 1 | 103.8 | 0.14 | 0.02 |
| Durian (2) | *Durio* sp. | | | | | 1 | 95.0 | | | 1 | 95.0 | 0.14 | 0.02 |
| Mergasing | *Kayea* sp. | | | | | | | 1 | 95.0 | 1 | 95.0 | 0.14 | 0.02 |
| Medang (14) | | | | | | | | 1 | 91.6 | 1 | 91.6 | 0.14 | 0.02 |
| Unknown 471 | | | | 1 | 84.9 | | | | | 1 | 84.9 | 0.14 | 0.02 |
| Rengas (3) | | | | | | | | 1 | 78.5 | 1 | 78.5 | 0.14 | 0.02 |
| Total | | 179 | 109,320.0 | 165 | 120,920.4 | 202 | 128,655.7 | 190 | 134,982.2 | 736 | 493,878.3 | 100.00 | 100.00 |

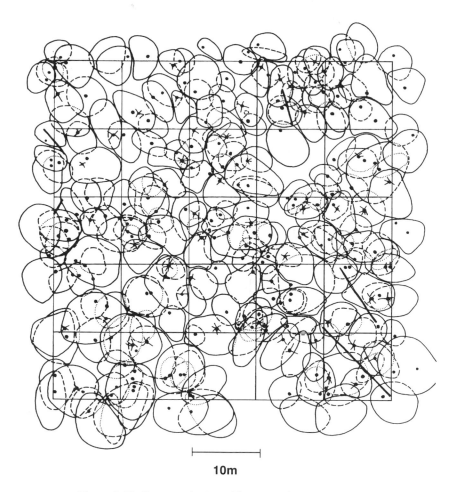

**10m**

Figure 2-22. Crown projection of *kapor paya* forest (50 × 50 m).

sional stage following logging or fire, as the dominant species, *Dryobalanops lowii* (*kapor paya*), is normally a large tree reaching 50 m.

PHYSIOGNOMY

The *Dryobalanops rappa*, which almost exclusively dominates the overstory, is readily identifiable by its straight trunks bearing a continuous canopy of hemispherical crowns. The trees are not particularly tall, around 30 m. No other species is notable in the overstory. Various species occupy the second story, and immediately below is the ground vegetation, which includes the *Salacca* palm. The forest floor has shallow litter and peat overlying sandy soil. In the dry season it is dry, but in the rainy season rainwater flows through the low ground (figs. 2-22 and 2-23).

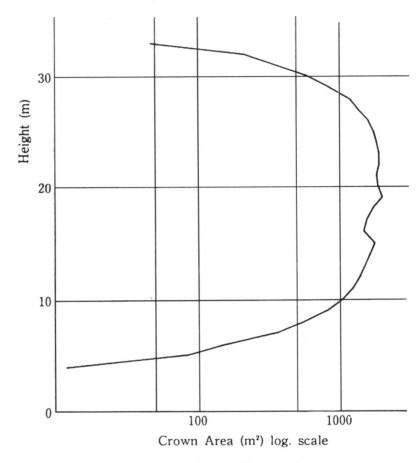

Figure 2-23. Stratification of *kapor paya* forest.

Climbers and epiphytes are scarce. Many trees can be seen that have regenerated through coppicing after forest fire. *Dryobalanops rappa* flowers every year in March. In 1986 flowering was abundant, and almost every tree in the overstory bore fruit.

TREE SPECIES COMPOSITION

The tree density is second highest after *ulat bulu,* which is probably a feature of secondary forest, but basal area, maximum height, and maximum diameter are small. The forest can be regarded as a regenerating forest about thirty years old.

*Dryobalanops rappa* is overwhelmingly predominant in numbers and basal area. However, the species number, at 28, was in the *ulat bulu* class. *Eugenia, Garcinia, Litsea, Palaquium,* and others were present as small trees. *Tetramerista* and other

Table 2-11

Tree Species Composition of *Kapor Paya* Forest

| VERNACULAR NAME | BOTANICAL NAME | FAMILY | SP-1 (50 × 50) | | TOTAL/HA | | % | |
|---|---|---|---|---|---|---|---|---|
| | | | No. | B.A. | No. | B.A. m² | No. | B.A. |
| Kapor paya | *Dryobalanops rappa* Becc. | Dipterocarpaceae | 138 | 79,094.1 | 552 | 316,376.4 | 62.44 | 86.21 |
| Ubah (1) | *Eugenia* sp. | Myrtaceae | 12 | 1,848.4 | 48 | 7,393.6 | 5.43 | 2.01 |
| Ubah (2) | *Eugenia* sp. | Myrtaceae | 9 | 1,200.2 | 36 | 4,800.8 | 4.07 | 1.31 |
| Kandis (2) | *Garcinia* sp. | Guttiferae | 8 | 1,944.6 | 32 | 7,778.4 | 3.62 | 2.12 |
| Medang (1) | | | 7 | 755.2 | 28 | 3,020.8 | 3.17 | 0.82 |
| Nyatoh (2) | | Sapotaceae | 6 | 930.0 | 24 | 3,720.0 | 2.71 | 1.01 |
| Kandis (1) | *Garcinia* sp. | Guttiferae | 5 | 950.5 | 20 | 3,802.0 | 2.26 | 1.04 |
| Rambutan hutan | *Nephelium* sp. | Sapindaceae | 4 | 596.1 | 16 | 2,384.4 | 1.81 | 0.65 |
| Nyatoh (1) | | Sapotaceae | 3 | 534.9 | 12 | 2,139.6 | 1.35 | 0.58 |
| Mengkulat | *Ilex hypoglauca* (Miq.) Loes. | Aquifoliaceae | 3 | 459.5 | 12 | 1,838.0 | 1.35 | 0.50 |
| Mengilas | *Parastemon urophyllum* A. DC. | Rosaeae | 3 | 280.3 | 12 | 1,121.2 | 1.35 | 0.31 |
| Resak air | *Vatica umbonata* (Hook. f.) Burck. | Dipterocarpaceae | 2 | 619.0 | 8 | 2,476.0 | 0.90 | 0.67 |
| Medang (2) | | | 2 | 284.2 | 8 | 1,136.8 | 0.90 | 0.31 |
| Merpisang (1) | *Polyalthia* sp. | Annonaceae | 2 | 209.9 | 8 | 839.6 | 0.90 | 0.23 |
| Medang (3) | | | 2 | 207.2 | 8 | 828.8 | 0.90 | 0.23 |
| Mentang kelait | | | 2 | 188.7 | 8 | 754.8 | 0.90 | 0.21 |
| Kedondong (2) | | | 2 | 173.3 | 8 | 693.2 | 0.90 | 0.19 |

*(continued on next page)*

Table 2-11   (continued)

| Vernacular name | Botanical name | Family | SP-1 (50 × 50) | | Total/ha | | % | |
|---|---|---|---|---|---|---|---|---|
| | | | No. | B.A. | No. | B.A. m² | No. | B.A. |
| Kedondong (1) | | | 1 | 213.7 | 4 | 854.8 | 0.45 | 0.23 |
| Merpisang (2) | *Polyalthia* sp. | Annonaceae | 1 | 193.5 | 4 | 774.0 | 0.45 | 0.21 |
| Rengas (1) | *Melanorrhoea* sp. | Anarcardiaceae | 1 | 179.0 | 4 | 716.0 | 0.45 | 0.20 |
| Unknown (1) | | | 1 | 176.6 | 4 | 706.4 | 0.45 | 0.19 |
| Amat | *Tetramerista glabra* Miq. | Tetramerisaceae | 1 | 156.1 | 4 | 624.4 | 0.45 | 0.17 |
| Pendarahan (1) | | | 1 | 120.7 | 4 | 482.8 | 0.45 | 0.13 |
| Pendarahan (2) | | | 1 | 98.5 | 4 | 394.0 | 0.45 | 0.11 |
| Nantungan | | | 1 | 86.5 | 4 | 346.0 | 0.45 | 0.09 |
| Medang (4) | | | 1 | 83.3 | 4 | 333.2 | 0.45 | 0.09 |
| Ubah (3) | *Eugenia* sp. | Myrtaceae | 1 | 83.3 | 4 | 333.2 | 0.45 | 0.09 |
| Bintangor | *Calophyllum* sp. | Gutiferae | 1 | 80.1 | 4 | 320.4 | 0.45 | 0.09 |
| Total | | | 221 | 91,747.4 | 884 | 366,989.6 | 100.00 | 100.00 |

### Table 2-12
### Typical Values of Sample Plots

| FOREST TYPE | PLOT SIZE (m²) | NO./HA | B.A./HA (m²) | SPECIES NO./PLOT | HMAX (m) | DMAX (cm) |
|---|---|---|---|---|---|---|
| Agathis | 100 × 50 | 610 | 38.4 | 33 | 47.0 | 92.7 |
| Alan batu | 100 × 50 | 522 | 42.6 | 24 | 57.5 | 140.0 |
| Alan bunga | 100 × 50 | 310 | 43.4 | 22 | 51.5 | 96.6 |
| Alan | 50 × 50 | 400 | 50.7 | 14 | 51.0 | 80.5 |
| Padang alan | 50 × 50 | 872 | 39.7 | 17 | 43.5 | 63.5 |
| Alan padang | 50 × 50 | 744 | 34.5 | 10 | 33.0 | 46.0 |
| Ulat bulu | 50 × 50 | 1172 | 32.6 | 26 | 30.5 | 46.5 |
| Mixed peat swamp (1) | 50 × 50 | 828 | 36.5 | 41 | 36.0 | 73.3 |
| Mixed peat swamp (2) | 100 × 100 | 645 | 33.9 | 58 | 47.0 | 140.0 |
| Mixed dipterocarp | 100 × 100 | 736 | 49.4 | 174 | 50.0 | 114.0 |
| Kapor paya | 50 × 50 | 884 | 36.7 | 28 | 33.5 | 47.5 |

trees of freshwater swamp forest and peat swamp forest were also found, as were *Vatica* and other Dipterocarpaceae (table 2-11).

## Overview

Table 2-12 shows typical values for the sample plots in each of the forest types described above. Of the eleven plots, nine are swamp forest. The reasons for this include the fact that, while mixed dipterocarp forests have been quite extensively surveyed, little apart from Anderson's work has been done in relation to swamp forests. In addition, swamp forests are important for Brunei's future forestry policy.

From the figures in the table, it is clear that while the swamp forests are limited in the number of species present, their sizes and volumes are considerable. The fact that peat swamp forest is now being exploited for lumber in Sarawak and Brunei is because there is such a high volume. The paucity of species, moreover, particularly the presence of *alan* as a single dominant species, is extremely advantageous.

The mixed dipterocarp forest here is not notably extensive, but the Andulau Forest Reserve is on the small side even for Brunei, and a more extensive forest lies in the interior of Labi. The results of the present survey might suggest peat swamp forest is more extensive, but it is probably necessary to investigate more survey plots before making a comparison.

The larger peat swamp forests all contained *alan*. It is obvious that the presence of this species greatly influences forest biomass. In the stratification profiles, too, *alan* predominates in the overstory. By comparison, no one species is particularly remarkable in the mixed dipterocarp forest, where dipterocarps are found from the overstory to the understory.

For forest management, selective felling at sufficiently long intervals should allow natural regeneration in mixed dipterocarp forest and other forest types where continuous regeneration takes place. Where exploitation approaches clear-felling, however, as with the *alan* of peat swamp forest, regeneration is highly problematic. This point needs to be tackled in earnest.

It is also probably desirable to develop methods for exploiting not only *alan* but other useful swamp forest species where comparatively large trees are present in large numbers, such as *Dryobalanops rappa, Diospyros, nantungan,* and *Copaifera palustris.*

The tapping of *Dyera lowii* for sap suggests a further possibility that is presently totally unexplored. Advanced methods of chemical component analysis should be applied in developing methods for extraction and utilization of trace quantities of useful substances.

The difficulty of approach must be overcome, for which it seems best to build fixed, permanent railway lines around the swamps. After felling and extraction, these can be used for experiments and research into forest management and silviculture. Japan has areas where forestry was previously established by the building of forest railways, and in the swamp zones, where road building is difficult, the railway probably remains more effective.

In my present survey, the number of plots was limited, and ideally this should be increased. There are many examples in various parts of the world where permanent plots have been established on scales from hundreds to thousands of hectares, with surveys repeated at regular intervals. In the future, the Brunei Forestry Department would do well to conduct surveys on a wider scale.

## 2. Flowering and Fruiting of the Main Tree Species in Virgin Forest

The flowering and fruiting habits of trees in tropical rain forests are not as simple as in temperate zones, where most species flower in spring and bear fruit in autumn. As is clear from several reports in the past, the periodicity of these habits depends not only on species but also on the individual tree, sometimes even the individual branch.

In the Malaysia area, Holttum, Koriba, Medway, Ng, Sasaki, Tamari, Chan, and others have conducted surveys to examine the question of phenology. These all relate to peninsular Malaysia and Singapore, and very little work has been done on Borneo.

In this section, I shall report the results of phenological surveys in the representative forest types around Sungai Liang.

## Survey Methods and Sites

In observing phenology, it is best to use observation towers and ladders that bridge between tree crowns, but because no preparations had been made on this occasion, all observations were made from the ground using binoculars. Observations, which are still continuing, were made once every two weeks, employing a team of two observers and one recorder. Here I summarize the results of two years' observations.

### ARBORETUM ANDULAU

An arboretum of virgin forest occupies a hill in what might be called the backyard of the Sungai Liang research center. Here a virgin stand of mixed dipterocarp forest is conserved, footpaths have been made, and species name-tags attached to some of the trees. A nonoverlapping route along paths through and around the arboretum was chosen so that observations could be completed in one day, along which 227 trees more than 5 cm in diameter were selected to give a reasonable representation of each story in the forest.

### COMPARTMENT 7 (ANDULAU FOREST RESERVE)

A stretch of virgin forest lies along the track leading to the forest where the Andulau permanent survey area was established. The forest is basically the same as in the arboretum, and it was selected partly to allow comparison with the arboretum.

### AGATHIS FOREST

In the *Agathis* forest at Badas, where a permanent survey area was established, the survey covered all 143 trees greater than 10 cm DBH in a plot 50 × 50 m. The main purpose here was to examine the phenology of *Agathis* and the associated *kerangas* forest.

### ALAN BUNGA FOREST

The *alan bunga* forest was a little removed from Sungai Liang, but because it was important to learn about the phenology of *alan,* the main forestry species of Brunei, a survey was conducted of all standing trees more than 10 cm DBH in the survey plot. The main purpose was to determine the periodicity of *alan* flowering, at the same time investigating the flowering of understory trees.

## Phenology by Forest Type

### Arboretum Andulau

The 227 trees surveyed here were the most numerous of any site, and the work took a full day. Many species are still to be identified, and only the main results were recorded.

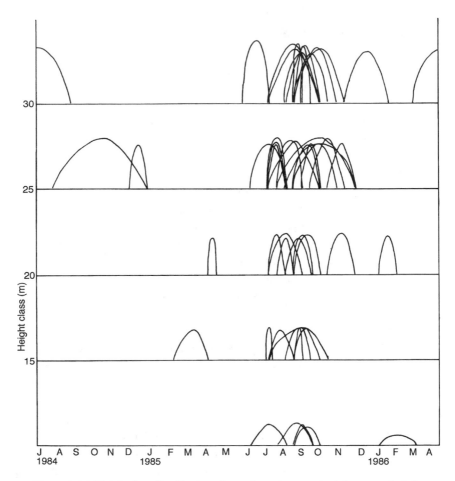

Figure 2-24. Height class distribution of trees flowering once in Arboretum Andulau.

Of the 227 trees, 53 flowered once, 5 twice, 3 three times, and 1 four times. Thirty-seven fruited once, 5 twice, and 3 three times. Leaf fall was observed in 21 trees once and in 4 twice.

Most of the trees that flowered once belonged to the families Dipterocarpaceae, Sapotaceae, Moraceae, Fagaceae, and Anacardiaceae.

The flowering times of trees that flowered and fruited once in the survey period, with the exception of *Cotylelobium melanoxylon* (127), were concentrated in the period from 15 June to 2 November 1985, particularly in the period from 27 July to 7 September, when 19 of the 24 trees flowered. Thus although flowering occurred in the plot year-round, it was most concentrated from late July to early September. Figure 2-24 shows the flowering seasons of trees that bloomed by height class (5-m intervals), from which it is also clear that flowering was concentrated in August.

Figure 2-25 shows the relation between the times of flowering and fruiting, of which three patterns can be discerned: fruiting beginning during flowering; fruiting beginning immediately after flowering; and fruiting beginning following an interval after flowering. Most common was the first pattern, shown by 18 of 30 trees.

The fruiting period was generally longer than flowering. It was particularly long in *Castanopsis* (2), *Calophyllum* (84), *Melanorrhoea* (4, 27), and *pendarahan* (42) (unidentified), and especially so with no. 42.

In the tropics, fruiting often begins before flowering ends, and examples have been recorded of individual branches of the same tree flowering and fruiting at different times.

Of 53 trees that flowered, 23 did not bear fruit, a sizable proportion. Genecology in the tropics remains an undeveloped field, but with phenological observations as a handhold, a deeper study of particular species can be undertaken.

The tall trees of the 30-m class virtually all belonged to the Dipterocarpaceae. Of the 227 trees surveyed, 56 were dipterocarps, of which 12 trees representing nine species flowered and 8 trees in seven species went on to bear fruit. The seven species were: *Anisoptera grossivenia* (72), *Shorea rubella* (53, 116), *Shorea scaberrima* (227), *Shorea ovata* (110), *Dryobalanops beccarii* (145), *Dipterocarpus globosus* (147), and *Cotylelobium melanoxylon* (127). Of these the *Cotylelobium melanoxylon* fruited from 22 September 1984 to 9 February 1985, while the others all fruited at slightly different times between 21 September 1985 and 22 February 1986. These phenological activities in 1985 can be considered to represent comparatively gregarious flowering and heavy fruiting of the dipterocarps in the wake of a drought in 1983.

Because these results derive from just under two years of observations, nothing can be said about the periodicity of flowering and fruiting, but they do indicate that flowering is concentrated in August for most tree species, followed by fruiting toward the end of the year, and from this, preparations can be made for seed collection and nursery construction.

In seed collection, it is important to collect from trees showing the best characters possible, to secure a basis for tree breeding and silviculture in the tropics. Because of the difficulty of collecting any seed, there has been a universal tendency to collect from any parent tree that is in fruit. Seeds are the basis of all silviculture, however, and without checking the genetic characters inherent in seeds, there can be no hope for better silviculture in the future. For this, a thorough examination is necessary, from a genetic viewpoint, of the characters of seedlings grown from the selected seeds.

Leaves were shed during the survey period once by 18 trees and twice by 5.

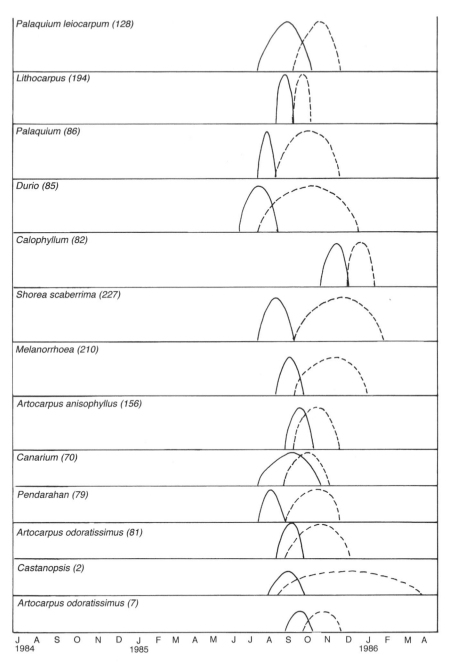

Figure 2-25. Relation between flowering and fruiting seasons in Arboretum Andulau.
Solid lines indicate flowering; broken lines indicate fruiting.

Plate 1. Mangrove forest at Temburong, Brunei. Compared with nearby
Malaysia, the natural forest is well preserved, with only small clearings
for charcoal burning.

Plate 2. Peat swamp forest at Badas, Brunei. The forest, dominated by *alan* *(Shorea albida)*, has a flat, densely crowded canopy.

Plate 3. *Alan* forest at Badas, Brunei. *Alan* covers an expanse of flat swampland. In the center is a water pipe.

Plate 4. *Alan bunga* forest at Badas, Brunei. *Mengkuwang (Pandanus andersonii)* covers the ground in this pure forest of *alan (Shorea albida)*.

Plate 5. Canopy of *alan bunga* forest at Badas, Brunei, seen from below. The small tree crowns are beautifully distributed.

Plate 6. Area defoliated by *ulat bulu* at Badas, Brunei. Defoliation resembles a volcanic crater covering a wide area.

Plate 7. Logging *alan* forest at Badas, Brunei. Workers from Sarawak haul out lumbe using a *kuda-kuda* sled.

Plate 8.
Land reclamation in mixed peat swamp forest along the Belait River, Brunei. The forest is cleared for planting pineapple, banana, and other crops.

Plate 9. Extensive swamp on the middle reach of the Sepik River, Papua New Guinea, with a vast expanse of freshwater swamp forest.

Plate 10. Fresh water swamp forest on the upper Belait River, Brunei. The rainy season transforms the landscape.

Plate 11. Planted sago at the edge of freshwater swamp forest, Belait River, Brunei.

Plate 12. Natural sago forest on the upper Sepik River, Papua New Guinea. Sago palms reaching 30 m are mixed with broad-leaf trees.

Plate 13. Mixed dipte-rocarp forest at Andulau, Brunei. Dipterocarps predominate, reaching heights of more than 50 m; most have straight trunks.

Plate 14. Mixed dipterocarp forest in the vicinity of Balikpapan, Kalimantan, Indonesia. Note the tall buttress roots.

Plate 15. Mixed dipterocarp forest at Labi, Brunei. Compared with peat swamp forest, the variations in canopy height are prominent.

Plate 16. Logging area in mixed dipterocarp forest at Andulau, Brunei. Felled areas are conspicuous.

Plate 17. Logged area in Balikpapan, Kalimantan, Indonesia. The remaining large trees are not cut because they have become hollow.

Plate 18. Canopy of mixed dipterocarp forest in Aceh, North Sumatra, Indonesia. Resembling a cauliflower, the canopy consists of overlapping layers with subtly different colors.

Plate 19. Area of repeated swiddening after clearing of mixed dipterocarp forest, Aceh, North Sumatra, Indonesia. Extensive grassland is formed, with pines on the ridges.

Plate 20. Oil palm cultivation in Sarawak, Malaysia. Virgin forest—visible in the distance—has been cleared to make way for a uniform expanse of plantations.

Plate 21. *Kerangas* forest, Badas, Brunei. Tall *Agathis* trees are dominant.

Plate 22. Montane forest at around 1,500 m, Mount Pangrango, West Java, Indonesia. This forest has the greatest abundance of epiphytes and creepers.

Plate 23. Moss forest at around 2,800 m, Mount Pangrango, West Java, Indonesia. Mosses grow to about twice the thickness of tree trunks, and the forest is perennially humid.

Plate 24. Monsoon forest at Sarakhet, northeast Thailand. Monsoon forest like this appears inland, to the north of the tropical rain forest, in areas, with a dry season of several months duration.

Plate 25. Savanna landscape resulting from human intervention in monsoon forest, Baluran, East Java, Indonesia.

Plate 26. Botanical garden at Bogor, West Java, Indonesia. A central establishment in botanical research in Southeast Asia, where ex-situ conservation of genetic resources has long been practiced.

For these 23 trees, abscission was most common from 9 March to 18 May 1985, which corresponds to the first dry spell after the rainy season. Next most common was from 25 August to 22 September 1984, and from 24 August to 7 September 1985, which correspond to the second dry spell of the year. Leaf fall thus occurs mainly at the drier times of the year, though a small amount of leaf fall was observed from 30 November to 14 December 1985, when rainfall was considerable. This again is a feature of the tropics: Leaf fall is concentrated in the dry seasons but also occurs at other times.

Trees that shed their leaves twice included *Calophyllum, Anisoptera grossivenia, Shorea scaberrima,* and *Koompassia malaccensis.*

## Compartment 7 (Andulau Forest Reserve)

This area lies within the Andulau Forest Reserve, 4.7 km east-southeast from the arboretum, and has a notable amount of remaining virgin forest. Here, 121 trees were observed.

Flowering of species other than dipterocarps formed a peak from 15 January to 26 March 1986, with a smaller one from 31 August to 25 October 1985. These were all trees that flowered once, while if those that flowered twice or three times are included, flowering was seen throughout the year.

Of the dipterocarps, those that flowered once were concentrated in the period from 31 August to 4 December 1985, with others flowering from 9 April to 21 May 1986 and from 12 September to 19 December 1984. If those that flowered twice or more are included, the peaks from 9 October to about 19 November 1985, and from 12 July to about 12 September 1984, increase. The dipterocarps thus can be regarded as flowering in the driest season in Brunei (fig. 2-26).

Fruiting occurred in the nondipterocarps with a slight delay after flowering and could be seen throughout the year, being most abundant from October 1985 to May 1986 (fig. 2-27).

Fruiting of the dipterocarps occurred at two times, from 11 September 1985 to 15 January 1986, and after 21 May 1986. Leaf fall showed two main peaks, from 12 to 31 July 1984, and from 19 November to 18 December 1985. It was also observed on 12 September, 19 June, and 18 July 1984.

Fruiting most commonly began during flowering. This was seen with *Shorea scaberrima, Calophyllum, Dyera lowii, Litsea, Lithocarpus, Blumeodendron,* and others. Species in which fruiting began immediately after flowering included *Dryobalanops beccarii, Dipterocarpus globosus, Dipterocarpus crinitus,* and *Canarium.* Fruiting after a delay was seen in *Canarium* (56), *Eugenia, Cratoxylum,* and others.

Flowering and fruiting of the dipterocarps occurred at almost the same time in the arboretum and in Compartment 7. For the nondipterocarps, however, many trees flowered and fruited outside of the peak seasons, and the same was also true of leaf fall. Apart from differences between species, this variation may

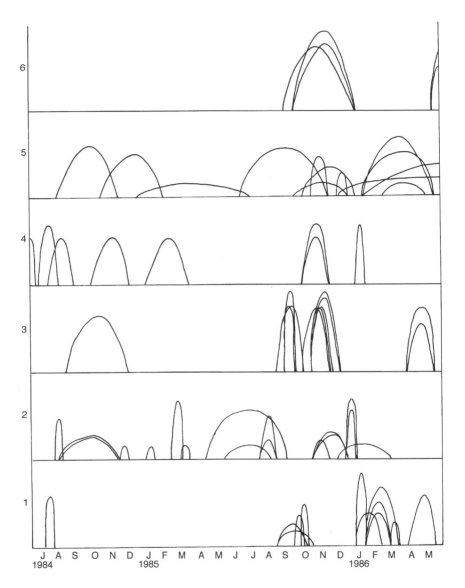

Figure 2-26. Phenology of trees in Compartment 7, Andulau Forest Reserve.
1. Nondipterocarps that flowered once in the survey period; 2. Nondipterocarps that
flowered twice or three times; 3. Dipterocarps that flowered once; 4. Dipterocarps that
flowered twice or more; 5. Nondipterocarps that fruited once or twice; 6. Dipterocarps
that fruited.

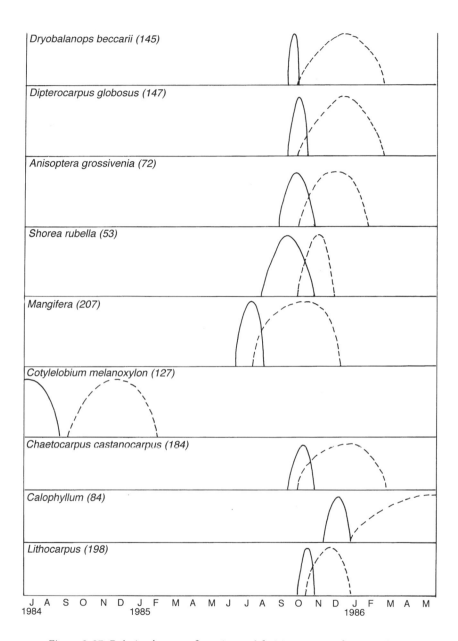

Figure 2-27. Relation between flowering and fruiting seasons of trees in Compartment 7, Andulau Forest Reserve. Solid lines indicate flowering; broken lines indicate fruiting.

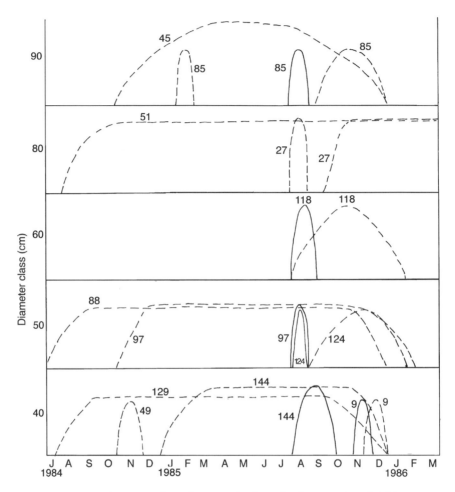

Figure 2-28. Phenology of *Agathis* in *Agathis* forest, by diameter class. Solid lines indicate male cones; broken lines indicate female cones.

have been because of differences in age, location, and stratum in individuals of the same species, and to differences in soil and other factors resulting from the fact that, although the arboretum and Compartment 7 lie in the same Andulau region, the latter occupies slightly higher ground farther from the sea.

## *Agathis* Forest

In the *Agathis* forest, cone development was observed in only 12 of 113 *Agathis* trees, while flowering and fruiting occurred in 16 of 20 *Cotylelobium burckii* trees. No other species flowered or fruited, and no leaf fall was observed.

*Agathis* produced male cones once in 6 trees and female cones once in 10 and

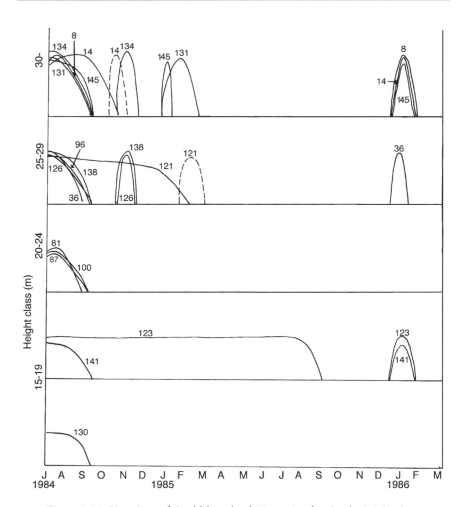

Figure 2-29. Phenology of *Cotylelobium burckii* in an *Agathis* plot, by height class.
Numbers indicate tree numbers. Solid lines indicate flowering; broken lines
indicate fruiting.

twice in 2 trees. Of *Cotylelobium burckii,* 6 trees flowered once, 9 twice, and 1
three times, while 2 trees fruited once (fig. 2-28).

Development of the male cones of *Agathis* was concentrated in the period
from 17 August to late September 1985. Female cones were seen throughout the
year, being abundant from August 1985 to around January 1986.

Flowering of *Cotylelobium burckii* occurred from July to late September 1984
and from 18 January to 1 March 1986, the former being in the dry season, the
latter at the end of the rainy season. The two trees that fruited did so in October-
November 1984 and February-March 1985 (fig. 2-29).

Examination of the phenology of *Agathis* reveals more vigorous reproductive activity with increasing diameter. In the smaller trees of 10-cm to 30-cm diameter classes, almost all trees showed virtually the same changes, with no cone formation. From 14 March to 24 June 1985, no development of new leaves or leaf buds was seen. The main leaf activity took place from 25 October 1984 to 2 March 1985, from 1 August 1985 to 4 January 1986, and from 1 February to 4 April 1986. These observations suggest that leaf activity in *Agathis* is intermittent.

Cones were produced in trees greater than 40 cm in diameter and can be summarized as follows for each diameter class.

*40-cm class:* Four of 11 trees produced female cones. The duration varied from tree to tree, the longest extending 74 weeks, from 6 August 1984 to 4 January 1986. Male cones were seen from 17 August to 26 October 1985 and from 23 November to 21 December 1985.

*50-cm class:* Three of six trees produced female cones. The longest period was 78 weeks, from 21 July 1984 to 18 January 1986, while the shortest was 14 weeks. Male cones were seen from 17 August to 14 September 1984.

*60-cm class:* One of two trees produced cones. Male cones appeared from 17 August to 18 September 1985 and female cones from the same period to 18 January 1986 over a period of 22 weeks.

*80-cm class:* Two of three trees produced cones, both almost perennially.

*90-cm class:* Both of two trees produced female cones. One did so for 64 weeks from 25 October 1984 to 18 January 1986, the other in two periods, for 6 weeks from 16 February to 30 March 1985, and for 13 weeks from 3 September to 7 December 1985. Male cones appeared from 17 August to 14 September 1985.

*Agathis* thus requires many years to begin cone production, corresponding in its position in the forest to its emergence from the understory into the direct sunlight of the middle and upper stories. The timing of cone emergence and maturation shows some degree of individual variation. In Brunei, cones generally mature in March and April, and this is supported by figure 2-28, which shows the cone-bearing period in most species ending in February or March.

Whitmore (1977) points out that although the taxonomy of *Agathis* remains confused, 13 species can be recognized, with three subspecies in the Southeast Asia region. Surprisingly few observations have been made of its phenology. One exception is *A. robusta,* in which, in the first year, male cones emerge in January and female cones in June and July, and pollination occurs in August and September. Fertilization occurs in September of the second year, and seeds fall in January of the third year, 18 months after the emergence of the female cone.

The age of first appearance of female cones in plantations has been recorded at 15 years in Central Java, 20 years in Queensland, and 20, 18, 12, 13, and 13 years in West Java (Whitmore, 1977). In Java, it was noted that good germination was obtained only with trees aged at least 25 years. In most plantations, mature female cones are produced once each year at about the same time over a period of two to three months. In Brunei, this centers on May to July, occurring occasionally in September and November.

In the present survey, where observations were made from the ground, some uncertainty remains about the exact development of the axillary male cones, although the growth and fall of female cones could be judged more accurately. The fact that cones were produced in trees of 40-cm diameter class and greater agrees with the finding in plantations that the germination rate is good in trees aged 25 years and more.

The emergence of male cones in August to September coincides with the dry season, and pollination probably occurs shortly thereafter. About 18 months later, in February to March, the mature female cones scatter their seeds, either on falling or while still on the tree. *Agathis* cones generally shatter on the tree, and in Brunei, too, seed has been collected on the tree by bagging cones before they matured. In the present survey, however, many ripe cones were seen on the forest floor in March and April, suggesting that shattering does not occur exclusively on the tree.

The most conspicuous tree in flower in the plot was *Cotylelobium burckii*. Almost no members of the 10-cm diameter class flowered, probably because they had not reached flowering age and also because they ranked in the understory. Trees of the 20-cm class and above flowered uniformly, and two trees shed their leaves completely. Two trees bore fruit. Fruiting and leaf fall occurred outside the period of low rainfall from 13 April to 3 September, while flowering occurred at the end of this period, in June and August, the driest time of the year, when flowering often occurs in the tropics. The fact that this species sheds its leaves, and the timing of budding, opening, and other leaf phenomena, suggest that *Cotylelobium burckii* can be regarded as deciduous. No. 131 shed all its leaves at the start of February 1985, and again extensively in July to August and from early November to mid-January 1986. No. 141 shed all its leaves in July 1984.

Some trees flowered twice, from July to September 1984 and from January to March 1985, of which the former period, corresponding to the latter part of the dry season, can generally be regarded as the flowering period. The present data also suggest that this species does not flower every year, and that a tree that flowers one year will not flower again at the same time the following year.

No. 14 bore fruit in November 1984 after a long period of flowering. No.

121 bore fruit from February to March 1985 after two continuous periods of flowering.

Trees other than *Agathis* and *Cotylelobium burckii* were mainly a group of understory trees, and their phenology showed no definite trend, varying from one species to another. The fact that neither flowering nor fruiting was observed indicates the extreme poverty of phenological activity in the shrub layer of *kerangas* forest.

### *Alan Bunga* Forest

Much remains unknown about the flowering of *alan*, the main species in this forest, which has been said to occur, for example, once in twenty-five years. Loggers interviewed in Brunei suggested once in seven or eight years, and someone who had observed the forest from a helicopter said the canopy had colored bright red three years previously.

In 1986, I was fortunate to observe a gregarious flowering and fruiting. One of the 34 *alan* trees in the plot had been windthrown, but the remaining 33 flowered and fruited. The earliest flowering was seen on 14 January 1986, while all others began on 11 February. Flowering ended on 8 April. Fruiting began in 32 trees on 24 March, before the end of flowering, and in the remaining tree on 8 April. Fruiting ended on 21 April in 25 trees, 5 May in 7 trees, and the remaining tree was still in fruit on 19 May. All of the *alan* trees were of the 30-m height class, making up the overstory, and the fact that they all flowered and fruited is a clear indication of gregarious flowering.

Other than *alan*, flowering was seen in *Eugenia* from 14 January to late February 1985, followed by fruiting from March to June. A second flowering occurred around April 1986. This species showed a lag between flowering and fruiting seasons.

*Lithocarpus* flowered from October 1985 to February 1986 and from September 1985 to February 1986. Both flowerings were of the rainy season type. Fruiting occurred only once in one tree, in the period of low rainfall in July 1985 (fig. 2-30).

In addition, flowering and fruiting were seen in two *Palaquium* trees in August 1984, and in one *Campnosperma auriculata* tree in August 1985, all in the period of low rainfall.

The period of *alan* flowering from February to early April 1986, corresponds to the transition from the period of highest rainfall to the first dry period of the year. The only other tree to flower in this period was one *Eugenia*, all the other *Eugenia* flowering slightly later in April. In the previous year, however, 11 *Eugenia* trees flowered simultaneously from January to February; and it is possible that in general this is a period of extensive flowering in *alan bunga* forest.

In the Arboretum Andulau, the flowering of dipterocarps was concentrated in

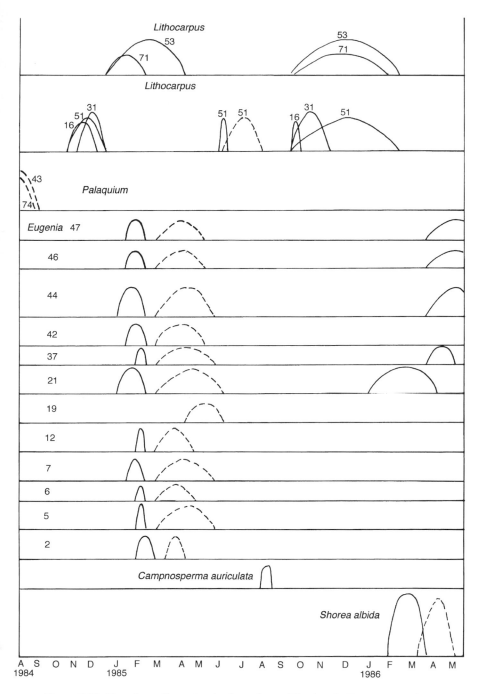

Figure 2-30. Phenology of trees in *alan bunga* forest. Numbers indicate tree numbers.
Solid lines indicate flowering; broken lines indicate fruiting.

the period from August to September 1985, while one tree flowered in April 1986. In Andulau K7, concerted flowering was seen from 31 August to 4 December 1985, while two trees flowered in April 1986 and one from January to March 1985. For *Cotylelobium burckii* in the *Agathis* forest, the periods were July and August 1985 and January to March 1986. Consequently, the flowering season of *alan bunga* appears to differ from that of dipterocarps growing on drier soil, and to match the second period of flowering of *Cotylelobium burckii*.

No leaf fall was seen in any tree.

## *Phenology*

Let us compare these phenological trends with the trends in rainfall data at Sungai Liang. As figure 2-31 shows, rainfall was generally low in March, June, and August. In February and March 1983, a major drought occurred in Borneo as the result of the El Niño phenomenon. In 1984, monthly rainfall was below 100 mm in February and March, and about 100 mm in August 1984. Rainfall was heavy from September 1984 to January 1985, and this pattern was repeated the following year.

Flowering in the tropics generally follows a prolonged dry season. Of the species discussed above, the flowering of *Cotylelobium burckii* correlates to some extent with the low rainfall of June and August shown in figure 2-31. The flowering of a considerable portion of the dipterocarps around September probably shows a similar correspondence. The gregarious flowering of *alan* from February 1986, however, followed several months of extremely high rainfall. Thus, even from the present data, it is impossible to substantiate this generalization about flowering.

Unlike Europe, no systematic study of Southeast Asia has been published that brings together phenological information on its various regions. The seasonal changes in various tree species of tropical rain forest have long been noted, though, and a considerable body of literature exists centering on Singapore, Malaysia, and Indonesia. Here I shall consider the diversity of phenology in tropical rain forest by introducing the classical research by Holttum, Koriba, and others in Singapore and the major reports of observations subsequently carried out in various parts of Malaysia.

Holttum (1931) investigated the phenology of trees in the Singapore Botanic Garden from 1927 to 1931. Many of the deciduous trees changed their leaves once a year, mainly with the change of season in February, but some in August. Four species changed leaves twice a year, in February and August. Others changed leaves at regular intervals of other than six or twelve months, regardless of season. Individual variations between members of the same species were also seen. Other species, like *Hevea braziliensis,* were extremely irregular. Small

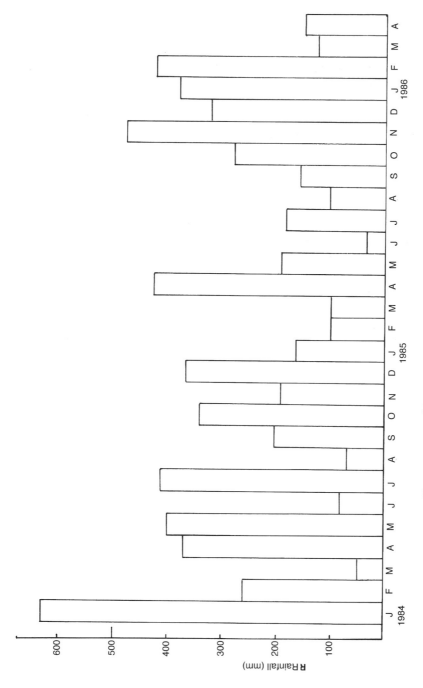

Figure 2-31. Rainfall at Sungai Liang, January 1984 to April 1986.

climatic variations combined with internal factors to trigger leaf fall and other phenomena.

In a later article based on almost ten years of observation, Holttum (1940) revised his earlier findings. He showed that leaf fall in deciduous species ranges from being totally independent of climate to being particularly sensitive. He divided species into those that change leaves once a year, those with a leaf cycle of more than one year, those with a leaf cycle of six to twelve months, and those with no fixed leaf cycle; and at the same time he discussed the relation between flowering and leaf cycle in deciduous and evergreen trees.

Of climatic variations, he noted that such events as a drop in air temperature of 6–8°C from a storm, or a cool day following a succession of hot days may stimulate a leaf change. In Singapore, climatic changes are most dramatic in January and February, while the driest times are February to March and July to August, and leaf phenomena are common in this second period.

Among species keeping their leaves for more than one year is *Koompassia malaccensis*. In the first observations this was thought to have a 12-month leaf cycle, but on extended observation the cycle was judged to be slightly longer (12.7 months). Particularly long cycles were found in *Heritiera elata* (20.5 months) and *H. macrophylla* (32.0 months). Short cycles were shown by *Sterculia* spp. among those with a 6- to 12-month cycle and *Dyera costulata* among those with an irregular period.

Flowering occurs at a certain point in the leaf cycle in deciduous trees, but in evergreen trees it is independent of this cycle and of the growth of new leaves. Leaf flush in evergreens occurs at intervals of from several months to one year or longer, but the timing is even more irregular than in deciduous trees. Normally, the period of old leaves is protracted, since they do not fall for some time after the emergence of the new. *Dyera costulata* shows an intermediate habit between deciduous and evergreen. The growth of new leaves in evergreens may be triggered by rain.

Of the *Koompassia malaccensis* trees incorporated into the botanic garden from the former forest, some that until 1930 had changed leaves in September did so in November in 1931, December in 1932, and not at all in 1933. Subsequent changes were recorded in March 1934, on 17 May 1936, 8 July 1938, 15 July 1939, and 12 May 1940, giving an average leaf cycle of 12.7 months with a range from 10.9 to 14.8 months.

Subsequently, Koriba made a detailed survey of the habits of trees in the Singapore Botanic Garden. He described the characters of deciduous trees, classified the modes of branch growth into evergrowing, manifold, intermittent, and deciduous types, and examined their relation to leaf activity, flowering, and fruiting. My work on the middle slopes of Java's Mount Pangrango, discussed in detail in chapter 3 in this volume, involving the use of litter traps to examine litter volumes and observation of seasonal changes in leaves, flowers, and fruits for various species, indicates that the difference between the deciduous type and the evergrowing and intermittent growth habits depends on variations in habitat among different strata in the forest.

Burgess (1972) has investigated data on flowering in Malaysia from 1925 to 1970. He reports that most dipterocarps flowered in cycles of from two to five years. *Shorea curtisii* and *S. platyclados* do not exceed five years. From 1960 to 1970, flowering peaked in May, being spread over the period from March to July. In the dry zone of northwest Malaysia, the peak was less distinct and tended to occur at the start of the year. The red *meranti* group mainly showed a peak in May over the eleven-year period. No particular correlation with climate was noted, but flowering appeared to occur at one time on portions of new growth formed several months earlier.

For future regeneration of dipterocarps, Burgess stresses the importance of learning their phenology because of the short period of viability of their seeds. Trees require twenty to thirty years to begin flowering, and research is required on the hormones and physiology involved in the initiation and opening of flower buds, and on the relation of leaf flushing.

The relation of drought to flowering is unclear, but Burgess suggests the possibility that some form of water stress is involved. *Shorea curtisii* is particularly sensitive to dry conditions, but the involvement of other factors must also be considered.

Medway (1972) studied the Ulu Gombak Forest Reserve (3°21′ N, 101°47′ E) in Selangor, Malaysia, observing the flowering, fruiting, and foliar activities in 61 canopy trees (45 species in 17 genera) at two-week intervals over a period of six years from July 1963 to July 1969. The site was a typical hill dipterocarp forest at an elevation of 515 m.

Synchronized flowering occurred in 5 of 10 species represented by two or more trees. Only the strangler *Ficus sumatrana* flowered in a regular succession of cycles, but since individual *F. sumatrana* flowered asynchronously at an average interval of four and a half months, this was clearly uncorrelated with climatic events. The remainder flowered at irregular intervals, often exceeding one year.

Ten species flowered annually from 1963 to 1969, and 6 bore fruit. In any given year, at least 20 species flowered and 12 set fruit. In the driest years of 1963 and 1968, gregarious flowering occurred and many trees produced fruit. Regardless of species, flowering from 1963 to 1969 was most vigorous from February to July, slackened somewhat from August to November, and was least active in December and January. Fruiting was similarly high from May to November, being particularly abundant from September to November. The community as a whole thus exhibited a fairly marked seasonality.

Leaf activities were also diverse. Three species showed continuous leaf growth, and 21 species showed an annually recurrent cycle, often involving two periods of activity. Five species showed irregular biannual cycles. Four species were annually deciduous, and 4 were deciduous at longer or less regular intervals. In at least 6 species, foliar activity was clearly related to the floral cycle, and in 5 others with annual floral cycles such a relationship was possible. In the remainder the two activities were independent. Nevertheless, the community overall showed an annual cycle of leaf growth, here with two seasonal peaks—the higher coinciding with the secondary wet season and the lower with the main rainy season.

Cockburn (1975) analyzed the flowering of dipterocarps in Sabah based on available data, including annual reports, research papers, records of illipe production, and herbarium observations, and found heavy flowering and seed production in 1973. As future tasks, he advocated research into grafting to produce flowers from immature trees and induction of flowering by use of giberellic acid and other chemical treatments, and detailed, long-term measurements of microclimate in the forest.

Ng (1977a) has summarized the results of monthly observations beginning in 1972 on the flowering of dipterocarps in the arboretum of the Forest Research Institute at Kepong, Malaysia. While 20 of the observed trees flowered in 1973, 15 in 1974, and 5 in 1975, 79 flowered in 1976. Early in 1976, a gregarious flowering was seen across the whole Malay Peninsula, the most extensive ever recorded. Other families also flowered, and in August to September 1976, mast fruiting occurred. Most notable were the families Euphorbiaceae, Polygalaceae, Moraceae, and Bombacaceae, in particular the genera *Baccaurea, Elateriospermum, Pimeleodendron, Xanthophyllum, Artocarpus,* and *Durio.* Fifteen of the fifty species of dipterocarps at Kepong flowered in 1976 for the first time since being planted, of which the oldest trees were 45 years old. The higher yield of mature seeds in 1976 compared to a normal year probably resulted from a higher rate of successful pollination.

While Burgess (1972) had pointed out the occurrence of peak flowering of dipterocarps in Malaysia in May, Ng noted peaks at Kepong in April in 1973, 1974, and 1976. Having surveyed earlier records, he concluded that flowering at Kepong peaks in March or April.

Ng suggested that flowering is stimulated by an increase in the number of hours of sunshine over the previous one or two months, in particular a sharp increase of two hours or more in mean daily sunshine. Such an increase is normally seen in January, February, or March. This factor appears to have triggered gregarious flowering in 1957, 1958, 1963, 1970, and 1976. In this context, Ng claims that drought, which is widely felt to be the stimulus of gregarious flowering, should not be measured by rainfall but as a long succession of hot sunny days following a cool period.

While many tree species were said to flower synchronously in the gregarious flowerings of 1957, 1958, 1963, 1970, and 1976 in Malaysia, Sasaki et al. (1979) observed various species of dipterocarps during the gregarious flowering of 1976 and concluded that synchrony was limited to certain species and to certain areas. They suggested that the timing of flowering is dependent on differences in altitude, species, location, topography, and climate, and that for breeding purposes, a large-scale survey of individual trees of selected species is required.

Putz (1979) conducted a phenological survey over four years, from March 1972 to February 1976, in the Sungai Buloh Forest Reserve, which lies 25 km northeast of the Malaysian capital of Kuala Lumpur at altitudes between 20 and 225 m. His survey covered 131 trees (35 families, 52 genera, 62 species), including Burseraceae, Leguminosae, Euphorbiaceae, Rubiaceae, Sapotaceae, Lauraceae, Myristicaceae, and Rhizophoraceae. The basal area for trees of more than 20 cm DBH in the 100-ha plot was 35 m²/ha. Observations were made from transects consisting of horizontal aluminum ladders supported by steel cables suspended between the tops of the largest trees, with an overall length of 363.5 m and a maximum unsupported span of 75 m. Several times each month, flowering, fruiting, and leaf production were recorded, and spectral and Colwell analyses were performed to evaluate phenological periodicities.

Flowering showed a lack of intraspecific synchrony, and the duration of flowering was variable between trees of the same species. For example, staminate trees of *Knema stenophylla* flowered continuously, while pistillate trees flowered intermittently. Seven of eight *Santiria laevigata* trees flowered asynchronously zero, one, or two times per year for a duration of about one month per flowering episode, while the eighth tree flowered in each month of 1973 and 1974. These differences showed no relation to tree size or crown position. Other species showing clear asynchrony included *Randia densiflora, Anisophyllea corneri,* and *Millettia atropurpurea.*

On average, 15 of the 62 species flowered per month. The constancy of flower production at the community level was attributed to a few species that flower continuously and others that flower for extended periods. Twelve of the 62 species flowered for more than 25 of the 49 months of observation. Of these, *Euodia glabra, Strombosia javanica,* and *Millettia atropurpurea* flowered continuously for periods of more than 12 months.

Some species did show regular periodic behavior, including five biennially flowering species, which flowered in the same months of 1972, 1974, and 1976: *Alangium ebenaceum, Actinodaphne oleifolia, Koompassia malaccensis, Intsia palembanica,* and *Ormosia venosa.* Four species flowered regularly every year, though not with complete regularity. *Alstonia angustiloba* flowered in March and/or April of 1972, 1974, 1975, and 1976 but, perhaps because it also flowered in December

1972, it did not flower in 1973. *Litsea castanea* flowered every year during
August to October, but also sometimes at other times of year. Six trees of *Elateriospermum tapos* all flowered annually in January and February, and sometimes
again in July, August, or September. And *Cryptocarya griffithii* flowered in the
first three months of the year for three of four years. These species that flower
synchronously are considered either to have endogenous rhythms or to respond to
photoperiod or climatic factors with a periodic cycle. However, as many as 66
percent of species did not flower for long periods or at regular intervals.

While no generalizations can be made about its seasonality, flowering often
follows a dry period. Significant negative correlations were found between
flowering and rainfall in the month, two months, and three months before
phenological observation. Wycherley (1973) notes that rapidly falling temperatures associated with rainstorms during or at the end of the dry season often
stimulate flowering.

Of the 62 species observed by Putz (1979), an average of 20.6 species were in
fruit in each month. More species fruited in the period from April to September,
but many differences were seen in the periods required for fruit maturation and
the durations for which fruit remained on the tree.

Species with fruits that remained on the tree for extended periods and that
contained seeds capable of germination were *Anisophyllea corneri, Barringtonia
pendula, Canarium littorale,* and *Millettia atropurpurea.* All of these have large
fruits, of which all but the last named are dispersed by animals.

Species that flowered and produced fruit continuously for periods of more
than twelve months included *Timonius wallichianum, Nauclea subdita, Santiria
laevigata, Dacryodes costata, Endospermum diadenum, Planchonella maingayi, Paropsia
vareciformis, Grewia laurifolia, Litsea castanea,* and *Myristica* sp.

All other species fruited intermittently. Of these, some fruited at intervals of
more than one year, while others fruited two or three times a year. The former
numbered 26 species, the latter 19 species, often with different trees or branches
taking part in fruiting. Three species were not observed in fruit.

Ninety percent of the species in the canopy had fleshy fruits, indicating dispersal by animals. In those with wind-dispersed seed, fruits tended to mature in
dry periods.

Leaf production at the community level continued without interruption
throughout the year. On the average, more than half the species were producing
new leaves each month, and no correlation was seen between leaf production and
climate. No periodicity was found by Colwell or spectral analysis. Peaks in leaf
production were difficult to discern, either for individuals or for species: Not
only was a range of growth modes seen from deciduous to evergreen and continuous to intermittent, but trees of the same species showed asynchrony. Also,

some trees produced leaves over the whole crown, while others did so on only a single branch.

Of the deciduous trees, none was leafless for more than two weeks. Those that lost all their leaves were *Ficus caulocarpa, Terminalia subspathulata, Intsia palembanica,* and *Irvingia malayana. F. caulocarpa* abscissed and produced new leaves during the period May to November, the others from January to March. *Alstonia angustiloba, Dracontomelum mangiferum,* and *Xanthophyllum obscurum* shed the bulk of their leaves before producing new foliage, and one of three trees of *Barringtonia pendula* behaved similarly.

In deciduous trees and those that gradually and completely replace old leaves by new over a period of a few weeks, only one generation of leaves was present on the tree at one time. In other species, up to three generations were present at one time. The number of leaf flushes per year also varied widely between species. *Xanthophyllum griffithii* and *Diospyros pendula* produced new leaves only twice in four years. At the other extreme, leaf production appeared continuous in some species: *Actinodaphne oleifolia, Knema stenophylla, Endospermum diadenum, Macaranga hypoleuca, Strombosia javanica, Nauclea subdita, Randia densiflora,* and *Timonius wallichianum.*

From the above results, it can be said that seasonality at the community level here is weak. Also, the species show a wide range of reactions to climate. For example, seed dormancy periods ranged from two to eighteen weeks and longer.

Aseasonal flowering and fruit production in a high and constant number of species has the advantage of maintaining stable populations of pollinators and dispersal agents throughout the year. Moreover, species that lack rigid timing mechanisms may benefit from being able to respond to local conditions of time and space.

The study areas of Putz (1979) and Medway (1972) are similar in the low proportions of species with annual and biennial cycles and the lack of intraspecific synchrony. Medway, however, observed distinctive, repeating phenological patterns. Despite specific variations, the community as a whole exhibited a regular seasonality, with single annual peaks of flowering and fruiting and a double peak of leaf production. No such patterns emerge in Putz' findings. Putz believes that this is probably attributable to the dominance of Medway's forest by dipterocarps.

Yap (1980) has summarized the phenology of *jelutong (Dyera costulata).* Production of latex from this species as a base for chewing gum has declined since its peak in the 1930s and 1940s, while logging of *jelutong* has increased markedly as its timber has become more widely used for light construction. Observations were made from 1974 to 1979 of 11 trees in the arboretum of the Forest Research Institute, Kepong. Of these, 3 trees flowered regularly from July to December, with a tendency toward synchronization. In 1978, the 3 arboretum trees and 34 of 45 other *jelutong* in the immediate vicinity flowered during the period August to October. Flowers opened during the night, and many white petals were seen on the ground the following morning. Flowering lasted for two to three weeks on a tree, and young fruits appeared two to three months later. Mature fruits were collected eight to nine months after the anthesis.

Chan (1980) studied the fruiting and seedling biology of dipterocarps in Pasoh, Malaysia. He found that initial fruit production was high, but that failure to produce mature fruits was marked in earlier flowering species. Shedding of immature fruits was particularly heavy in the first two weeks after flowering. Seed dispersal in dipterocarps is by wind, with the seed first falling vertically for several meters before spreading its wings and beginning to spin rapidly. This slows its fall, allowing it to drift away from the mother tree before reaching the ground. In the absence of wind, it lands near the tree. Under normal wind conditions, mature fruit could be dispersed some 50 m from the mother tree, but the majority land within 20 m, their dispersal impeded by the density of the canopy. When the mother tree is on a slope or ridge, however, a gust of wind can carry a seed a great distance. Burgess (1975) records a distance of up to 80 m for *Shorea curtisii.* Janzen (1970) argued that the survival rate should increase with distance from the mother tree, but Liew and Wong (1973) and Daljeet Singh (1976) found a decrease in fruit number with increasing distance.

The work by Chan and Appanah (1980) on the flowering of dipterocarps involved a large-scale survey of 100 adult trees of the red *meranti* group (section Muticae), represented by *Shorea leprosula, S. macroptera, S. acuminata, S. lepidota, S. parvifolia,* and *S. dasyphylla,* during an episode of gregarious flowering in Pasoh, Malaysia. In going beyond simple phenological observation, this opened a new field of research.

All the 100 selected trees flowered. *Shorea macroptera* flowered first, followed in succession by *S. dasyphylla, S. lepidota, S. parvifolia, S. acuminata,* and *S. leprosula.* The flowering times were staggered, with some overlapping between the later phases of an earlier species and the initial phases of the species following, but their blooming peaks were distinct.

The duration of bloom was short, ranging from two weeks *(S. macroptera, S. dasyphylla, S. lepidota)* to three and a half weeks *(S. leprosula),* with the later flowering species tending to bloom for a longer duration. With *S. leprosula,* the flowering intensity was estimated to be from 63,000 to 4 million flowers per tree.

As part of a collaborative study of reproductive biology, Yap (1982) conducted phenological observations in the Pasoh Forest Reserve of 130 trees of *Xerospermum intermedium* for five years from 1972 to 1977. Several trees of other species, including *Baccaurea, Nephelium, Lansium, Durio, Myristica,* and *Knema,* were also observed for three years in the same period. Rainfall, temperature, and duration of bright sunshine were recorded daily.

*Xerospermum intermedium,* an understory tree that produces male and female flowers on different trees, bloomed annually following a dry spell. Flowering was prolonged, lasting three to four months, and apparently was unaffected by the sharp rise in the number of hours of bright sunshine in December 1975, which preceded the gregarious flowering of dipterocarps in 1976. Of the other species, those that reached the main canopy flowered together with the dipterocarps. Aoki et al. (1975) have reported that 60 percent of incoming solar radiation is absorbed in the top 5 m of the canopy, so variation in sunshine above the canopy may have less influence on the understory.

Appanah and Chan (1982) describe methods employed for studying the reproductive biology of *Shorea leprosula* and *Xerospermum intermedium,* giving details of methods of tree climbing and bagging of flowers.

Tamari and Jacalne (1984) investigated the fruit dispersal of *Shorea contorta* in the Makiling forest of Los Baños, the Philippines, in 1979 and found that the majority of viable fruits were dispersed within 30 m of the mother tree. Having surveyed the literature, they conclude that, except in special circumstances, dispersal within 30 m of the mother tree is usual in virgin forest. In a phenological study of the same forest, Tamari and Domingo (1979) found that dipterocarps flowered over widely extended seasons, with a main season running from February to May and a scanty flowering season from August to December, when flowering and fruiting are rare. Fruit maturation occurs from May to October, coinciding with the wet season in Makiling. This forest lies in the dry climatic region of the Philippines, but flowering and fruiting habits of dipterocarps there resemble those in the wet region. In the Philippines in general, dipterocarps flower over widely extended seasons with peaks in April in the dry region, May in the intermediate region, and May and June in the wet climatic region. Peak fruiting seasons differ by two to three months between the wet and dry regions, and flowering and fruiting seasons are of longer duration in the wet and intermediate regions than in the dry region.

In Malaysia, Tamari (1976) reported some outstanding work covering the phenology, seed properties, and seed storage tests of dipterocarps in the vicinity of Kuala Lumpur. He found two annual flowering seasons, which coincided with the pronounced rainy seasons, although some variation may be caused by abnormal wet or dry seasons. The interval between flowering and fruiting in most dipterocarps was from two to five months. Loss of flowers and young fruits due to physiological factors, insects, and other natural causes exceeded 90 percent. His figure 3 clearly shows the coincidence of flowering with rainy seasons, although it is also possible to interpret the data as showing flowering being triggered by spells of low rainfall two to three months earlier, in February and July.

Wong (1983) investigated the phenology of understory trees in virgin and regenerating forests in Pasoh. Reproductive activity was more vigorous in the virgin forest than the regenerating forest, and the seasonality of phenological activity was less pronounced in the understory than in the canopy vegetation.

The foregoing review of literature on phenology centers on Southeast Asia; for other regions of the world there are the following works: on tropical America, Janzen (1970, 1974), Daubenmire (1972), Frankie et al. (1974b), Gentry (1974), Croat (1975), Opler et al. (1976, 1980), Borchert (1980), Mueller-Dombois et al. (1981), Reich and Borchert (1982), and many others. Also notable are the works of Wium-Anderson and Christensen (1978) on Thai mangroves; of Gill (1971) on flower morphology; of Shimizu (1983) on Japan; of Wycherley (1973) on Micronesia; and of Hopkins (1968) and Malaisse (1974) on Africa.

General works include Lieth (1974), Frankie et al. (1974a), and Baker et al. (1983). In Indonesia, there are the early works of Coster (1923, 1926) and, more recently, Okimori (1987). Ashton et al. (1988) also discuss the gregarious flowering of dipterocarps.

Borchert (1980) observed leaf fall, flowering, and shoot emergence of *Erythrina poeppigiana* in Costa Rica and found a clear lack of synchrony among trees in the same population, which he attributed to strong endogenous control of tree development. He also advocated the progression of phenological study to the analysis of many trees of a single species rather than the general observation of many different species.

From observations of *Tabebuia neochrysantha,* Reich and Borchert (1982) found a close correlation between phenology and water stress, which was reflected in shrinkage of stem diameter.

From the above survey of the literature, it is apparent that while dipterocarps show a clear synchrony, nothing definite can be said of other trees. As dipterocarps constitute the most important group of species in Southeast Asia, they have been most extensively studied, and there is abundant literature on their gregarious flowering. The causes of this gregarious flowering remain obscure, however, and far more data need to be collected for analysis, together with more precise and more localized meteorological information.

Nevertheless, the general tendency for increasing seasonality with higher canopy positions is readily understandable. Koriba attributes seasonality to the interaction of exogenous and endogenous factors, and suggests that hormonal action is among the major endogenous factors. While little work has been done on the hormones involved in growth, flowering, and fruiting in tropical trees, much can be expected of these studies in the future.

Ashton (1964b) has summarized our knowledge of the dipterocarps of Brunei, but much remains unknown about their phenology. Beyond a partial elucidation of their taxonomy and ecology, it is probably true to say that the characteristics of the various tree species are unexplored. Of course, the same is true for temperate countries, where a thorough knowledge is limited to a handful of representative species. In Japan, these would extend to *Cryptomeria japonica, Chamaecyparis obtusa,* and *Pinus densiflora.*

The large diversity of tree species in the tropics has further delayed such investigations. Problems should be cleared up for a limited number of species before turning to the wider field.

In my own survey, I examined the phenology of a limited number of representative species. Such representative species or communities in the various forest types include the dipterocarps in mixed dipterocarp forest, *Agathis* and *Cotylelobium burckii* in *kerangas* forest, and *Shorea albida* in *alan* forest. The survey revealed a difference in flowering season between dipterocarps growing on dry land and those growing on peat swamp, with those in *kerangas* forest being intermediate between the two, and that understory trees in *kerangas* forest are extremely inactive biologically. In relation to canopy position, the understory trees tended to show weaker seasonality. Another finding, as will be described, is the weak seasonality of girth growth.

The present survey was the first such attempt in Brunei and the first such experience for the observers, and in the future more detailed observation will be required. Nevertheless, the results were interesting, and I was fortunate to observe the flowering and fruiting of *alan.* Because, as noted at the start of this section, flowering and fruiting habits in the tropics are irregular, it is essential to study these habits closely for each of the main tree species and even individual trees in order to practice effective silviculture.

Apart from the *alan bunga* forest, the forests chosen for study were typical forest types within easy reach of Sungai Liang. In addition, it will be necessary to establish a new plot in the *alan* forest near Badas, and to study the flowering and fruiting habits of representative species over the long term.

In such a study, trees of superior phenotype should be chosen for observation. This is an iron rule that has been recorded in the FAO *Genetic Resources Information* publications and elsewhere, but at present it is not observed very strictly. Unlike temperate zones, pure stands in the tropics are few, and choice tends to be limited to the specimens encountered.

In Brunei, at least as far as *alan* and *Agathis* are concerned, considerable numbers of trees are present, and selection of superior phenotypes should be possible. In practice, *Agathis* seeds are collected even if the parent trees are not outstanding. With *alan,* seed preservation is difficult, so cuttings are propagated by use of small mist boxes. Along with the expansion and refinement of this technique, other methods of propagation should be examined, based on seedlings from the one thousand seeds collected after the present flowering and fruiting.

For the dipterocarps, with Arboretum Andulau and the Andulau Forest

Reserve as a basis, about ten superior trees each of a limited number of major species should be chosen, and efforts should be directed toward grafting and cutting propagation and seed preservation. It has frequently been reported that many species take part in gregarious flowering and fruiting in seed years, and at such times it is necessary to prepare nurseries for planting and raising seedlings on a large scale.

While even-aged plantations of fast-growing imported species may serve as stopgaps for the supply of timber and pulp material, long-term stability is provided by the indigenous species, which have long been valued for their timber quality and other properties. These find a ready market and are resistant to pests and diseases. Also, because their preferred habitats are known, silviculture is simple.

*Agathis* and the various dipterocarps described here, including *alan*, constitute the representative forest communities of this area, and their regeneration in virgin and logged forests requires full-scale study.

In his monograph on *Agathis,* Whitmore (1977) describes detailed surveys of past literature and present conditions in the field, and for the future he notes the need to explore the feasibility of silviculture to take advantage of the wide ecological adaptability, wide geographic range, timber quality, and other favorable features of this genus. Future tasks include refinement of seed preservation and vegetative propagation techniques, establishment of taxonomic criteria, and investigation of ecotypes.

To judge from the forest at Badas, Brunei, it appears feasible to establish plantations of *Agathis,* in particular, plantations of superior clones from parents of superior phenotype. The present forest is small, far smaller than the five thousand trees that Whitmore claims as the minimum effective number for genetic resource conservation; but by selecting and propagating superior phenotypes from the forest, it should be possible to develop a population of superior strains.

In Brunei, small-scale line planting of *Agathis* has been carried out at Ulu Badas, the Labi Hills Forest Reserve, and the Andulau Forest Reserve. All are too small for experimentation, and regular maintenance and measurements have not been conducted.

Subsequent work on *Agathis* includes successful propagation by means of leaf cuttings (Smits, 1983) and growth analyses by Whitmore and Bowen (1983), who found that seedlings can be successfully planted into small canopy gaps. For full-scale silviculture, a comprehensive system must be established encompassing phenological study of the Badas forest, seed collection and storage, raising of seedlings, and maintenance and growth monitoring after transplantation. In this process, it will be necessary to tackle and solve problems one by one.

## 3. Growth Modes of the Main Tree Species in Virgin Forest

### *Survey Methods and Sites*

Permanent plots were established to carry out periodic surveys and to observe continuously the growth, death, and regeneration of constituent tree communities. Such data are particularly sparse for the tropics and present a serious obstacle to studies of growth volume and growth prediction for tropical trees, which lack annual rings. Besides the regular annual measurement of diameter growth, mentioned earlier, at the positions marked with a white band, growth was measured in detail by use of a dial gauge at the time of the phenological survey.

The annual measurement of girth growth was made with a diameter tape placed over the white band painted around each tree. These data will be reported later, when measurements have been accumulated for several years.

The dial gauge is a modification of the one developed in America by Liming (1957), consisting of an aluminum dendrometer band with a gauge readable to 0.05 mm. This method has been employed in Java, Sumatra, Thailand, and Japan with consistently good results.

The gauges used were made in Japan and assembled in the field. The position of attachment either above or below the DBH band (1.3 m aboveground) was decided after examination of the irregularity of the trunk and bark. Before beginning the survey, it was confirmed that the gauges had been stably attached, with no slippage or misplacement of the spring. Suitable trees of various species were chosen in the four phenological survey plots, and measurements were taken at two-week intervals parallel with the phenological survey.

### *Growth Modes of Main Tree Species*
### Arboretum Andulau

Dial gauges were attached to 49 trees in the arboretum, but for such reasons as faulty attachment, disturbance by small animals and people after attachment, and malfunction because of resin or irregularities in the bark, proper values were not obtained from 35 of them. Of the 14 that functioned properly, half were attached to dipterocarps.

The best growth was shown by a dipterocarp, *Shorea ovata*, for which the results are plotted in figure 2-32. This tree had a diameter of 19.3 cm, a height of 17.0 m, and a height to first branch of 12.5 m. Its growth of 16.8 mm was the highest recorded in the survey period. The fluctuations in the course of growth shown in the figure involve steady growth from September 1984 to February 1985, a period of almost no growth in March, and a spurt in April followed by slower growth until August. September saw another growth spurt, followed by slower growth again from October to March 1986, with another spurt in the

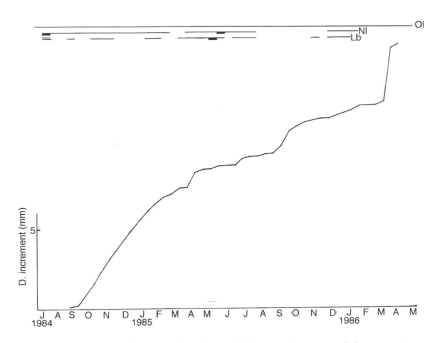

Figure 2-32. Relation between phenology and diameter increment of *Shorea ovata* in
Arboretum Andulau. *Lb:* Leaf buds; *Nl:* New leaves; *Ol:* Old leaves.

latter part of March. Phenological activity in the period included remarkable production of leaf buds and new leaves in May 1985 but no other major
events, although leaf buds and new leaves continued to be produced with considerable frequency. The period from August to November 1985 saw a complete hiatus in such activity. This tree can thus be regarded as belonging to
Koriba's intermittent growth type. The growth spurts of September 1985 and
March 1986 can probably be regarded as coinciding with the rest periods in
foliar activity.

The three trees of *Shorea acuta* (fig. 2-33) were all small, with diameters of
13.4, 21.6, and 31.2 cm, and occupied a similar canopy position to the previous species. As the figure shows, the overall growth was not particularly large.
No. 200 and no. 8 showed similar phenological activity, and their growth rates
tended to increase slightly during lulls in foliar activity. This tendency was
less pronounced in no. 46, which showed more frequent new leaf activity, but
growth did quicken somewhat in September 1985 and from January to March
1986.

Next is *Dryobalanops beccarii*, which is in the same size class. Its leaf growth
seems to be of the continuous type, with leaf buds and new leaves present
throughout the year. The relation of growth with phenological activity is thus

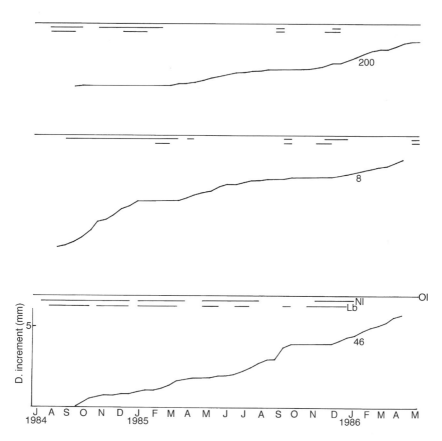

Figure 2-33. Relation between phenology and diameter increment of *Shorea acuta* in Arboretum Andulau. *Lb:* Leaf buds; *Nl:* New leaves; *Ol:* Old leaves.

obscure, although growth appears to be fastest around the time of interrupted foliar activity in May 1985 (fig. 2-34).

On the sample tree of *Anisoptera grossivenia,* the coil came off during the survey, and only half of the expected data were collected. Although no plot was made, a growth spurt occurred after leaf development. In the period before the new leaves, however, from February to April 1985, growth was slow despite the interruption of foliar activity.

Finally, the *Copaifera palustris* tree (also shown in fig. 2-34) flowered and fruited, and growth tended to slow somewhat at this time, while from July to September 1985 it accelerated to some extent.

For the dipterocarps as a whole, such phenological activities as leaf bud production and leaf opening, flowering, and fruiting showed a fairly close correlation with stem diameter growth, although not always.

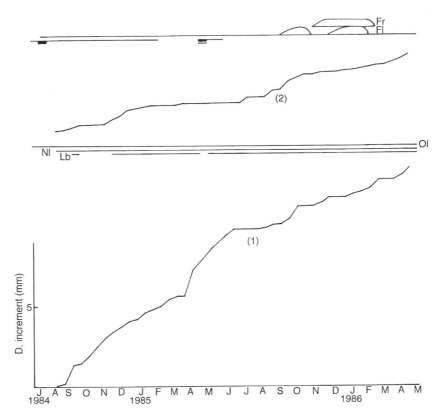

Figure 2-34. Relation between phenology and diameter increment of *Dryobalanops beccarii* (1) and *Sindora leiocarpa* (2) in Arboretum Andulau. *Lb:* Leaf buds; *Nl:* New leaves; *Ol:* Old leaves; *Fl:* Flowers; *Fr:* Fruit.

The results for the nondipterocarps are plotted in figure 2-35 and reveal a relationship between phenology and diameter growth similar to that observed in dipterocarps.

*Artocarpus odoratissimus* showed a slowing of growth during episodes of flowering and fruiting, and slight increases in growth rate during intermissions in new leaf activity, as in October 1984 and May 1985. In *Litsea,* growth rate increased when new leaf activities stopped after October 1985. Two *Aquilaria malaccensis* trees showed the same tendency for growth rate to increase when foliar activities were in abeyance (data not shown). *Melanorrhoea* showed a similar but less marked tendency, while in *Calophyllum,* growth increased more during intermissions in new leaf activities. The same was true of *Koompassia malaccensis* (data not shown).

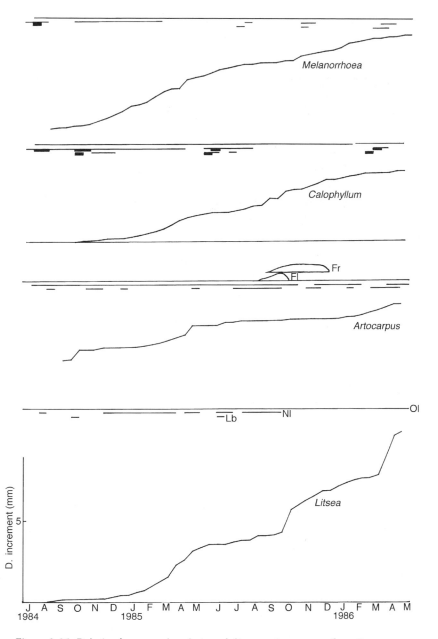

Figure 2-35. Relation between phenology and diameter increment of nondipterocarps in Arboretum Andulau. *Lb:* Leaf buds; *Nl:* New leaves; *Ol:* Old leaves; *Fl:* Flowers; *Fr:* Fruit.

## Andulau Forest Reserve

Forty dial gauges were set here, but accurate data were obtained only for 5 dipterocarps and 21 nondipterocarps.

Results for the dipterocarps are plotted in figure 2-36. *Shorea kunstleri* showed somewhat faster growth during intermissions in new leaf activity. *Dipterocarpus crinitus* showed continuous leaf production, with a tendency for faster growth during somewhat reduced leaf activity, though the tendency was not marked. In *Shorea* spp., where growth occurred in the intermissions in new leaf activity from January to April 1985 and March 1986, the tendency was clear. Finally, the two trees of *Shorea acuta* produced new buds and leaves with considerable frequency, and growth was again higher during the intermissions. Consequently, the dipterocarps here showed the same correlation as those in the arboretum.

The results for the 21 nondipterocarp trees, shown in figure 2-37, can be summarized as follows:

(a) Trees that flowered and fruited and in which growth slowed during those times and also during new leaf production, but which conversely showed clearly enhanced growth during intermissions in new leaf activity:

| | | |
|---|---|---|
| *Lithocarpus* sp. 38 | (D = 15.9) | Intermittent growth |
| *Calophyllum* sp. 55 | (D = 32.4) | Intermittent growth |
| *Cratoxylum* sp. 10 | (D = 10.5) | |

(b) Trees that flowered and fruited but in which the relationship with growth was unclear:

| | | |
|---|---|---|
| Pendarahan 73 | (D = 14.6) | Intermittent growth |
| Pendarahan 116 | (D = 28.3) | Intermittent growth |

(c) Trees in which growth declined during new leaf production and increased during intermissions:

| | | |
|---|---|---|
| *Xylopia* sp. 79 | (D = 17.1) | Continuous growth |
| *Alstonia* sp. 25 | (D = 15.9) | Intermittent growth |
| *Canarium* sp. 95 | (D = 15.8) | Intermittent growth |

(d) Trees that showed no clear correlation between diameter growth and leaf production or other phenological activity but showed enhanced growth at certain times of intermission:

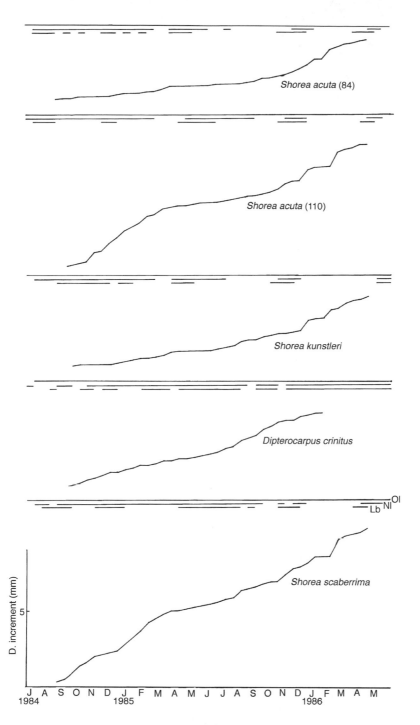

Figure 2-36. Relation between phenology and diameter increment of dipterocarps in Compartment 7, Andulau Forest Reserve. *Lb:* Leaf buds; *Nl:* New leaves; *Ol:* Old leaves

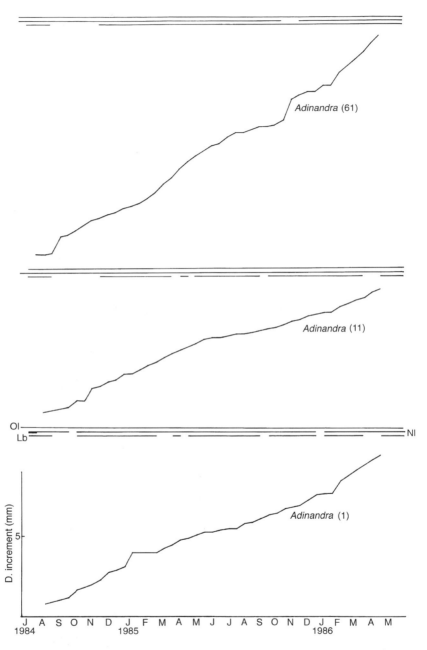

Figure 2-37 (1)–(6). Relation between phenology and diameter increment of nondipterocarps in Compartment 7, Andulau Forest Reserve. *Lb:* Leaf buds; *Nl:* New leaves; *Ol:* Old leaves; *Fl:* Flowers; *Fr:* Fruit.

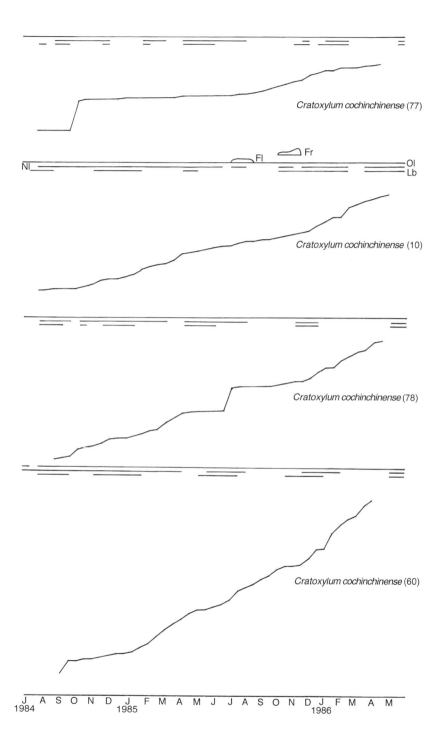

*Cratoxylum cochinchinense* (77)

Fl Fr

Nl Ol Lb

*Cratoxylum cochinchinense* (10)

*Cratoxylum cochinchinense* (78)

*Cratoxylum cochinchinense* (60)

J A S O N D J F M A M J J A S O N D J F M A M
1984       1985       1986

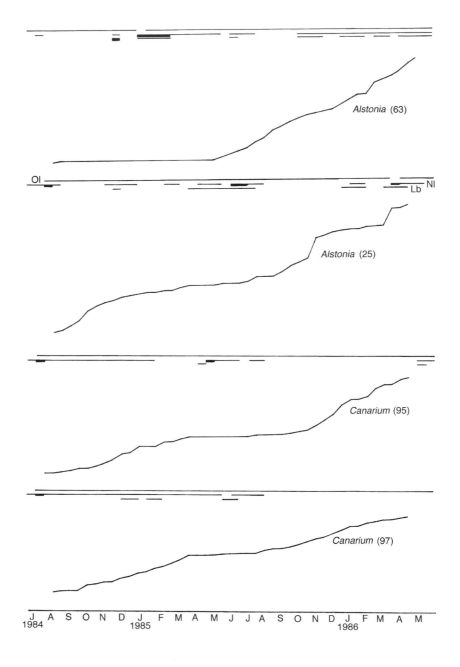

Alstonia (63)

Ol ———————— Nl
Lb

Alstonia (25)

Canarium (95)

Canarium (97)

J A S O N D J F M A M J J A S O N D J F M A M
1984          1985                          1986

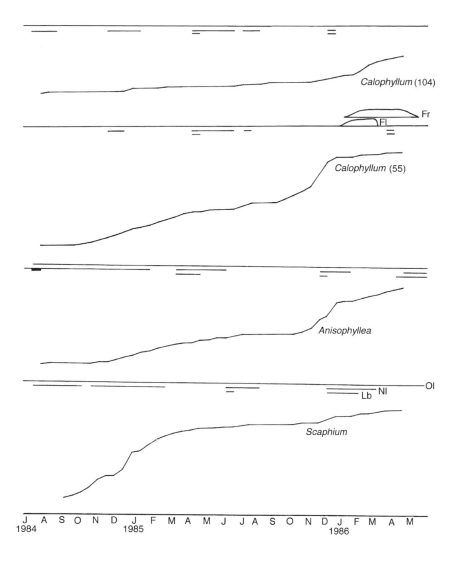

*Calophyllum* (104)

Fr

Fl

*Calophyllum* (55)

*Anisophyllea*

Ol

Lb  Nl

*Scaphium*

J A S O N D J F M A M J J A S O N D J F M A M
1984        1985                    1986

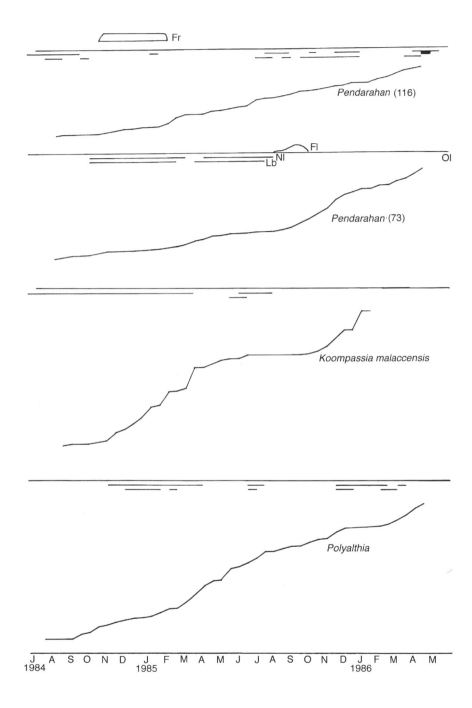

Fr

Pendarahan (116)

Fl
Lb NI                                                                 Ol

Pendarahan (73)

Koompassia malaccensis

Polyalthia

J A S O N D J F M A M J J A S O N D J F M A M
1984          1985                              1986

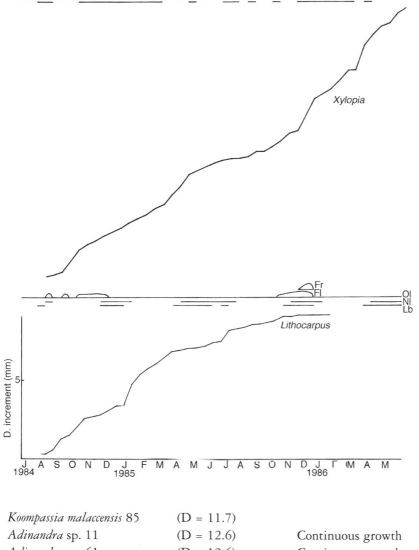

| *Koompassia malaccensis* 85 | (D = 11.7) | |
| *Adinandra* sp. 11 | (D = 12.6) | Continuous growth |
| *Adinandra* sp. 61 | (D = 12.6) | Continuous growth |
| *Calophyllum* sp. 104 | | Intermittent growth |
| *Polyalthia* sp. 67 | (D = 24.4) | Intermittent growth |

(e) Trees in which girth growth clearly correlated with periods of new leaf activity:

| *Anisophyllea* sp. 118 | (D = 15.9) | Intermittent growth |
| *Scaphium* sp. 41 | (D = 29.5) | Intermittent growth |
| *Alstonia* sp. 63 | (D = 16.0) | Intermittent growth |

| *Canarium* sp. 97 | (D = 12.4) | Intermittent growth |
| *Adinandra* sp. 1 | (D = 16.1) | Continuous growth |
| *Cratoxylum cochinchinense* 77 | (D = 12.2) | Continuous-type intermittent growth |
| *Cratoxylum cochinchinense* 78 | (D = 16.5) | Continuous-type intermittent growth |
| *Cratoxylum cochinchinense* 60 | (D = 13.7) | Continuous-type intermittent growth |

The results for the nondipterocarps were various, and it cannot be said that diameter growth correlated with periods of new leaf activity.

### *Agathis* Forest

Dial gauges were attached to all *Agathis* trees in a 50 x 50 m plot in the *Agathis* forest and to many other trees in the plot. However, minimal girth growth was recorded on many trees here. Further investigation will be required to determine whether this reflects the true situation or was because of inept attachment, the failure of the aluminum band to function due to exudation of resin, or other factors.

The *Agathis* trees showing good growth, nos. 118, 124, 29, and 24, displayed no such tendency as the dipterocarps in mixed dipterocarp forest. All showed the same pattern of girth growth, and their phenologies, too, were similar. There was no tendency for diameter growth to parallel cone formation or resting periods in new leaf activity; rather, girth growth was reduced at these times. The same tendency was also seen generally in the *Agathis* trees with lower growth volumes. Figure 2-38 shows that diameter growth occurred from August 1984 to around February 1985, then became extremely slight. After August 1985 the growth rate again increased and remained high until April 1986. The growth rate thus declined in periods of low rainfall and increased in periods of high rainfall, when new leaf activities also were seen.

Of species other than *Agathis*, *Cotylelobium burckii* showed the most pronounced flowering and fruiting activities, but these had no particular correlation with the movement of the dial gauge. Like *Agathis*, diameter growth was faster in periods of heavy rainfall than low rainfall. In *Eugenia* (28) and *Hopea pentanervia* (35), diameter growth was extremely small, but these appear to show the same tendency (fig. 2-39).

### *Alan Bunga* Forest

Dial gauges were attached to 12 *alan* trees and 12 trees of other species and functioned well in all but 3 trees of each group with low growth.

In *alan* tree no. 8, the diameter growth rate decreased or even became nega-

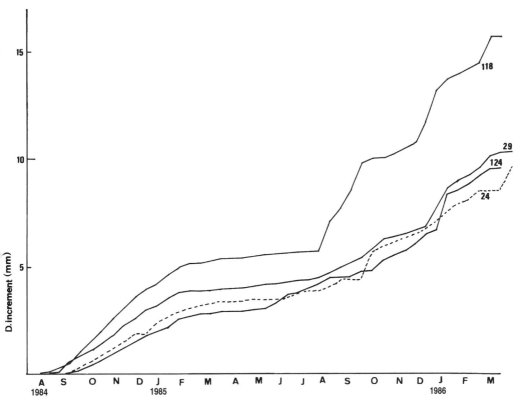

Figure 2-38. Diameter increment of *Agathis* in *Agathis* forest.

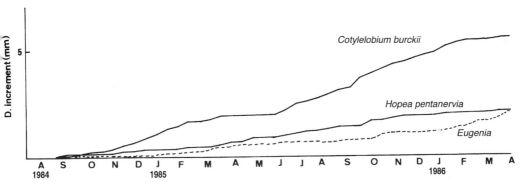

Figure 2-39. Diameter increment of tree species other than *Agathis* in *Agathis* forest.

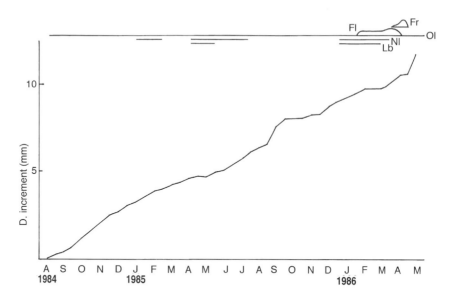

Figure 2-40. Relation between phenology and diameter increment of *alan* in *alan bunga* forest. *Lb:* Leaf buds; *Nl:* New leaves; *Ol:* Old leaves; *Fl:* Flowers; *Fr:* Fruit.

tive at times of leaf bud formation and leaf opening. A similar tendency was also observed during flowering and fruiting, with faster growth resuming after the fall of fruit (fig. 2-40). These tendencies were observed clearly only in this tree; other trees showed them only slightly.

Of the species other than *alan, Lithocarpus sundaicus* showed similar diameter growth to *alan* but no clear correlation between this and phenological phenomena. *Eugenia* was similar (fig. 2-41). Both showed similar growth patterns, with faster growth in periods of low rainfall.

## Phenology and Growth Modes

Several points merit attention in connection with the dial gauge measurements just described. These include (1) the functioning of the gauge after attachment, (2) the correlation between diameter growth and phenological events in dipterocarps and nondipterocarps in mixed dipterocarp forest, (3) the modes of diameter growth and leaf growth in periods of high rainfall in *Agathis* forest, and (4) the increase in growth in periods of low rainfall in *alan* forest.

### Functioning of the Dial Gauge

The aluminum dendrometer employed in the present survey was developed in the United States and has recently been used by Japanese researchers in various

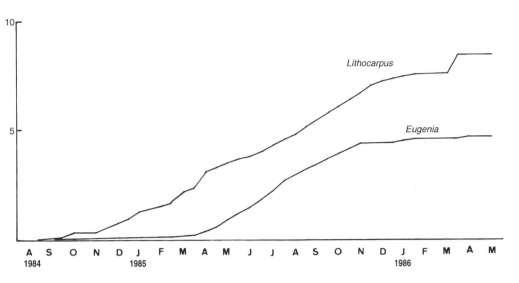

Figure 2-41. Diameter increment of tree species other than *alan* in *alan bunga* forest.

parts of Japan and Indonesia. It had functioned satisfactorily when used in Sumatra, except when the spring was disturbed by small animals.

The fact that many gauges malfunctioned in the Arboretum Andulau was in part because of disturbance by people and small animals and in part because this was where gauges were first installed, and some were installed improperly—for example, too loosely. For large trees, two aluminum bands had to be joined, but this caused no problems. The problem with *Agathis* was that if the surface of the bark was shaved too deeply in attaching the band, resin was exuded and the band stuck to the trunk. In the mixed dipterocarp and *alan* forests, in several instances the roots of climbers appressed the band. This problem commonly arose with the thinner bands used for the more slender trees. The growth of epiphytes around the bark is remarkably fast in the tropics, and many species besides *Agathis* exuded resin. The use of this kind of gauge thus requires the utmost care.

## Seasonality of Dipterocarps and Aseasonality of Nondipterocarps

In the mixed dipterocarp forests of the Arboretum Andulau and the Andulau Forest Reserve, a fixed relationship was discerned in the dipterocarps between stem diameter growth and phenological events such as new leaf activity, flowering, and fruiting in dipterocarps. In nondipterocarps, although some show a similar tendency, many do not, and no such fixed relationship was seen.

The point to be noted here is that flowering and fruiting in the dipterocarps occur at intervals of several years rather than annually. It seems reasonable to expect that, in the lead-up to flowering, leaf buds decrease to the extent that flower buds are formed, and that the energy requirement for flowering and fruiting leads to a reduction of diameter growth. Further, so far as the flowering and fruiting of dipterocarps is of the type where all branches participate synchronously, a considerable portion of photosynthetic materials needs to be diverted from diameter growth. Similarly, where leaf growth showed clear correlation with diameter growth, leaf growth was of the intermittent type, and it appears that during leaf opening, energy was diverted to it from diameter growth.

The nondipterocarps did not show the clear tendency seen with the dipterocarps. The latter all belong to the same family and show similar characters, including gregarious flowering, occupation of the upper canopy positions, and intermittent leaf growth; but the latter represent several families, occupy various canopy positions, and have various habits of leaf growth, from evergrowing to intermittent. It is only natural, therefore, that the correlation with diameter growth is unclear.

## Diameter Growth of *Agathis* in Periods of High Rainfall

The increased rate of diameter growth and relative abundance of new leaf development at times of high rainfall in *Agathis* forest should be examined in the context of locational conditions. The *kerangas* forest where *Agathis* grows stands on white podzolic soil that drains almost as rapidly as sand, and is unusual under a rain forest climate in presenting overall a strongly xerophytic appearance. Consequently, the effect of rainfall volume on growth is greater here than in any other forest type, and productive activities are naturally limited at times of low rainfall. Leaf development takes place at times of high rainfall, photosynthetic activity increases accordingly, and stem diameter growth also accelerates. This seems to be true not only of *Agathis* but of other species growing under the same conditions.

## Good Growth in *Alan* Forest in Periods of Low Rainfall

The decline in diameter growth at times of leaf budding, flowering, and fruiting in *alan bunga* forest resembles that in mixed dipterocarp forest. What is important here is that the rate of diameter growth increases sharply at times of low rainfall. Interestingly, this is completely opposite to the tendency seen in *Agathis* forest. It can be interpreted as a result of the fact that diameter growth is always somewhat repressed by the perhumid conditions of the peat swamp forest. At times of low rainfall, when conditions become slightly drier, the repression is released, and diameter growth approaches what it would be under a more normal environment. Thus, although the tree species that grow in swamp forest are better adapted to the wet conditions than those growing on dry soil, the swamp conditions nevertheless are a major factor inhibiting growth.

The small changes in stem diameters measured by the dial gauge showed several tendencies, which can be summarized as follows:

1. Normal moisture conditions—mixed dipterocarp forest
*Upper story:* intermittent leaf growth, flowering, and fruiting accompanied by cessation of diameter growth; dipterocarps.
*Middle and lower stories:* continuous to intermittent leaf growth, no clear correlation with diameter growth; nondipterocarps.
2. Dry conditions—*Agathis* forest
Improvement of growth conditions by rain; diameter growth in rainy season.
3. Perhumid conditions—*alan* forest
Improvement of growth conditions in dry season: diameter growth in dry season.

In reviewing the literature on phenology, I noted the finding by Wong (1983) of the apparent aseasonality of phenological events in the understory, which contrasts with the marked seasonality reported elsewhere for dipterocarps in the overstory. This accords with the general notion that species occupying the top of the canopy are more sensitive to seasonal changes than those in the understory, where the climate is modified. Richards (1952) cites several examples of this.

Here I would like to examine briefly the characteristics a tree species needs to reach the overstory, based on the phenological data and dial gauge measurements obtained in this study.

For periodicity of branch growth, Koriba (1958) distinguished evergrowing and intermittent modes. I shall consider them in chapter 3 in this volume from the standpoint of species ecology. Here I shall consider what factors most greatly influence canopy position.

From his analysis of tropical deciduous tree species, Koriba (1958) concluded that evergreen species are better adapted to climatic conditions. Though deciduous trees must shed their leaves to survive a dry spell or winter, evergreens can withstand these conditions. Compared with the temperate zones, harsh conditions for tree growth appear not to arise in the tropics; but, in fact, it is known that such tropical occurrences as a sudden drop in temperature following a storm or a succession of hot sunny days after a period of rain can promote leaf abscission or flowering and fruiting. Evergreens are predominant in the tropics, but an appreciable proportion of tropical trees are deciduous. Thus, under the relatively stable climate of the tropical rain forest, a substantial number of trees can be found that are sensitive to small climatic changes.

Height growth can be broadly classified into evergrowing and intermittent types. The former habit is characterized by continuous growth with no resting period, the latter by the interruption of growth at certain definite intervals. There is no simple answer to which type is advantageous, but at least for pioneer vegetation the evergrowing habit appears to be favorable. Such species, however, do not usually attain a large size. Having reached a certain stage through the rapid turnover of thin, large leaves, they cease growing. The evergrowing habit may have further advantages in vigorously regenerating secondary forest and in gaps in primary forest, allowing the tree quickly to reach the canopy and to secure a better environment than other species left behind in the understory. Virgin forest contains a high proportion of evergrowing species.

What, then, are the advantages of the intermittent habit, which features a so-called resting period? While the evergrowing species are constantly cycling metabolic products to new growth, the period of cessation of leaf growth in the intermittent species allows this energy to be directed to expansion and storage in other parts. It is during this period that diameter growth occurs, and it is also conceivable that photosynthetic materials are stored to provide energy for the next new leaves or flowering and fruiting. Compared with the evergrowing type, the intermittent type can be regarded, in a sense, as exercising an effective restraint to provide protection against various environmental conditions. In other words, for tolerance, the intermittent habit is superior to the evergrowing.

The question of dominance in the forest can be approached from various angles; I shall examine first the generalization of reproduction. The gregarious flowering of dipterocarps described in the earlier section on phenology undoubtedly plays a major role in dominance, at least for these species. For the individual tree, successful reproduction involves the completion of a continuous process by which flower buds form on the majority of branches, flowering and fruiting proceed smoothly, and a crop of mature seeds establish themselves on the ground. Clearly, this process has a better chance of success if reproductive activities are carried out on a large scale.

How does the phenomenon of gregarious flowering relate to growth type? It may be that the evergrowing type tends also to flower and fruit continuously, while the intermittent type tends also to have the habit of gregarious flowering. If this is so, then the intermittent type, with its higher success of reproduction, must be capable of retaining a dominant position in virgin forest over a long period.

Of the vegetation in Singapore surveyed by Koriba (1958), the only dipterocarps to display the evergrowing habit belonged to the genus *Dryobalanops;* all others were the intermittent type or deciduous. Of the overstory species in tropical rain forest, the great majority are probably of the intermittent type. This type is more sensitive to climatic change and shows a wider range of adaptation to climate; but at the same time, as Koriba points out, endogenous factors should also be considered. Unfortunately, at least as far as tropical species are concerned, no investigation appears to have been made into the conditions and mechanisms of the action of the hormones involved in growth, flowering, and other physiological processes, and this question is not one to concern us here. Suffice it to note that gregarious flowering is undoubtedly advantageous to dominant species.

Another factor in dominance is tree size—the absolute sizes that different species will attain under the same conditions, and their relative growth rates. Gregarious flowering would not guarantee the dominance of the dipterocarps if trees were small; the dominance of this family derives from the fact that individuals grow to more than 1 m in diameter and 40 m high to dominate the overstory of the canopy. What is important here is the process of reaching the overstory, namely, the difference in growth between dipterocarps and the surrounding trees, from seedling and sapling to giant tree. No detailed data on this process are available; but growth rates probably differ considerably between species and ultimately determine whether a species will become dominant. Consequently, once a seedling established after a gregarious flowering has escaped insect damage and other adversity to become a small tree, the second barrier to its becoming a medium-sized tree is difference in growth rate. Then it must compete with large trees in the vicinity, including their allelopathic influences, before becoming a dominant tree. Having attained the overstory, it begins vigorous photosynthetic activity, consolidates its position as a dominant species, and repeats the cycle of gregarious flowering and fruiting.

As Koriba (1958) also states, almost all tree species probably were of the ever-growing type in geologic times, developing the different habits that are found today as a result of various environmental changes.

Other important characters determining dominance are tolerance to dryness and wetness. The relationship between periods of activity and volume of rainfall seen in *Agathis* and *alan* shows that although these species have the greatest tolerance in their respective locations, the conditions there work as limiting factors, and when these conditions are mitigated, their activities increase. Intermediate between the two are the locational conditions of mixed dipterocarp forest.

Another question concerns the numbers of dominant species. The greater the limiting factors in a particular location, the smaller the number of species that will be found there; and as the conditions are alleviated, the number of species will increase. Where species number decreases, those with greater tolerance to the limiting conditions become dominant; and the greater their range of tolerance, the wider the gap becomes between them and other species.

An example is *alan* in peat swamp forest. *Alan* shows a gregarious flowering habit and rapid growth. Although its growth process has not been studied in detail, the fact that the overstory of *alan* stands far above other species attests to this rapid growth. A major factor in this growth is its double-layered root system. A network of feeding roots ramifying over the lower buttresses absorbs nutrients over a wide area, and decomposition within the false forest floor formed by this network increases the amount of available nutrients. Below the buttresses, a system of large, long roots develops, sometimes reaching the deepest layers of the peat. Furukawa (1988) has stated that a narrow tidal range and the presence of standing water for prolonged periods further promotes the dominance of species like *alan,* which have giant buttresses. A large, spreading root system naturally is effective in the poor conditions of peat swamp in providing physical support for such giant trees.

*Agathis* lacks these remarkable features, and the factors that have produced the even-aged forest are unclear. The thick root system in *Agathis* forest suggests, however, that this species established itself in *kerangas* forest on the long accumulation of litter on white sand, and through its superior regeneration it gradually displaced other species. Another factor could well be a negative allelopathic effect of chemical components of *Agathis* resin on other tree species.

The gregarious flowering of *alan* took place in the middle of the rainy season. On the other hand, many trees of the mixed dipterocarp forest flower at the end of the dry season. This difference, I believe, is the result of the same kind of stress. For the swamp-dwelling *alan,* rainwater coming on top of the standing water already present exacerbates its growth conditions, and this produces qualitatively the same kind of stress that dry conditions from lack of rainfall produce in species growing on dry land. In both, when the stress becomes fairly extreme, the response is gregarious flowering.

# 3.

# *Forest Ecology of Mount Pangrango*

## 1. Location of Mount Pangrango

I first set foot in Indonesia on 1 October 1969. In Thailand, which I saw in 1965, researchers from Osaka Municipal University and Kyoto University had already carried out forest surveys in cooperation with Chulalongkorn University and Kasetsart University. If possible, I wanted to work on an area where Japanese researchers had not conducted forest surveys, and when I applied to the Center for Southeast Asian Studies to do research overseas, I chose Indonesia. From the information I obtained in Japan, I could not grasp clearly the state of the forests in Indonesia; and so I departed, and having looked around the survey areas of my predecessors in Thailand and Malaysia, I entered Indonesia.

My destination was what was then known as Lembaga Biologi Nasional (National Biological Institute). Having gathered information on the forests in Indonesia, I judged that it would be financially impossible to study the outer islands and therefore chose a virgin forest in Java as accessible as possible. As I will describe later, the area of forest in Java is small, and not much virgin forest remains. The only candidate areas in West Java were Mount Pangrango and Ujung Kulon. Most of the other regions were secondary forest, rice lands, tea gardens, rubber plantations, or dry fields. Ujung Kulon is a peninsula at the western extremity of West Java, two days' travel from Jakarta. It was inconvenient for periodic measurements, and I therefore decided to concentrate my efforts on Mount Pangrango. A search of the literature on the botany of Mount Pangrango revealed more than two hundred published reports (van Steenis and van Steenis–Kruseman, 1953). Most of the surveys had been carried out in the Dutch period, and many were written in Dutch or German. The majority focused on taxonomy, morphology, or physiology; only two works, Seifriz (1923) and Meijer (1959), addressed forest ecology. The former dealt with vertical distribution zones, differentiated by the dominants at each altitude, and the latter was a survey of plant sociology at around 1,400 m.

In dealing with montane forest vegetation, I wanted to chart its changes with altitude in as much detail as possible. Vertical distribution has been discussed by many writers. Representative works include Imanishi's (1937) vertical distribution zones of the Japan alps, Kira's (1949) theory of distribution zones of major tree species of Japanese origin according to warmth index, and Whittaker's (1956) analysis of the Great Smoky Mountains by environmental gradient analysis. Of these, the former two concern distribution of tree species, that is, whether a particular species is present. Whittaker dealt with the quantitative distribution of species by environmental gradient of the mountains. Change in altitude results in change in the tree species present and also in the forest structure. This is particularly notable in the tropics, where Wyatt-Smith (1964) has reported an example in Malaysia. Whittaker (1970), too, has pointed this out.

Several methods have been applied in an attempt to portray forest structure by stratification, including the crown depth diagram (Ogawa et al., 1965), analysis by distribution density function (Hozumi et al., 1968), and observation from the ground (Ogino, 1974).

At Mount Pangrango, I established nine plots in an altitude range of 1,400 m, from the foothills at 1,600 m to the summit at 3,000 m, and observed the constituent species and stratification at each altitude to gain a picture of the distributions of individual species. My main objective was to attempt to clarify the ecological significance of each species by examining which part of the canopy it occupies. I conducted surveys for one year, from January 1969 to January 1970, and for one month in October 1976.

## Overview of the Survey Area
### Geography of Indonesia

The Indonesian archipelago has often been likened to an "emerald necklace adorning the equator." It stretches from 95° E to 135° E and from 6° N to 11° S. Its span of 4,800 km from Sumatra in the west to Irian Jaya in the east is equal to the distance from Kamchatka to Taiwan, and the 1,600 km from north to south is roughly equivalent to the distance between the northern and southern borders of the United States (Watanabe, 1959). Its more than three hundred islands, large and small, lie on a geologic structure composed of three basic units: the Sunda Shelf, the Sahul Shelf, and the young mountains located between the two continental shelves.

The Sunda Shelf is a stable block at the southern edge of the Asian continental mass that has retained its integrity as a land mass through long geologic periods. It is the basement underlying Borneo, eastern Sumatra, northern Java, and the Malay Peninsula. The Sahul Shelf is a northward extension of the Australian continental mass, again with a stable structure. Its northernmost edge lies south of the main range running east-west through central New Guinea.

The area sandwiched between the Sunda and Sahul Shelves is described as the product of comparatively recent orogenic movements in or after the Tertiary period. Fault and fold systems form complex patterns that appear as arcs of islands.

The Burma-Java loop stretches south from Burma and emerges in Sumatra, Java, the Kei Islands, and Seram. On the Indian Ocean side, this arc is paralleled across a deep oceanic trough by the Nias-Mentawai Islands and the islandless submarine ridge south of Java (Dobby, 1961). Work on the theory of plate tectonics has produced a detailed explanation of land formation in this area (Hamilton, 1979), but the process is far from simple and will not be explored further here.

The second feature influencing the landform of Indonesia is volcanic activity. At this meeting point of the circum-Pacific and Mediterranean volcanic belts, volcanoes are truly plentiful. Many volcanoes lie in the area sandwiched between the Sunda and Sahul shelves; Indonesia overall has more than one hundred active and three hundred dormant volcanoes.

The largest land area in Indonesia is Kalimantan (occupying the southern region of Borneo island). This is followed by Sumatra. In population, the Java-Madura region overwhelmingly has the highest concentration. Though small in area, its population density is extremely high at 499 people/km². The second highest density occurs in the Bali–Nusa Tenggara region, with 79 people/km². This is followed by Sulawesi with 40 people/km² and Sumatra with 35 people/km², while Kalimantan has 8 people/km². Where population density is high, the area of forest continues to decrease. In Kalimantan, forest covers 60 percent of the land area, in Sulawesi 53 percent, and in Java-Madura only 23 percent.

Kalimantan, Sumatra, and elsewhere support tropical rain forests dominated by dipterocarps, thought to contain the greatest diversity of species and the largest trees in any of the world's three tropical zones. From Sulawesi to the Moluccas and New Guinea, the land is characterized by rain forests with no particular dominant species group, but a paucity of surveys means few details are known. The most extensively surveyed region is Java, with forest types ranging from rain forest to monsoon forest. The flora has been described in detail by Backer and van den Brink (1963–1968) and van Steenis et al. (1972).

## Geography of Java

The area of Java and Madura, 132,187 km², amounts to only 6.94 percent of Indonesia as a whole. Statistics for 1963 show forest covered 23 percent of the land area in this region, far less than the national average of 64 percent. It is limited to the peninsulas of Ujung Kulon and Baluran at the eastern and western extremes, offshore islands such as Nusa Kambangan and Pulau Seribu, the survey area of Mount Pangrango, and other mountain areas such as Mount Slamet.

On the densely populated island of Java, the ubiquitous landscape is paddy fields backed by graceful, soaring, conical volcanoes. It is totally different from the lush green world of the outer areas, which are covered by tropical rain forest. The volcanoes that characterize Java's landscape became active in the Tertiary period, increased in activity during the Quaternary, and now make the island one of the world's foremost volcanic areas. Almost all mountains in Java more than 2,000 m high are volcanic. The highest peak is Mount Semeru (3,676 m) in East Java; the highest in West Java is the survey area of Mount Pangrango (3,019 m).

The cool plateaus of Bandung, Garut, Dieng, Malang, and others are nearly all surrounded by volcanoes. These highlands have been summer resorts since colonial times, and they are important for the production of fruit, vegetables, tea, and other crops. West Java has many tea and orchard estates at altitudes of 1,200 to 1,500 m, which were established under the Dutch. These, together with rice fields and vegetable gardens, cover the land completely.

Thus, practically no primary forest remains at altitudes below 1,500 m. In the mountains south of Bandung, shifting cultivation extends to almost 2,000 m, and the smoke of cultivators' fires can be seen everywhere in the dry season. The exhaustive stripping of forest cover as a result of population growth inevitably has caused soil erosion on a large scale and has led to severe socioeconomic problems.

## Climate of Java

Java has a monsoon climate, and the prevailing wind direction is nearly constant depending on the season. Temperatures are high and almost constant throughout the year. Jakarta has an average temperature of 26.6°C and an annual variation of 1.1°C, showing no remarkable seasonal differences. The climate could be described as tropical marine. Rainfall is governed by the two seasonal winds, which alternate roughly every six months. From March to September the prevailing winds are generally southeasterly or easterly, and from September to March they are mainly from the west or northwest. Strictly speaking, March to April and October to December are transitional periods, and the winds are sometimes easterly, sometimes westerly.

Both seasonal winds bring rain to Java. Rainfall volumes in each season, however, vary with location. West Java receives rain during the westerly and west-northwesterly winds; southern Java during the southerly and southeasterly winds. The rainy season therefore comes at different times in different places. The westerly and west-northwesterly winds tend to bring a greater volume of rainfall, and the areas where it falls are thus wetter. Rainfall volumes decrease progressively from the west of Java to the east.

In Jakarta, the 1,014 mm of rainfall during the rainy season from December to March account for 56 percent of the annual total. Monthly rainfall in the dry season from June to September, centering on July and August, rarely exceeds 100 mm. In the mountains of West Java, annual rainfall may exceed 4,000 mm.

The central part of Central Java has a massif of mountains in the 3,000-m class, and in some places rainfall is very high. In East Java, some areas receive less than 1,000 mm of rainfall annually and are markedly dry. Here, the seasonal variation in rainfall is pronounced, producing distinct dry and rainy seasons.

Walter (1955) defined dry season by plotting monthly rainfall and temperature lines on the same graph, taking one division on the vertical axis to represent 1°C of temperature and 2 mm of rainfall. Months when the rainfall line fell below the temperature line were taken as dry-season months.

Figure 3-1 shows temperature and rainfall charts prepared for various localities in Indonesia by Walter's method. Citajam in West Java has an annual average rainfall of 3,469 mm and no dry season; Klaten in Central Java has 1,639 mm of rainfall and a dry season of three months; and Pasuruan in East Java has 1,284 mm of rainfall and a dry season of four months. Average temperatures are the same for all three locations, at 26–27°C. It is clear that rainfall in Java decreases from west to east as the dry season becomes progressively longer. This tendency to increasing dryness extends farther eastward from Java.

## Vegetation of Java

In a climate characterized by the two features of almost invariably high temperature throughout the year and rainfall volume governed by the monsoons, the natural flora of Java, at least in the lowland, consists of various types of forest, from closed forests of tall trees to sparse, open forest. In the whole of Java, only fragments remain of what can be regarded as primary vegetation. Nevertheless, it is in Java that the flora and vegetation have been relatively well researched.

Briefly, the flora of Java is described as containing elements both of Asia and Australia. Species unique to Java are few, and the mountain flora includes many elements common with Sumatra. Dipterocarps, however, the Asian elements characteristic of the Malay Peninsula, Borneo, and Sumatra, are relatively few. *Eucalyptus* species typical of the Australian element are absent; only *Podocarpus* occurs, and in small amounts.

Van Steenis (1965) divided the vegetation of Java into the following eleven types: submerged littoral vegetation; mangrove; beach formation; lowland swamp forests; hydrophytic vegetation; rheophytic vegetation; mixed lowland and hill rain forest on dry land; montane everwet rain forest; mountain swamps and lakes; subalpine vegetation; and monsoon forest.

Besides the minor types with extremely limited distribution along the coasts and rivers or in the mountains, the major vegetation types in Java include coast mangrove and the *Barringtonia* association; swamp forest with such genera as *Alstonia* and *Gluta;* mixed rain forest and monsoon forest on lowland and hills; and montane rain forest and subalpine vegetation. The areas of coastal forest and swamp forest are extremely limited, in contrast with the wide expanses found in Sumatra and Kalimantan. Lowland rain forest originally may have become established in the high rainfall regime of West Java, but today, because of extensive

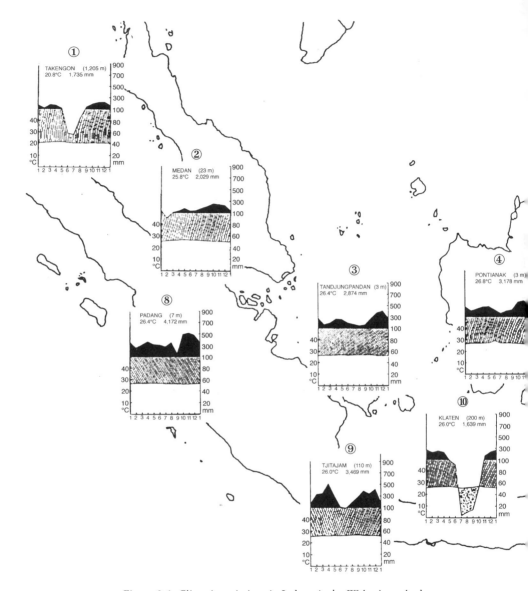

Figure 3-1. Climatic variations in Indonesia, by Walter's method.

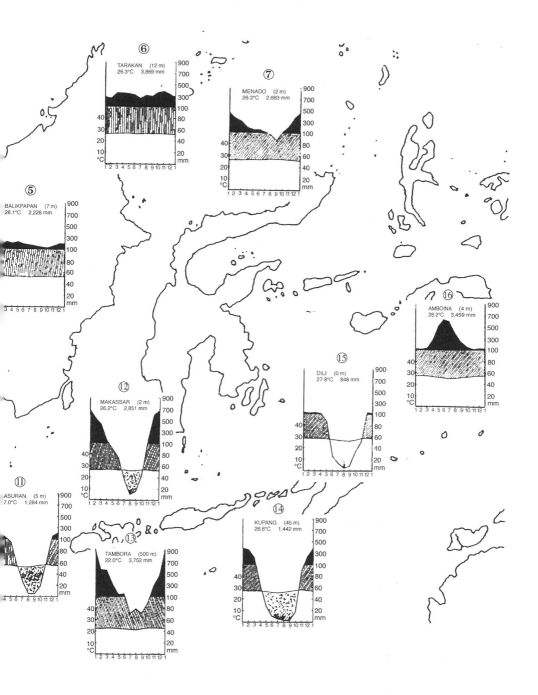

human intervention, virtually none remains. The small areas left in forest reserves at Ujung Kulon and elsewhere suggest that it was a well-developed, closed forest of tall trees up to 50 m high, with a canopy composed of five stories. It differs from the forests of Sumatra and Kalimantan, which contain an admixture of dipterocarps, in lacking a layer of giant emergent crowns and in containing a small admixture of deciduous species. It is very similar in appearance to the forest type that Ogawa (1974) termed tropical evergreen seasonal forest. The Pasoh Forest Reserve in the center of the Malay Peninsula has typical tropical rain forest, but farther north, in South Thailand, the forest resembles the lowland forest of West Java. This area can be regarded as a forest type with seasonal rainfall distribution, deviating slightly from the typical rain forest climate.

East Java, with its distinct dry season, is a world of the so-called monsoon forest. Teak and other deciduous species such as *Butea, Acacia, Albizia,* and *Lagerstroemia* are notable. Forest structure simplifies to two strata, and few trees reach above 20 m high. Many species have thorns and thick, corklike bark, affording resistance to drought and fire. In areas severely disturbed by humans in the past —in Baluran, for example—grassland has developed. Some of the grasslands in East Java have even become conservation areas. In places where dry-season deciduous species dominate, teak has been planted. Although Indonesia does not produce high-quality timber like Thailand and Burma, teak silviculture is an important element of forestry in Java.

In the mountains, forests dominated by Fagaceae appear first, then, at higher altitudes, Ericaceae become dominant. Van Steenis terms the former montane everwet rain forest and the latter subalpine vegetation. Compared with lowland forests, the canopy structure of montane forests is extremely simple, but the canopy is closed, and trees more than 30 m high are common. The species composition is considerably diverse, and climbers and epiphytes are abundant. From the higher altitudes of the montane zone into the subalpine zone, in Java as elsewhere in Southeast Asia, so-called moss forest occurs, where tree trunks, branches, and the forest floor develop a thick covering of moss.

The vegetation of Java may be summarized as follows: In the lowland, the changing water regime from west to east gives rise to a series of forest types from lowland rain forest (tropical evergreen seasonal forest) to monsoon forest; and vertically, lowland rain forest gives way to montane forest, which in turn is replaced by subalpine vegetation.

### *Forests of Mount Pangrango*

The area chosen for survey was the northeastern slope of the Gede-Pangrango massif, which has two volcanic peaks, Mount Pangrango (3,019 m) and Mount Gede (2,967 m). On Mount Gede, volcanic activity has destroyed vegetation

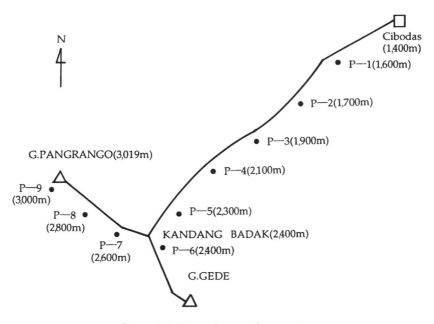

Figure 3-2. Schematic map of survey sites.

above 2,500 m, but on Pangrango, apart from a localized area of grassland near the summit, the entire upper slope is forested.

Below 1,400 m, the massif is extensively covered by tea estates, settlements, and fields, and none of the primary vegetation appears to have survived. On the middle slopes of the massif, at 1,400 m, is the botanic garden at Cibodas. Figure 3-2 shows a schematic view of the area.

Located about 40 km south of Bogor (about 100 km south of Jakarta), the Cibodas Botanic Garden is one branch of the Bogor Botanic Garden, which forms part of the Bogor National Biological Research Institute, the center of biological research in Indonesia. This area, including the twin peaks of Gede-Pangrango, was designated a nature reserve in 1925. Although there have been several changes in the responsible authority since then, the fact that the area has been maintained as a site for tropical montane forest research owes much to a consistent policy of strict conservation. The slopes below 1,400 m have, as mentioned, a totally artificial landscape, but above that the vegetation remains intact.

Between 1,400 and 1,500 m, *Altingia excelsa* (Indonesian: *rasamala*) dominates. The biggest individuals of this tree reach higher than 60 m, emerging above the continuous canopy and dominating the overstory. Above about 1,500 m, *Schima wallichii* and species of Fagaceae and Lauraceae appear, and above about 1,600 m, these species become abundant.

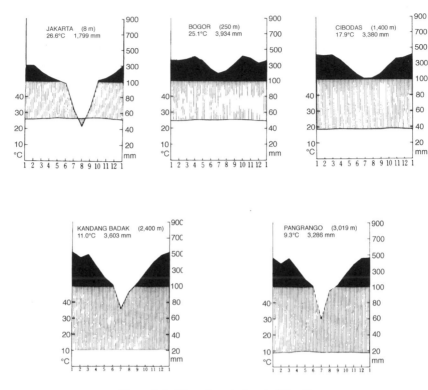

Figure 3-3. Climatic variations in West Java.

At around 1,800 m, *Podocarpus imbricatus* is notable. This is the area that van Steenis termed montane forest. At the saddle (2,400 m) between Gede and Pangrango, the average tree height is around 20 m, and the canopy structure becomes simpler. *Myrsine affinis, Eurya obovata, Vaccinium varingiaefolium,* and other species begin to appear. Approaching the summit of Pangrango, the quantity of moss increases, and the area takes on the appearance of moss forest. Near the summit, stunted species of Ericaceae become dominant. *Anaphalis javanica* grows in localized bare patches.

Figure 3-3 shows Walter's temperature-rainfall plots for West Java from the coast to the mountains. Jakarta has a distinct dry season in August, but there is no clear dry season in Bogor or Cibodas. Annual rainfall in Jakarta is 1,799 mm, while in Bogor and Cibodas it exceeds 3,000 mm. Temperature fluctuations are small. The average temperatures are 26.6°C for Jakarta, 25.1°C for Bogor, and 17.9°C for Cibodas, decreasing to the south, but none of the locations shows a marked seasonality.

Table 3-1

Climate at the Summit of Mount Pangrango

| | Average Temp. (°c) | Average Relative Humidity (%) | Monthly Rainfall (mm) | Days of Rainfall > 0.1 mm | Insolation Period (hr) | Insolation Rate |
|---|---|---|---|---|---|---|
| January | 8.9 | 91 | 461 | 29 | 80 | 27 |
| February | 9.0 | 91 | 386 | 26 | 51 | 28 |
| March | 9.0 | 91 | 451 | 30 | 88 | 22 |
| April | 9.6 | 88 | 329 | 28 | 99 | 33 |
| May | 9.9 | 82 | 191 | 25 | 161 | 41 |
| June | 9.5 | 78 | 109 | 20 | 172 | 50 |
| July | 9.1 | 70 | 60 | 16 | 219 | 69 |
| August | 9.1 | 72 | 95 | 17 | 186 | 67 |
| September | 9.1 | 76 | 114 | 20 | 172 | 53 |
| October | 9.2 | 84 | 259 | 26 | 131 | 39 |
| November | 9.2 | 88 | 376 | 27 | 110 | 30 |
| December | 8.9 | 88 | 455 | 28 | 73 | 29 |
| Annual | 9.3 | 83 | 3,286 | 291 | 1,522 | 41 |
| Period | 1912–1938 | 1912–1938 | 1912–1936 | 1912–1936 | 1912–1936 | 1912–1936 |
| No. of years | (27) | (27) | (16) | (16) | (16) | (10) |

Meteorologic records covering the twenty-seven-year period from 1912 to 1938 are available for elevations of 2,400 m and 3,019 m on Mount Pangrango (table 3-1). Temperature averaged 11°C at 2,400 m, decreasing to 9.3°C at 3,019 m. Annual rainfall volumes were, respectively, 3,603 mm and 3,286 mm, not significantly different from levels at Cibodas. Even in the drier months of July and August, the average number of rainy days was high, and humidity remained above 70 percent. The summit received fewer hours of sunshine.

## 2. Altitudinal Changes in Species Composition and Stratification

### Survey Method

Survey plots were located along the mountain path from 1,500 m to the summit of Mount Pangrango in the Gede-Pangrango Nature Reserve, which was managed by the Cibodas Botanic Garden (Cibodas branch of the National Botanical Institute) until 1979, when management was transferred to the Ministry of Agriculture and Forestry. In selecting plots, I sought primary forest occupying, as far as possible, gentle slopes that faced the same direction. As shown in table 3-2, however, the slopes and orientations of the plots were considerably scattered. Plots 1, 6, and 9 were established in February 1969, the remainder in October 1976.

Though the total areas of the plots differ, each was divided into quadrats of 10 × 10 m, and the trees of 10 cm and greater in DBH were identified by species, and their DBH, height, and height to lowest live branch were recorded. The same items were surveyed within the quadrats indicated in the table for trees of 4.5 cm up to 10 cm DBH and for those of less than 4.5 cm DBH and more than 1.3 m high. For ground vegetation, quadrats of 1 × 1 m were set up in several places, and the species identity, species number, individual number, plant height, and so on were recorded. Diameters were measured with a diameter tape for the larger trees and with dial calipers or slide calipers for the smaller ones. Heights were measured with a Weisse hypsometer or, for small trees, a measuring rod.

The survey results for each plot were summarized for the number of individuals and relative dominance of each tree species present. These appear in the tables in the appendix.

The following is the result of a survey of changes in species composition and stratification with altitude and of the characteristics of each species.

## *Altitudinal Changes in Species Composition*
### Vertical Distribution Zones in Mountains of Southeast Asia

Seifriz (1923) distinguished the following vertical distribution zones on the Gede-Pangrango massif by dominant species.

| | |
|---|---|
| 1,400–1,670 m | *Rasamala* subzone |
| 1,670–2,120 m | *Podocarpus* subzone |
| 2,120–2,400 m | Herbaceous subzone |

Above 2,400 m, only Mount Gede is mentioned, divided into a *Vaccinium* subzone and an Edelweiss subzone.

Okutomi (1977) divided the vertical distribution on the eastern ridge of Mount Kinabalu (4,101 m), the highest peak in Borneo, as follows:

| | |
|---|---|
| 1,200–1,700 m | Tropical low montane rain forest |
| 1,700–2,900 m | Tropical high montane rain forest |
| 2,900–3,700 m | Tropical subalpine vegetation |
| 3,700–4,101 m | Tropical alpine stunted forest |

### Table 3-2
### Sample Plots

|  |  | P-1 | P-2 | P-3 | P-4 | P-5 | P-6 | P-7 | P-8 | P-9 |
|---|---|---|---|---|---|---|---|---|---|---|
| Elevation (m) |  | 1,600 | 1,700 | 1,900 | 2,100 | 2,300 | 2,400 | 2,600 | 2,800 | 3,000 |
| Slope |  | 3° | 10° | 14° | 10° | 20° | 20° | 12° | 32° | 7° |
| Direction |  | N40 °W | W | N10 °W | N5 °W | N10 °W | N | N70 °W | N40 °E | N55 °W |
| Plot size | $D \geq 10$ | 1ha | 1,000 m² | 1,000 m² | 1,000 m² | 1,000 m² | 400 m² | 400 m² | 400 m² | 400 m² |
|  | $4.5 \leq D < 10$ | 1ha | 1,000 m² | 1,000 m² | 400 m² | 100 m² | 400 m² | 400 m² | 400 m² | 400 m² |
|  | $H > 1.3, D < 4.5$ | 1,000 m² | 100 m² | 100 m² | 100 m² | 25 m² | 100 m² | 25 m² | 100 m² | 100 m² |
|  | Ground vegetation | 10 m² | 100 m² | 100 m² | 100 m² | 25 m² | 5 m² | 25 m² | 25 m² | 5 m² |

In the mountains of peninsular Malaysia, Wyatt-Smith (1964) made the following divisions:

| | |
|---|---|
| 0–300 m | Lowland dipterocarp forests |
| 300–760 m | Hill dipterocarp forests |
| 760–1,300 m | Upper dipterocarp forests |
| 1,300–1,700 m | Montane oak forests |
| Above 1,700 m | Montane Ericaceous forests |

On Mount Wilhelmena, New Guinea, Brass (1941) made the following divisions:

| | |
|---|---|
| 480–2,350 m | Mid-mountain forest |
| 850–3,100 m | Beech forest |
| 1,500–3,000 m | Mossy forest |
| 3,000–4,050 m | Subalpine forest |

Van Steenis et al. (1972), having investigated altitudes of marked change in species composition by finding the upper and lower limits of distribution of various species, divided the Malesia region (including the Malay Peninsula, Sumatra, Java, Borneo, Sulawesi, and New Guinea) into the following altitudinal zones:

| | |
|---|---|
| 1–1,000 m | Tropical zone |
| 1,000–1,500 m | Submontane zone |
| 1,500–2,400 m | Montane zone |
| 2,400–4,000 m | Subalpine zone |
| 4,000–4,500 m | Alpine zone |
| 4,500–5,000 m | Nival zone |

Comparison of the above classifications reveals as a common feature of the montane forests of Southeast Asia the presence of forest zones dominated by Fagaceae and Ericaceae. The former is the dominant community in what van Steenis et al. called the montane zone, while the latter is dominant in the sub-alpine zone. Both are further characterized by the admixture of conifers such as *Podocarpus* and *Dacrydium.* The development of so-called mossy forest in the upper montane and subalpine zones also characterizes Southeast Asian forest.

## Altitudinal Distribution of Various Species

The number of individuals and relative dominance of each tree species represented by individuals greater than 10 cm DBH in each of the plots are summarized in table 3-4. The total species in the nine plots was 81, of which 57 were found in plot 1. The most frequently found species were *Acronodia punctata* (seven of nine plots), *Schima wallichii* (six plots), *Polyosma ilicifolia* (six plots), *Myrsine affinis* (four plots), and *Eurya obovata* (four plots).

Some species appeared not in high frequency but in separate rather than adjacent plots. Since these could be expected to appear continuously but by chance were not included in the chosen plots, they should be regarded as the next most frequent to the above species. They include *Astronia spectabilis, Geniostoma arboreum, Glochidion macrocarpum, Leptospermum flavescens, Lithocarpus elegans, Lithocarpus tijsmannii, Macropanax undulatus, Neolitsea javanica, Photinia notoniana, Symplocos laurina, Syzygium gracile, Vaccinium laurifolium,* and *Viburnum coriaceum.*

Forty-one species, accounting for 51 percent of the total, appeared in only one of the nine plots.

For the 30 species that appeared in two or more plots, the vertical distributions are shown in figure 3-4. The solid lines show the range of altitudes at which species were recorded in the survey plots or confirmed by observation outside the plots, and the broken lines show the range of distribution given by Backer and van den Brink (1963–1968). The vertical distribution zones distinguished by van Steenis et al. (1972) and Seifriz (1923) are included.

The distributions of various species cover wide ranges and overlap each other, but overall the species composition can be seen to change gradually. It is difficult, however, to recognize a particular elevation where a striking change in species composition takes place. I therefore adopted the method that van Steenis et al. used in dividing the distribution zones of the Malesia region and examined the upper and lower limits of distribution for individual species. Table 3-3 lists the numbers of species having their upper or lower limit of distribution at each given altitude.

At 2,400 m, five species have their upper limit of distribution and two species their lower limit, giving a total of seven species. At 2,600 m, the total is five species. Thus the altitude of 2,400 m or 2,600 m can be taken as that at which species numbers change on Mount Pangrango.

Species occurring in high frequency have a wide range of vertical distribution but not always a high degree of dominance. The dominance of each species at each altitude is represented by its relative dominance (proportion of basal area) (table 3-4).

## Table 3-3
### Distribution of Tree Species by Altitude, by van Steenis' Method

| ELEVATION (m) | 1,700 | 1,900 | 2,100 | 2,300 | 2,400 | 2,600 | 2,800 |
|---|---|---|---|---|---|---|---|
| Upper limit of species | 2 | 1 | 3 | 4 | 5 | 4 | 0 |
| Lower limit of species | 0 | 1 | 1 | 0 | 2 | 1 | 1 |
| Total | 2 | 2 | 4 | 4 | 7 | 5 | 1 |

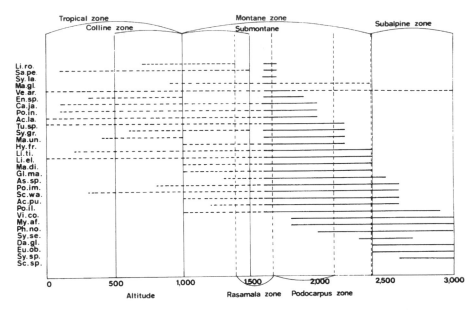

Figure 3-4. Vertical distribution of main tree species. The dotted line indicates the distribution range in Java, taken from Backer and van den Brink Bakhuizen, 1963–1968; the solid line is the distribution determined in the present study. The zonation at top was proposed by van Steenis et al., 1972; below, by Seifriz, 1923. Note that many of the species show overlapping altitudinal ranges and that no clear zonations are discernible.

| | | | |
|---|---|---|---|
| Ac. la. | : *Acer laurinum* | Ma. un. | : *Macropanax undulatus* |
| Ac. pu. | : *Acronodia punctata* | My. af. | : *Myrsine affinis* |
| Ar. ja. | : *Ardisia javanica* | Ne. ja. | : *Neolitsea javanica* |
| As. sp. | : *Astronia spectabilis* | Pe. ri. | : *Persea rimosa* |
| Ca. ja. | : *Castanopsis javanica* | Ph. no. | : *Photinia notoniana* |
| Da. gl. | : *Daphniphyllum glaucescens* | Po. il. | : *Polyosma ilicifolia* |
| De. fr. | : *Decaspermum fruticosum* | Po. im. | : *Podocarpus imbricatus* |
| | var. *polymorphum* | Po. in. | : *Polyosma integrifolia* |
| En. sp. | : *Engelhardia spicata* | Sa. pe. | : *Saurauia pendula* |
| Eu. ob. | : *Eurya obovata* | Sc. sp. | : *Schefflera sp.* |
| Fi. ri. | : *Ficus ribes* | Sc. wa. | : *Schima wallichii* |
| Fl. ru. | : *Flacourtia rukam* | Sy. fa. | : *Symplocos fasciculata* |
| Gl. ma. | : *Glochidion macrocarpum* | Sy. gr. | : *Syzygium gracile* |
| Hy.fr. | : *Hypobathrum frutescens* | Sy. la. | : *Symplocos laurifolia* |
| Le. fl. | : *Leptospermum flavescens* | Sy. se. | : *Symplocos sessilifolia* |
| Li. el. | : *Lithocarpus elegans* | Sy. sp. | : *Symplocos sp.* |
| Li. ps. | : *Lithocarpus pseudomoluccus* | Tu. sp. | : *Turpinia sphaerocarpa* |
| Li. ro. | : *Lithocarpus rotundatus* | Va. va. | : *Vaccinium varingiaefolium* |
| Li. ti. | : *Lithocarpus tijsmannii* | Ve. ar. | : *Vernonia arborea* |
| Ma. di. | : *Macropanax dispermus* | Vi. co. | : *Viburnum coriaceum* |
| Ma. gl. | : *Magnolia glauca* | | |

The most frequently occurring species, *Acronodia punctata,* has a relative dominance of 0.22 percent in plot 1 (1,600 m), by no means high when compared with *Schima wallichii* or *Castanopsis javanica.* In no plot up to plot 5 (2,300 m) does the relative dominance exceed 5 percent. In plot 6 (2,400 m), this rises to 22.7 percent, and together with *Eurya obovata,* it has the highest dominance. But in plot 7 (2,600 m), relative dominance falls to 7.8 percent, and the dominants become *Polyosma ilicifolia* and *Myrsine affinis.* Thus while *Acronodia punctata* has the widest distribution range, it has a high degree of dominance with respect to other species only in the vicinity of plot 6, at 2,400 m.

*Schima wallichii* is another species that occurs frequently. At 1,600 m, its relative dominance of 25 percent was the highest in this plot. This increased with altitude to 44 percent at 2,100 m in plot 4. In plot 6 (2,400 m), however, it declined sharply. No other species shows such a high relative dominance over a wide range of altitudes up to plot 4 at 2,100 m.

*Myrsine affinis* is a tall tree found at the higher altitudes of plot 5 (2,300 m) and above. It showed a high relative dominance over its entire range of distribution, with a maximum value of 42 percent in plot 8 (2,800 m). It is clearly a dominant species at high altitudes. *Eurya obovata* is another high-altitude species that makes its appearance in plot 5 (2,300 m) and can also be called a dominant at high altitudes.

*Astronia spectabilis* has a relatively wide distribution from plot 1 (1,600 m) to plot 5 (2,300 m). Its relative dominance is low in the two lowest plots but increases with altitude to reach 19 percent in plot 5, the highest of any species there. *Podocarpus imbricatus,* which is represented by a single individual in plot 1 (1,600 m), has a relative dominance of around 35 percent, the highest of any species, in plot 2 (1,700 m) and plot 3 (1,900 m). In table 3-4 this species is recorded only up to plot 4 (2,100 m), but its distribution has been noted up to around 2,600 m (fig. 3-4). At higher altitudes, however, its numbers decline, and its optimal range of distribution should be regarded as 1,700–1,900 m.

From the foregoing discussion, it is clear that species with wide ranges of distribution fall into several types through their relative dominance with respect to other species occurring at the same altitudes: for example, those with constantly high values and those with high values over a narrow range. The dominant species in each plot in terms of relative dominance are: plot 1 to plot 4, *Schima wallichii* and *Podocarpus imbricatus;* plot 5, *Astronia spectabilis;* and plot 6 to plot 9, *Polyosma ilicifolia, Acronodia punctata, Myrsine affinis, Eurya obovata,* and others.

Relative dominance allows quantitative comparison of species within a plot but not between plots. Thus we turn to the changes of basal area with altitude.

Figure 3-5a shows the variation with altitude of basal areas of trees more than 10 cm DBH in each plot. At 1,600 m the basal area is 52 m²/ha, while at the summit it is 36 m²/ha. The variation between these points is not a gradual de-

Table 3-4

Relative Dominance by Plot of All Tree Species Observed

| SPECIES | P-1 N | P-1 R.D. (%) | P-2 N | P-2 R.D. (%) | P-3 N | P-3 R.D. (%) |
|---|---|---|---|---|---|---|
| *Schima wallichii* spp. *noronhae* (Reinw. ex Bl.) Bloemb. | 47 | 25.12 | 47 | 25.12 | 80 | 32.31 |
| *Saurauia pendula* Bl. | 46 | 2.24 | 20 | 0.31 | | |
| *Castanopsis javanica* (Bl.) DC. | 37 | 21.01 | 50 | 3.40 | 10 | 2.82 |
| *Persea rimosa* (Bl.) Kosterm. | 30 | 7.61 | | | | |
| *Turpinia sphaerocarpa* Hassk. | 24 | 1.23 | 60 | 1.57 | | |
| *Lithocarpus pseudomoluccus* (Bl.) Rehd. | 22 | 4.83 | | | | |
| *Decaspermum fruticosum* var. *polymorphum* (Bl.) Bakh. f. | 20 | 0.76 | | | | |
| *Vernonia arborea* Buch.-Ham. | 19 | 4.64 | 20 | 7.58 | | |
| *Symplocos fasciculata* Zoll. | 16 | 0.49 | 10 | 0.13 | | |
| *Polyosma integrifolia* Bl. | 16 | 1.37 | 60 | 4.62 | | |
| *Polyosma ilicifolia* Bl. | 15 | 1.20 | | | 30 | 1.81 |
| *Ficus ribes* Reinw. ex Bl. | 10 | 0.29 | | | | |
| *Macropanax dispermus* (Bl.) O.K. | 7 | 1.23 | 10 | 0.31 | 40 | 1.70 |
| *Flacourtia rukam* Z. & M. | 6 | 0.37 | | | | |
| *Syzygium antisepticum* (Bl.) Merry & Perry | 6 | 2.36 | | | | |
| *Astronia spectabilis* Bl. | 5 | 0.21 | 20 | 1.61 | | |
| *Castanopsis argentea* (Bl.) DC. | 5 | 0.11 | | | | |
| *Lithocarpus rotundatus* (Bl.) A. Camus | 5 | 5.75 | 20 | 1.54 | | |
| *Mischocarpus fuscescens* Bl. | 5 | 0.36 | | | | |
| *Saurauia blumiana* Benn. | 5 | 0.25 | | | | |
| *Villebrunea rubescens* (Bl.) Bl. | 5 | 0.14 | | | | |
| *Antidesma tetrandrum* Bl. | 4 | 0.11 | | | | |
| *Lithocarpus indutus* (Bl.) Rehd. | 4 | 1.45 | | | | |
| *Pygeum parviflorum* Teysm. et Benn. | 4 | 1.17 | | | | |
| *Syzygium rostratum* (Bl.) DC. | 4 | 0.30 | | | | |
| *Viburnum sambucinum* Bl. | 4 | 0.25 | | | | |
| *Acronychia laurifolia* Bl. | 3 | 0.58 | 10 | 0.16 | 10 | 0.14 |
| *Casearia tuberculata* Bl. | 3 | 0.50 | | | | |
| *Eurya acuminata* DC. | 3 | 0.10 | | | | |
| *Lithocarpus elegans* (Bl.) Hatus. ex Soepadmo | 3 | 0.81 | | | | |
| *Litsea resinosa* Bl. | 3 | 0.68 | | | | |

| P-4 | | P-5 | | P-6 | | P-7 | | P-8 | | P-9 | |
|---|---|---|---|---|---|---|---|---|---|---|---|
| N | R.D. (%) | N | R.D. (%) | N | R.D. (%) | N | R.D. (%) | N | R.D. (%) | N | R.D. (%) |
| 150 | 42.55 | 290 | 43.63 | 53 | 5.88 | | | | | | |
| 260 | 16.28 | 340 | 16.25 | 399 | 18.84 | 750 | 26.50 | | | | |
| 40 | 5.75 | 50 | 19.04 | | | | | | | | |
| 30 | 0.67 | 80 | 15.60 | | | | | | | | |

Table 3-4 *(continued)*

| Species | P-1 N | P-1 R.D. (%) | P-2 N | P-2 R.D. (%) | P-3 N | P-3 R.D. (%) |
|---|---|---|---|---|---|---|
| *Manglietia glauca* Bl. | 3 | 0.16 | 10 | 0.19 | | |
| *Michelia montana* Bl. | 3 | 0.70 | | | | |
| *Platea latifolia* Bl. | 3 | 3.62 | | | | |
| *Acronodia punctata* Bl. | 2 | 0.22 | 10 | 0.90 | 70 | 4.64 |
| *Dysoxylum alliaceum* Bl. | 2 | 0.14 | | | | |
| *Elaeocarpus stipularis* Bl. | 2 | 0.05 | | | | |
| *Engelhardia spicata* Lech. ex Bl. | 2 | 1.96 | 30 | 5.73 | 20 | 1.17 |
| *Hypobathrum frutescens* Bl. | 2 | 0.04 | 10 | 0.19 | | |
| *Litsea mappacea* (Bl.) Boerl. | 2 | 0.13 | | | | |
| *Pithecellobium clypearia* (Jack) Bth. | 2 | 0.33 | | | | |
| *Pyrenaria serrata* Bl. | 2 | 0.03 | | | | |
| *Saurauia reinwardtiana* Bl. | 2 | 0.03 | | | | |
| *Cinnamomum parthenoxylon* (Jack) Meissn. | 1 | 1.61 | | | | |
| *Cinnamomum sintoc* Bl. | 1 | 0.31 | | | | |
| *Glochidion macrocarpum* Bl. | 1 | 0.26 | | | | |
| *Gordonia excelsa* (Bl.) Bl. | 1 | 1.79 | | | | |
| *Ilex cymosa* Bl. | 1 | 0.07 | | | | |
| *Lithocarpus tijsmannii* (Bl.) Rehd. | 1 | 0.48 | | | | |
| *Macropanax undualatus* (Wall. ex G. Don) Seem | 1 | 0.02 | | | 40 | 1.20 |
| *Meliosma ferruginea* Bl. | 1 | 0.02 | | | | |
| *Podocarpus imbricatus* Bl. | 1 | 0.38 | 30 | 37.74 | 30 | 34.56 |
| *Prunus arborea* (Bl.) Kalkman | 1 | 0.02 | | | | |
| *Saurauia nudiflora* DC. | 1 | 0.03 | | | | |
| *Tarenna laxiflora* (Bl.) K. & V. | 1 | 0.03 | | | | |
| *Viburnum coriaceum* Bl. | 1 | 0.02 | | | | |
| *Viburnum lutescens* Bl. | 1 | 0.02 | | | | |
| *Turpinia montana* (Bl.) Kurz | | | 10 | 0.29 | | |
| *Myrsine hasseltii* Bl. ex Scheff. | | | 10 | 0.56 | | |
| *Syzygium gracile* (Korth.) Amsh. | | | 10 | 0.64 | | |
| *Omalanthus populneus* (Geisel) Pax | | | 10 | 0.23 | | |
| *Neolitsea javanica* (Bl.) Back. | | | | | 60 | 5.44 |
| *Ardisia javanica* DC. | | | | | 10 | 0.14 |
| *Acer laurinum* Hassk. | | | | | 40 | 0.91 |

| P-4 | | P-5 | | P-6 | | P-7 | | P-8 | | P-9 | |
|---|---|---|---|---|---|---|---|---|---|---|---|
| N | R.D. (%) | N | R.D. (%) | N | R.D. (%) | N | R.D. (%) | N | R.D. (%) | N | R.D. (%) |
| 20 | 3.20 | 80 | 4.81 | 345 | 22.72 | 225 | 7.75 | | | | |
| 10 | 1.42 | | | | | | | | | | |
| 10 | 0.23 | 10 | 1.17 | | | | | | | | |
| 70 | 1.81 | | | | | | | | | | |
| 30 | 1.80 | | | | | | | | | | |
| 30 | 6.23 | | | | | | | | | | |
| | | | | | | | | 75 | 7.46 | | |
| | | 20 | 0.74 | | | | | | | | |
| | | 20 | 1.57 | | | | | | | | |
| 50 | 12.50 | | | | | | | | | | |

## Table 3-4  (continued)

| SPECIES | P-1 | | P-2 | | P-3 | |
|---|---|---|---|---|---|---|
| | N | R.D. (%) | N | R.D. (%) | N | R.D. (%) |
| *Prunus* sp. | | | | | 20 | 2.08 |
| *Lindera polyantha* (Bl.) Boerl. | | | | | 10 | 0.47 |
| *Symplocos laurina* (Retz.) Wall. | | | | | 20 | 0.38 |
| *Weinmannia blumei* Planch. | | | | | | |
| *Geniostoma arboreum* (Reinw.) O.K. | | | | | | |
| *Leptospermum flavescens* J. E. Smith | | | | | | |
| *Myrsine affinis* DC. | | | | | | |
| *Eurya obovata* (Bl.) Korth. | | | | | | |
| *Meliosma nervosa* K. & V. | | | | | | |
| *Daphniphyllum glaucescens* Bl. | | | | | | |
| *Symplocos sessilifolia* (Bl.) Gurke | | | | | | |
| *Schefflera* sp. | | | | | | |
| *Photinia notoniana* W. & A. | | | | | | |
| *Vaccinium laurifolium* (Bl.) Miq. | | | | | | |
| *Schefflera rugosa* (Bl.) Harms | | | | | | |
| *Symplocos* sp. | | | | | | |
| *Vaccinium varingiaefolium* (Bl.) Miq. | | | | | | |

cline but, as the figure shows, a curve with two peaks. The first and highest peak occurs in plot 3 at 1,900 m, with a value of 71 m²/ha. This declines to 46 m²/ha at 2,300 m; but with increasing altitude the basal area again increases, reaching 70 m²/ha at 2,600 m, almost as high as at 1,900 m. This again declines to 34–35 m²/ha at 2,800–3,000 m. Plot 7 at 2,600 m happened to include a large specimen of *Leptospermum flavescens,* which accounted for a considerable proportion of the basal area at this altitude and is one of the factors in the high value overall. As the table in the appendix shows, however, the other constituent species also contributed appreciably.

Figure 3-5b shows the variations with altitude in the basal areas of species with relatively wide distributions. The tendency noted at the forest level is also apparent in the quantitative changes of the constituent species. Of the ten species in the figure, only *Astronia spectabilis* peaks at 2,300 m, while the others show their respective peaks above or below this elevation. Notable are *Schima wallichii* and *Podocarpus imbricatus* below 2,300 m, and *Myrsine affinis* and *Eurya obovata* above 2,300 m.

| P-4 | | P-5 | | P-6 | | P-7 | | P-8 | | P-9 | |
|---|---|---|---|---|---|---|---|---|---|---|---|
| N | R.D. (%) | N | R.D. (%) | N | R.D. (%) | N | R.D. (%) | N | R.D. (%) | N | R.D. (%) |
| 10 | 0.97 | | | | | | | 25 | 0.62 | | |
| 10 | 3.18 | | | | | | | | | | |
| 10 | 0.15 | | | 50 | 0.90 | | | | | | |
| 10 | 2.17 | | | | | 50 | 20.10 | | | | |
| | | 250 | 12.60 | 399 | 22.39 | 775 | 29.93 | 400 | 41.61 | 1158 | 20.99 |
| | | 180 | 16.34 | 80 | 4.27 | 50 | 1.24 | 150 | 19.59 | 860 | 22.47 |
| | | 10 | 0.19 | | | | | | | | |
| | | | | 50 | 7.09 | 75 | 1.98 | | | | |
| | | | | 80 | 15.90 | 350 | 11.99 | 100 | 4.95 | 202 | 4.54 |
| | | | | | | 25 | 0.50 | 325 | 24.26 | 50 | 0.74 |
| | | | | 27 | 1.40 | | | 25 | 1.52 | 50 | 1.03 |
| | | | | 25 | 0.61 | | | | | 25 | 0.31 |
| | | | | | | | | | | 50 | 1.00 |
| | | | | | | | | | | 50 | 1.36 |
| | | | | | | | | | | 1425 | 47.56 |

*Schima wallichii* shows its maximum value, and the highest of all species, in plot 3 (1,900 m), and in plot 4 (2,100 m) it is again higher than other species, clearly the dominant species at these altitudes. *Podocarpus imbricatus* has its maximum values in plot 2 (1,700 m) and plot 3 (1,900 m). Together with *Schima wallichii,* its basal area far exceeds those of other species. At above 2,200 m, the basal areas of these two species decline.

*Castanopsis javanica, Engelhardtia spicata,* and other species similarly show their respective maxima over a relatively narrow range of altitudes. The distribution of basal areas shows a boundary at plot 5 (2,300 m), above and below which forest compositions are distinct, and it can be said that the dominant species change at this altitude. This is borne out by observation of the forest, which changes in appearance at around this altitude.

We have seen that at 2,400 m and 2,600 m there are fairly marked changes in species numbers. As to relative dominance, the dominant species of the lower regions on the mountain are replaced by those of the upper regions in the vicinity of plot 5 (2,300 m); while for basal area, plot 5 shows an appreciably lower

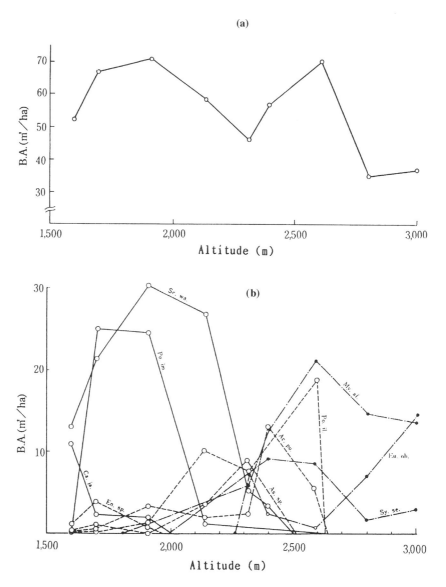

Figure 3-5. Altitudinal variation of basal area (a) by plot and (b) by tree species.
NOTE: The abbreviations are the same as in figure 3-4.

value than the plots immediately above and below. Together, these findings point to a boundary at around 2,300 meters, where species composition changes. Van Steenis et al. took the altitude of 2,400 meters as the boundary between the montane and subalpine zones, but on Mount Pangrango this appears to correspond to the altitude of 2,300 m.

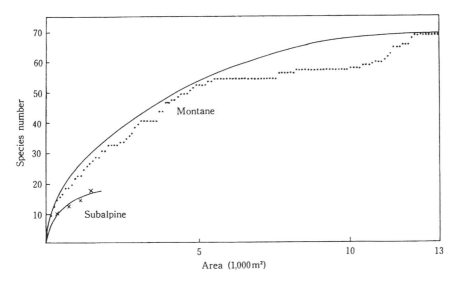

Figure 3-6. Species number-area curve.

The vertical distribution zones on Mount Pangrango thus can be considered to consist of a montane zone dominated by *Schima wallichii* and *Podocarpus imbricatus,* and a subalpine zone dominated by *Polyosma ilicifolia, Acronodia punctata, Eurya obovata, Myrsine affinis,* and others. Between the two we can see a transition zone.

## Species Number–Area Curve

For each of these three zones on Mount Pangrango showing different vertical distributions of species, I examined the relation between survey area and species number. Plots 1 to 4 were taken as the montane zone, plot 5 as the transitional zone, and plots 6 to 9 as the subalpine zone. The results are illustrated in figure 3-6.

The curve for the montane zone shows that species number increases with survey area while the area is still small, reaching 50 species at around 0.5 ha, and above this the curve begins to flatten, approaching a constant value at above 0.7 ha. The maximum value at 1.3 ha is 69 species. Ellenberg (1956) drew a line from the origin to the point where the curve becomes flat, and found the point where a line parallel to this is tangential to the curve. He called the abscissa of this point the *minimum area,* regarding it as the minimum unit of area necessary for a community survey. For the curve in figure 3-6, the point of minimum area lies near 0.2 ha. In the present survey, only plot 1, at 1 ha, meets this criterion; all other plots were 0.1 ha. Thus in discussing species composition, plots 2 to 4 individually may be too small.

One feature of the species number–area curve that clearly shows the difference between the montane and subalpine zones is the difference in species numbers. Extrapolation of the curves suggests that the upper limits may be 70 species for the montane zone and 20 species for the subalpine zone. One further difference is between the large area required for the species number to reach saturation in the montane zone and the small area in the subalpine zone.

## *Stratification and Altitude*

### Stratification in the Montane Zone on Mount Pangrango

I turn next to the vertical arrangement of individual tree crowns in the forest of the montane zone on Mount Pangrango. The crowns composing the canopy in plot 1 were divided according to the criteria described hereafter, and following the 1976 survey I drew crown projection diagrams. Between the surveys of 1969 and 1976, localized changes had taken place because of the growth or death of trees. Gaps in the canopy after the death of a tree would have been filled by the growth of surrounding trees. But despite such localized changes, the overall stratification of the forest can be termed stable over a sufficiently large area. The trees constituting the canopy can be divided into the following layers by comparing the vertical arrangements of their crowns.

$I_0$: Crowns emergent above the canopy, with the greater part of each crown receiving direct sunlight.

I: Crowns forming a continuous canopy, with the greater part of each crown receiving direct sunlight.

II: Crowns with the greater part lying below $I_0$ or I, but with a smaller part reaching the top of the continuous canopy and receiving direct sunlight.

III: Crowns lying below the three upper layers and receiving no direct sunlight.

Crowns in the survey plot were classified into one of the above four types, and their widths were measured in four directions. Figure 3-7 shows crown projection diagrams for each of the four types. Crown areas were measured from these diagrams, and the results are summarized in table 3-5 to show the stratification in the plot.

Layer $I_0$ contained only 32 trees, each with fairly large crown areas compared with the other layers, and with no overlapping of crowns. The average tree height in this layer was 34.7 m, and the average crown depth was 13 m.

Layer I consisted of 56 trees, with an average height of 29.6 m and an average crown depth of 11.4 m. Some crown overlapping occurred in this layer, but since the crown area amounted to only 48 percent of the ground area, it appears that this layer alone did not form a continuous canopy. Examination of the vertical relationship between layers $I_0$ and I in the plot overall shows that they overlap

Table 3-5
Overview of Stratification in Plot 1

| Layer | Maximum Height H (m) | Average Height H (m) | Number of Trees N (no./ha) | Crown Area CA (m²) | Average Crown Area CA (m²) | Average Crown Diameter CR (m) | Crown Volume CB (m³) | Average Crown Volume CB (m³) | Average Crown Depth CD (m) |
|---|---|---|---|---|---|---|---|---|---|
| $I_0$ | 41.0 | 34.7 | 32 (8.1%) | 3409.00 (20.9%) | 106.5 | 5.8 | 44318.02 (28.0%) | 1384.94 | 13.0 |
| I | 38.0 | 29.6 | 56 (14.2%) | 4828.25 (29.6%) | 86.2 | 5.2 | 54968.06 (34.7%) | 981.57 | 11.4 |
| II | 36.0 | 19.9 | 134 (34.1%) | 5025.90 (30.8%) | 37.5 | 3.5 | 42524.45 (26.8%) | 317.35 | 8.5 |
| III | 34.5 | 13.8 | 171 (43.5%) | 3038.25 (18.6%) | 17.8 | 1.8 | 16736.32 (10.6%) | 97.87 | 5.5 |

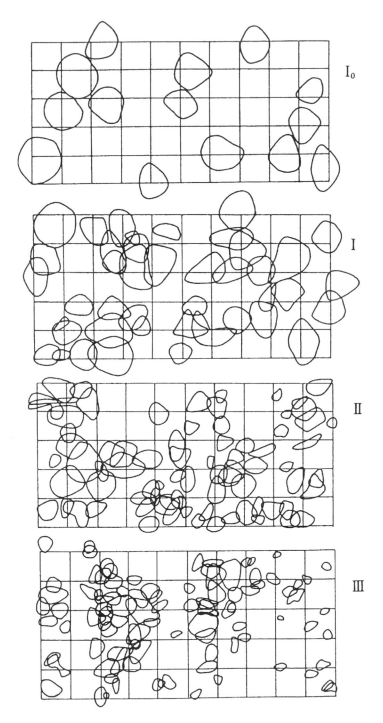

Figure 3-7 (1). Crown projection of plot 1 (trees 1–247).

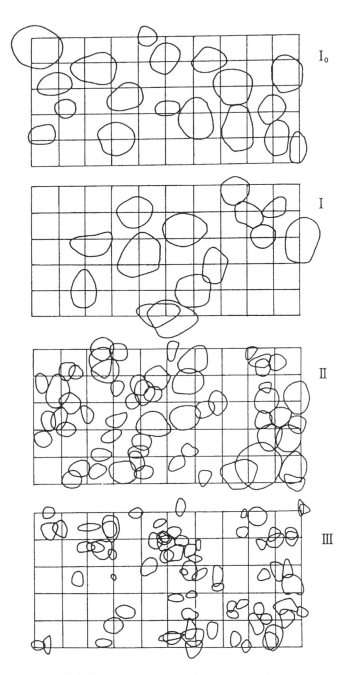

Figure 3-7 (2). Crown projection of plot 1 (trees 248–432).

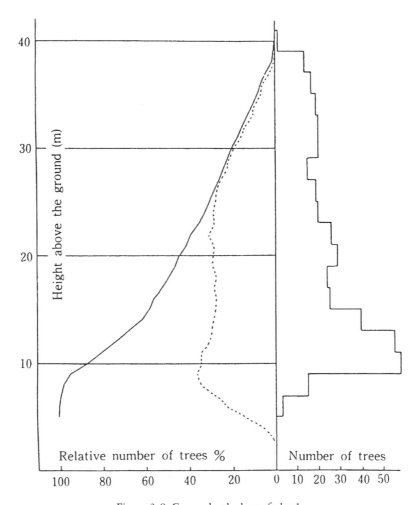

Figure 3-8. Crown depth chart of plot 1.

each other by more than half of their respective crown depths. So although it is possible to assign an individual tree to a particular layer, the overall distinction between the two layers is far from clear.

Layer II has many more trees. The 134 trees have an average height of 19.9 m and an average crown depth of 8.5 m. It overlaps layer I in the vertical plane to some extent, but the two layers are quite distinct.

In layer III, the number of trees rises again, to 171. The average height and crown depth are 13.8 m and 5.5 m, respectively. The average crown area is small, and the total crown area amounts to no more than 30 percent of the ground area. The projection diagrams reveal that although crowns in the upper layers are distributed irregularly, those in layer III show a degree of clustering. Layer III overlaps layer II in the vertical plane, but the two are fairly distinct.

The above four layers can be distinguished for individual trees in relation to their neighbors, but for the forest overall, the vertical overlap between layers $I_0$ and I prevents differentiation between them. Thus three layers can be distinguished: $I_0$–I, II, and III.

Ogawa et al. (1965) attempted to depict stratification with a crown depth diagram, on which they plotted frequencies of tree heights and heights to lowest branch. A similar diagram for plot 1 is shown in figure 3-8. The diagram has a crown frequency curve (dotted line) and a cumulative height curve (solid line). Stratification should be discernible from the pattern of the crown frequency curve, but in plot 1, at no height does either curve show a clear point of variation. If the minima which are discernible on the crown frequency curve are taken to be boundaries between layers, then three layers can be distinguished. Layer I stretches from 41 to 18 m, layer II from 18 to 10 m, and layer III from 10 to 6 m. Again, layers $I_0$ and I cannot be distinguished. Each of the three discernible layers overlaps the adjacent layer or layers to a greater or lesser extent, forming a deep canopy layer extending on average more than 26 m in the vertical plane.

The vertical distribution of crown area is depicted in figure 3-9. The vertical axis shows aboveground height (m), and the horizontal axis shows the aggregate crown projection area at each height. The crowns form a vertical continuum from 5 m aboveground to the highest tree at 41 m. At 24 m, the aggregate crown area reaches more than 10,000 m². Thus slightly above the midpoint of the canopy depth, almost exactly in the middle of layer I, the canopy can be said to be closed. This space is occupied mainly by trees of layer I, but in part by some trees belonging to other layers.

Stratification was determined by observing the position of individual crowns in the canopy in terms of the light reception of the crowns. This method allows the crown position of a tree to be defined in relation to the forest canopy, but it cannot predict to which layer a tree with a certain crown height above the ground will belong. Some trees in the $I_0$ layer are relatively low, while some in layer II are taller.

In the 1-ha plot 1 as a whole, the canopy is closed at a height of 24 m, but this does not express accurately the stratification of this forest. It is probably more accurate to say that the closed forest layer undulates, being sometimes higher and sometimes lower, and above this is the irregular $I_0$ layer. A view over the top of the forest would reveal a strikingly uneven canopy.

## Stratification and Species Composition

Let us examine whether the species appearing in each of the layers ($I_0$ to III) that make up the stratified structure show any notable characteristics.

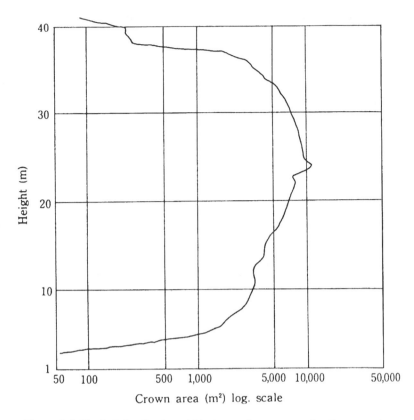

Figure 3-9. Vertical distribution within layers of crown areas for all tree species
in plot 1.

Table 3-6 lists the 56 species found in plot 1 and shows the proportion of the
crown area in each layer occupied by the species appearing there. Species occupying more than the average proportion of the total crown area for the species appearing in a given layer, and which were observed on the plot to be dominant, are
regarded as dominants in that layer and are marked with an asterisk in the table.

Inspection of species numbers and dominant species reveals that, of the 12
species in layer $I_0$, *Schima wallichii* and *Castanopsis javanica* occupy a large proportion of the crown area.

In layer I there are 20 species. The dominants are *Castanopsis javanica, Schima
wallichii, Vernonia arborea,* and *Lithocarpus rotundatus.*

Layer II has 36 species, of which the dominants include *Schima wallichii,
Castanopsis javanica, Persea rimosa, Lithocarpus pseudomoluccus, Saurauia pendula,
Polyosma integrifolia, Syzygium antisepticum, Polyosma ilicifolia, Vernonia arborea,
Michelia montana, Mischocarpus fuscescens,* and *Macropanax dispermus.*

Table 3-6

Stratification of All Tree Species in Plot 1

| SPECIES NAME | $I_0$ | | I | | II | | III | | TOTAL | |
|---|---|---|---|---|---|---|---|---|---|---|
| | C.A. | C.B. | C.A. | C.B. | C.A. | C.B. | C.A. | C.B. | C.A. | C.B. |
| Schima wallichii | 27.4* | 30.6 | 20.6* | 23.7 | 14.5* | 16.2 | 3.0* | 4.9 | 16.9 | 21.7 |
| Sauraria pendula | | | | | 5.7* | 3.4 | 18.3* | 15.3 | 5.2 | 2.5 |
| Castanopsis javanica | 18.9* | 19.7 | 26.0* | 26.8 | 8.8* | 9.4 | 7.2* | 9.7 | 15.7 | 18.3 |
| Persea rimosa | 7.0 | 6.4 | 4.4 | 3.6 | 8.5* | 8.5 | 6.5* | 8.3 | 6.6 | 6.2 |
| Turpinia sphaerocarpa | | | | | 2.7 | 1.5 | 7.6* | 7.0 | 2.3 | 1.1 |
| Litbocarpus pseudomoluccus | 6.2 | 6.3 | 4.9 | 6.3 | 7.3* | 6.5 | 3.6* | 2.8 | 5.7 | 6.0 |
| Decaspermum fruticosum | | | | | 1.6 | 1.5 | 7.5* | 8.9 | 1.9 | 1.3 |
| Vernonia arborea | 4.5 | 4.0 | 9.1* | 7.0 | 3.8* | 2.4 | 1.2 | 1.2 | 5.1 | 4.3 |
| Symplocos faciculata | | | 0.3 | 0.2 | 1.4 | 0.8 | 6.2* | 3.8 | 1.7 | 0.7 |
| Polyosma integrifolia | | | 2.3 | 2.3 | 5.7* | 5.1 | 4.1* | 5.1 | 3.2 | 2.7 |
| Polyosma ilicifolia | | | 3.7 | 2.8 | 3.9* | 3.5 | 3.9* | 4.5 | 3.0 | 2.4 |
| Ficus ribes | | | | | 1.5 | 1.0 | 3.3* | 3.2 | 1.1 | 0.6 |
| Macropanax dispermus | | | 1.8 | 2.3 | 3.1* | 3.8 | 2.1 | 2.1 | 1.9 | 2.0 |
| Flacourtia rukam | | | | | 1.7 | 1.9 | 0.6 | 0.7 | 0.7 | 0.6 |
| Syzygium antisepticum | 1.8 | 1.8 | 2.3 | 2.0 | 4.9* | 7.4 | 1.1 | 2.5 | 2.7 | 3.4 |
| Astronia spectabilis | | | | | 0.2 | 0.8 | | | 0.0 | 0.2 |
| Castanopsis argentea | | | | | 0.5 | 0.2 | 1.4 | 0.9 | 0.4 | 0.2 |

(continued on next page)

Table 3-6  (continued)

| SPECIES NAME | I₀ | | I | | II | | III | | TOTAL | |
|---|---|---|---|---|---|---|---|---|---|---|
| | C.A. | C.B. | C.A. | C.B. | C.A. | C.B. | C.A. | C.B. | C.A. | C.B. |
| Litbocarpus rotundatus | 9.4 | 10.0 | 7.9* | 7.6 | | | | | 4.3 | 5.4 |
| Miscbocarpus fuscescens | | | | | 3.3* | 4.1 | 1.9 | 1.4 | 1.4 | 1.2 |
| Sauvauia blumiana | | | | | 0.5 | 0.2 | 2.7 | 2.3 | 0.6 | 0.3 |
| Villebrunea rubescens | | | | | | | 2.3 | 1.7 | 0.4 | 0.2 |
| Antidesma tetrandrum | | | | | 0.5 | 0.2 | 1.8 | 1.3 | 0.5 | 0.2 |
| Litbocarpus indutus | 7.7 | 6.4 | | | 0.8 | 1.2 | | | 1.8 | 2.1 |
| Pygeum parviflorum | | | 3.9 | 4.9 | | | | | 1.1 | 1.7 |
| Syzygium rostratum | | | | | 1.7 | 1.6 | 1.5 | 1.2 | 0.8 | 0.6 |
| Viburnum sambucinum | | | 0.3 | 0.1 | | | 2.2 | 1.6 | 0.5 | 0.2 |
| Acronychia laurifolia | | | 2.3 | 2.0 | 2.3 | 2.6 | | | 1.4 | 1.4 |
| Cesearia tuberculata | | | 2.1 | 0.9 | | | 0.8 | 0.9 | 0.8 | 0.4 |
| Eurya acuminata | | | | | 0.6 | 0.3 | 1.5 | 1.4 | 0.5 | 0.2 |
| Litbocarpus elegans | | | 1.7 | 2.0 | | | 1.0 | 0.6 | 0.7 | 0.8 |
| Litsea resinosa | | | | | 2.1 | 1.9 | | | 0.6 | 0.5 |
| Manglietia glauca | | | | | | | 2.1 | 3.0 | 0.4 | 0.3 |
| Michelia montana | | | | | 4.1* | 5.3 | | | 1.3 | 1.4 |
| Platea latifolia | 3.6 | 3.6 | 1.9 | 1.7 | 0.8 | 1.5 | | | 1.6 | 2.0 |
| Acronodia punctata | | | | | 0.8 | 0.9 | | | 0.2 | 0.3 |

(continued on next page)

Table 3-6  *(continued)*

| SPECIES NAME | $I_0$ C.A. | $I_0$ C.B. | I C.A. | I C.B. | II C.A. | II C.B. | III C.A. | III C.B. | TOTAL C.A. | TOTAL C.B. |
|---|---|---|---|---|---|---|---|---|---|---|
| *Dysoxylum alliaceum* | | | | | 0.6 | 0.5 | 0.7 | 0.3 | 0.3 | 0.2 |
| *Elaeocarpus stipularis* | | | | | 0.4 | 0.2 | 0.1 | 0.2 | 0.1 | 0.1 |
| *Engelhardia spicata* | 7.6 | 6.6 | | | | | | | 1.6 | 1.9 |
| *Hypobathrum frutescens* | | | | | | | 0.9 | 0.5 | 0.2 | 0.0 |
| *Litsea mappacea* | | | 0.7 | 0.3 | | | 0.3 | 0.2 | 0.3 | 0.1 |
| *Pithecellobium clypearia* | | | | | 1.7 | 1.5 | | | 0.5 | 0.4 |
| *Pyrenaria serrata* | | | | | | | 0.1 | 0.1 | 0.0 | 0.0 |
| *Saurauia reinwardtiana* | | | | | | | 0.4 | 0.4 | 0.1 | 0.0 |
| *Cinnamomum sintoc* | | | 2.4 | 2.5 | | | | | 0.7 | 0.8 |
| *Glochidion macrocarpum* | | | 1.5 | 1.1 | | | | | 0.5 | 0.4 |
| *Gordonia excelsa* | 2.6 | 2.4 | | | | | | | 0.5 | 0.7 |
| *Ilex cymosa* | | | | | 0.6 | 0.6 | 0.8 | 0.9 | 0.3 | 0.3 |
| *Lithocarpus tijsmannii* | 3.1 | 2.1 | | | | | | | 0.6 | 0.6 |
| *Macropanax undulatus* | | | | | | | 0.3 | 0.2 | 0.0 | 0.0 |
| *Meliosma ferruginea* | | | | | | | 0.1 | 0.0 | 0.0 | 0.0 |
| *Podocarpus imbricatus* | | | | | 1.8 | 2.1 | | | 0.5 | 0.6 |
| *Pygeum arboreum* | | | | | | | 0.5 | 0.4 | 0.1 | 0.0 |
| *Saurauia nudiflora* | | | | | 0.5 | 0.4 | | | 0.2 | 0.1 |

*(continued on next page)*

Table 3-6  *(continued)*

| Species Name | I$_0$ | | I | | II | | III | | Total | |
|---|---|---|---|---|---|---|---|---|---|---|
| | C.A. | C.B. | C.A. | C.B. | C.A. | C.B. | C.A. | C.B. | C.A. | C.B. |
| *Tarenna laxiflora* | | | | | 0.7 | 0.5 | | | 0.2 | 0.1 |
| *Viburnum coriaceum* | | | | | | | 0.5 | 0.5 | 0.1 | 0.1 |
| *Viburnum lutescens* | | | | | 0.7 | 0.4 | | | 0.2 | 0.1 |
| Total | 100.0 | 100.0 | 100.0 | 100.0 | 100.0 | 100.0 | 100.0 | 100.0 | 100.0 | 100.0 |

CA: Crown area (%)    CB: Crown volume (%)

Layer III has 37 species. The dominants include *Saurauia pendula, Turpinia sphaerocarpa, Decaspermum fruticosum, Symplocos fasciculata, Persea rimosa, Polyosma integrifolia, Polyosma ilicifolia, Ficus ribes, Lithocarpus pseudomoluccus, Schima wallichii,* and *Castanopsis javanica.*

Overall, the number of species increases from the higher layers $I_0$ and I to the lower II and III.

As mentioned earlier, plot I is in a montane zone dominated by *Schima wallichii* and *Castanopsis javanica.* These species are dominant not only in layers $I_0$ and I but also in the lower layers II and III.

Four other species appear in all layers: *Persea rimosa, Lithocarpus pseudomoluccus, Vernonia arborea,* and *Syzygium antisepticum.* However, their dominance is restricted to layers I and II, II and III, or II only. Many of the other species that appear in layers II and III are concentrated in one or both of these two layers, and all are limited to the lower part of the canopy. Worthy of note is the appearance in layers II and III of dominants of the subalpine zone, like *Polyosma ilicifolia* and *Acronodia punctata,* and dominants of the transition zone, like *Astronia spectabilis,* all species whose center of distribution is at higher altitudes.

Figure 3-10 illustrates the vertical distribution of crown areas of representative species that are dominant in one or more of the layers $I_0$ to III.

At heights aboveground of 16 m and over, *Castanopsis javanica* and *Schima wallichii* have large crown areas. Over wide areas of the forest canopy, no other overstory species have comparable crown areas. *Lithocarpus pseudomoluccus* and *Persea rimosa,* though having smaller crown areas, also occupy the overstory. *Vernonia arborea, Lithocarpus rotundatus,* and other species show similar patterns.

Below 20 m, where the overstory recedes, *Turpinia sphaerocarpa* and *Saurauia pendula* appear. These are typical understory species with maximum crown areas 7–9 m aboveground. Other species with similar distribution patterns include *Symplocos fasciculata, Ficus ribes, Decaspermum fruticosum,* and *Macropanax dispermus. Polyosma ilicifolia* is a dominant species of the subalpine zone. In plot 1 it appeared below about 25 m aboveground, corresponding to the lower part of layer I. Its maximum distribution was 15–20 m aboveground, below which it rapidly receded. It is characterized by having an overstory type of distribution without really appearing in the overstory.

Examination of the vertical distribution of crown areas reveals that different species appear, respectively, in the overstory and understory of the canopy, displaying a vertical segregation of habitats according to species. This segregation occurs about 11 m aboveground, right at the boundary between layers II and III. In other words, examination of the species composition of the stratified canopy allows differentiation of a species group occupying layers $I_0$ to II and another group occupying layer III. This could also be seen as a division into a

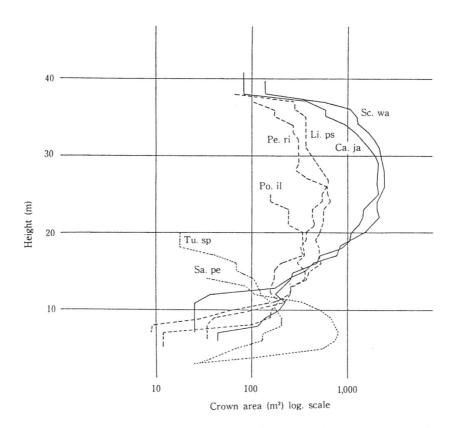

Figure 3-10. Vertical distribution within layers of crown areas for representative tree species in plot 1. NOTE: The abbreviations are the same as in figure 3-4.

layer occupied by an overstory species group and a layer occupied by an understory species group.

Layers $I_0$, I, and II show no segregation of habitats by species. In terms of light conditions, we would expect a marked difference between the spaces occupied by layer $I_0$ and the lower part of layer II. Nevertheless, it appears that the species that are dominant over a broad range from layer $I_0$ to layer II, like *Schima wallichii*, must be tolerant of a broad range of environmental conditions. In this sense, the overstory species group can be considered to have a wide adaptability to the environment.

Layers $I_0$ to II show a stratified structure that arose through segregation of habitats by an overstory species group with a wide ecological adaptability to the environment, but layer III is a space that can be filled only by the understory species group. Viewed this way, it becomes apparent that the stratification into

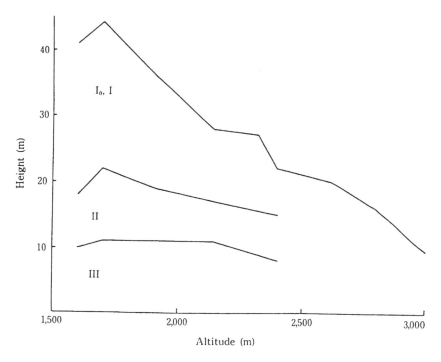

Figure 3-11. Variations in layers by altitude.

the four layers $I_0$ to III also reflects an ecologically significant difference between the upper layers $I_0$ to II and the lower layer III.

## Changes in Stratification with Altitude

In the plots other than plot 1, it was not possible to draw crown projection diagrams. Here, stratification was examined by the crown depth diagram method of Ogawa et al. based on measurements of tree height and height to first branch. Let us see how stratification changes with altitude. Figure 3-11 shows a plot of the changes with altitude of maximum tree heights and the boundaries between layers I, II, and III.

Maximum tree height in plot 1 was 41 m, while near the summit it was 9.5 m. In plot 2 (1,700 m), a value of 44 m was recorded, but overall the maximum height decreased with altitude. In plot 1, layers $I_0$, I, II, and III were distinguished. At altitudes up to 2,400 m, three layers could be discerned, but above this no stratification was discernible. Thus, in the subalpine zone the stratification seen in the montane zone gave way to a single layer.

While the changes in the maximum tree height are marked, the average

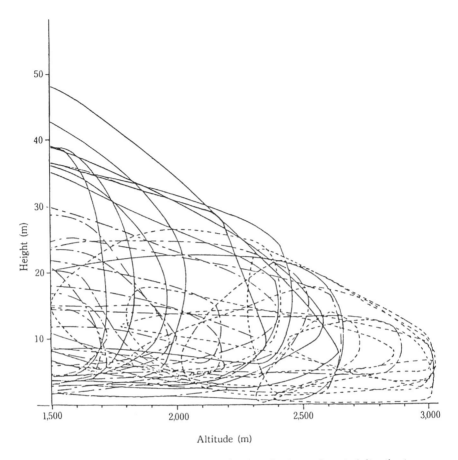

Figure 3-12. Distribution of tree species by stratification and vertical distribution.

heights of layers II and III do not vary greatly. The explanation for this is probably that while the maximum height is determined by such external factors as
temperature and rainfall, stratification is related to how plants react to the environment within the forest. It is clear that stratification becomes simpler with
increasing altitude.

Figure 3-12 is a distribution diagram obtained by connecting the points in
successive plots representing maximum tree height and minimum height to
lowest branch for species that appeared in at least two plots with more than 10 cm
DBH and that were dominant in one of the layers where they appeared.

This diagram shows not only the range of distribution of each species but also
its crown position in the canopy and how this changes with altitude. Although
the diagram overall is extremely crowded, examination of the crown positions of

Table 3-7

Classification of Tree Species by Stratification and Vertical Distribution

| | |
|---|---|
| Type A | *Schima wallichii, Podocarpus imbricatus, Myrsine affinis, Eurya obovata, Castanopsis javanica, Engelhardtia spicata, Cinnamomum sintoc, Vernonia arborea, Acer laurinum, Leptospermum flavescens, Vaccinium varingiaefolium* |
| Type B | *Viburnum coriaceum, Saurauia pendula, Saurauia reinwardtiana, Decaspermum fruticosum, Ficus ribes, Turpinia sphaerocarpa, Ardisia javanica, Vaccinium laurifolium* |
| Type C | *Acronodia punctata, Astronia spectabilis, Polyosma ilicifolia* |

these species within the canopy and how these change with altitude revealed the following three types of species:

A. Species constantly occupying the overstory, regardless of altitude
B. Species constantly occupying the understory, regardless of altitude
C. Species whose crown position changes with change in altitude (understory at low altitude, overstory at high altitude)

Table 3-7 shows the classification of tree species into these three types, and figure 3-13 shows plots for representative species. *Schima wallichii* and *Podocarpus imbricatus* are chosen for the A type in the montane zone. They appear constantly in the overstory at altitudes from 1,600 m to about 2,600 m. In the subalpine zone, *Myrsine affinis* and *Eurya obovata* dominate the overstory from 2,400 m to 3,000 m.

The representative species of type B is *Viburnum coriaceum*. From 1,500 m to 2,800 m it appears constantly in the understory. Although its relative dominance is far from high, its area of distribution shows the greatest range of altitude.

Type C is represented by *Acronodia punctata*. At around 1,600 m its crown depth is small and centers on layer II, but with increasing altitude its crown depth also increases, and at around 2,400 m it appears in the overstory.

## *Summary*

In species composition, a change in the dominant species group occurs at 2,300 m. The species group dominant at lower altitudes includes *Schima wallichii, Castanopsis javanica,* and *Podocarpus imbricatus,* and the upper limit of its distribution coincides closely with the upper limit of what van Steenis termed the montane zone. The upper limit of this group thus can be taken as the upper limit of the montane zone.

At higher altitudes, the dominant species group includes *Myrsine affinis* and *Eurya obovata.* I have again followed van Steenis in terming the area of distribution of this group the subalpine zone.

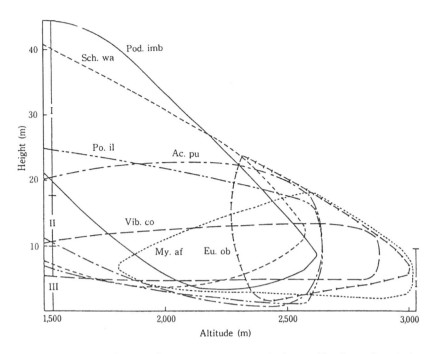

Figure 3-13. Distribution of representative tree species by stratification and vertical
distribution. NOTE: The abbreviations are the same as in figure 3-4.

Between the montane and subalpine zones is a transitional zone where *Acronodia punctata, Astronia spectabilis,* and other species are dominant. The altitude
range of this transitional zone is barely more than 100 m, almost negligible compared with the other two zones. As to the change in basal area with altitude, this
narrow range is one in which the montane species decline and the subalpine
species increase, and neither group is dominant.

In the montane zone, the relative positions of individual crowns could be
differentiated into four vertical layers, $I_0$, I, II, and III, but when the horizontal
spread was also considered, it became clear that the stratification consisted of
three layers, I, II, and III, since layers $I_0$ and I could not be distinguished. Examination of the species composition of the canopy, namely, the pattern of appearance of the species groups occupying each layer, revealed two groups of species in
the montane zone: *Schima wallichii, Castanopsis javanica,* and other overstory
species occupying layers $I_0$, I, and II; and *Saurauia pendula, Turpinia sphaerocarpa,*
and other dominant species of layer III. With increasing altitude in the montane
zone, maximum tree height decreased, but the triple-layered structure of the
canopy was maintained. In the subalpine zone, the maximum tree height continued to decrease, and the canopy became single layered. Overall, three types of

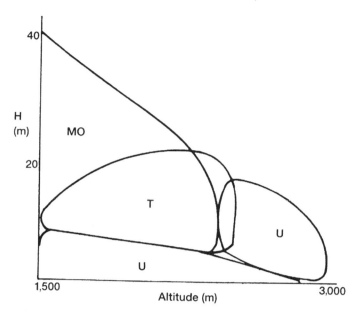

Figure 3-14. Vertical distribution of Mount Pangrango forest by stratification.
*MO:* Montane overstory species; *T:* Transitional zone species; *U:* Understory species;
*SA:* Subalpine zone species.

species were recognized: (A) those constantly occupying the overstory; (B) those constantly occupying the understory; and (C) those whose crown position changes with altitude. This is shown schematically in figure 3-14. Species with type A characteristics are the montane zone overstory species group and the subalpine zone species group.

It is noteworthy that the transition from the montane zone to the subalpine zone takes place in a narrow transitional zone, where the replacement of the dominant species is accompanied by a change in the stratification. This is not to say that the range of the dominant species group of the montane zone, including *Schima wallichii, Castanopsis javanica,* and *Podocarpus imbricatus,* does not overlap to some extent with the dominant species group of the subalpine zone, including *Myrsine affinis* and *Eurya obovata.* Indeed, it is not difficult to find places where the two groups are intermixed at the same altitudes. It is probably more accurate, however, to regard such areas not as areas of overlap of the two domains but as buffer zones where transitional species such as *Astronia spectabilis* and *Acronodia punctata* appear. In other words, the replacement of one species group by another should be regarded as the result not of a drastic change but simply of the fact that each has its dominant domain at a different altitude.

The determinant of these domains is probably temperature conditions. The average annual temperature near plot 6 at 2,400 m is 11°C, and the warmth index is 72°C-month. At the boundary between the montane and subalpine zones at 2,300 m, the warmth index is roughly 80°C-month. Interestingly, this value is close to the value of 85°C-month set by Kira (1949) for the boundary between the laurel forest zone and the temperate deciduous forest zone in Japan. Even if it is not possible to compare the temperature domain of the montane dominant species group with that of the coniferous forest zone species, it can be stated that the distribution of these montane species centers on places with a high warmth index. The work of Backer and van den Brink (1963–1968) and observation of other areas in Java show that these species are widely distributed from the area observed here down to the lowland.

The dominant species group in the subalpine zone, on the other hand, appears to have its center of distribution in a lower temperature domain, and this is supported by the fact that the maximum basal area occurs in plot 7, above 2,300 m. Nevertheless, the temperature domain of these species is far from wide. In plot 9 at 3,000 m, a new dominant species appeared in *Vaccinium varingiaefolium* (table 3-4). Van Steenis (1965) states that Ericaceae dominate the upper part of the subalpine zone, and it may be possible to regard the appearance of *Vaccinium* in plot 9 as a sign of the replacement of *Myrsine affinis* and *Eurya obovata*. To confirm this point, it will be necessary to survey the distribution of forest zones in a mountain area at higher altitude.

The dominant species group of the montane zone occupies the overstory of the forest canopy. We have seen that the term *overstory* covers substantial differences in light conditions between layer $I_0$ above and layer II below. And the fact that these species occupy stratified habitats with a wide range of light conditions seems to be evidence that they possess considerable adaptability to light. Seedlings of *Schima wallichii* can be seen regenerating in gaps in the forest left by fallen trees and other places where the canopy is open. *Castanopsis javanica* regenerates through coppicing. These species cannot produce saplings under a completely closed canopy, but where light conditions are somewhat better, they regenerate. The stratification of crowns in the overstory may represent different stages of growth of individual trees or groups of trees resulting from such localized regeneration.

In the subalpine zone, the simplification of the forest structure suggests that the adaptability of these species to light conditions is narrower than for the montane zone species. However, because the maximum tree heights are considerably lower and the space available for development of stratification is thus greatly constricted, it may not be possible to attribute the single-layer structure to differences in the reactions of species to light conditions.

The dominant understory species of the montane zone remained in the understory irrespective of altitude. Because temperature conditions in the understory are to some extent protected from the influences of extreme climatic fluctuations by the overstory canopy, the distribution of the understory species, even for those with a wide range, appears instead to be determined by their reaction to light conditions. The fact that these species always appear only in the understory strongly suggests their shade-tolerant nature.

The species in the transitional zone appear either in the overstory or the understory, depending on altitude. As we have seen, this species group appears in such a way as to fill the space between the montane and subalpine species groups. It is extremely limited in distribution, clearly lacking the kind of wide range that corresponds to a climatic zone. These species must possess a certain degree of adaptability to light conditions. Thus it is natural that they should show shade-tolerant behavior when they appear in the montane zone, which is the area of distribution of other species.

From this standpoint, the species groups found on Mount Pangrango can be summarized as follows:

MONTANE OVERSTORY SPECIES

*Schima wallichii, Podocarpus imbricatus, Castanopsis javanica, Vernonia arborea, Engelhardtia spicata, Cinnamomum sintoc, Persea rimosa, Glochidion macrocarpum, Polyosma integrifolia, Manglietia glauca, Lithocarpus elegans, Lithocarpus tijsmannii,* and *Lithocarpus rotundatus.*

MONTANE UNDERSTORY SPECIES

*Saurauia pendula, Turpinia sphaerocarpa, Viburnum coriaceum, Acronychia laurifolia, Macropanax undulatus, Macropanax dispermus,* and *Hypobathrum frutescens.*

SUBALPINE ZONE SPECIES

*Myrsine affinis, Eurya obovata, Symplocos* sp., *Symplocos sessilifolia, Schefflera* sp., *Photinia notoniana,* and *Daphniphyllum glaucescens.*

TRANSITIONAL ZONE SPECIES

*Acronodia punctata*, Astronia spectabilis, and *Polyosma ilicifolia.*

## 3. Seasonal Variations in Litter Fall

### Survey Method

In plot 1 (1,600 m) of the montane zone, a survey was conducted of the quantity of fallen material. Litter traps were set, and the material that fell into them was inspected and divided into various elements: leaves, branches, flowers, fruits, insect remains, insect feces, and so on. To determine the quantities of these and their seasonal changes, the variations in each element were examined over the course of one year. Because of the possibility that seasonality varies from species to species within each element, the fallen material was divided as finely as possible by species to see the seasonality of each species. At the same time, an attempt was made to determine what factors give rise to seasonality in each species.

Very few measurements of fallen material have been made for tropical forests, and for the mountain regions of Southeast Asia the only reported study is by Edwards and Grubb (1977) for a montane forest in New Guinea. Even for lowland, there is only the study by Ogawa (1974).

The traps used in the present study had a frame of boards 20 cm high enclosing a square basal area of 1 x 1 m, across the bottom of which was stretched plastic-coated wire netting of 1-mm mesh. Thirty-seven traps were placed at random in the 1-ha plot. Measurements were conducted for one year beginning 29 April 1969. Litter was collected from the traps once weekly and taken to the laboratory at Cibodas Botanic Garden where, after being dried naturally for several days, the leaves, flowers, fruits, and other plant materials were sorted as finely as possible by species. Branches, mosses and lichens, insect remains, feces, and other materials were simply sorted by element. Materials that could not be classified either by species or element were left as "others." The separated materials from each trap were sent to Japan and subjected to drying at 80°C for forty-eight hours in an air dryer. Nothing was lost in transit between Indonesia and Japan, and nothing appeared to have decomposed during transit. The classified materials were weighed to the nearest 0.01 gram.

## Total Annual Litter Fall

The quantity of litter collected each week was the quantity for 37 m², and the cumulative quantity for the year was the annual total litter quantity per 37 m², which was equivalent on a per-hectare basis to 5.96 ton/ha·yr. The value and standard deviation with a 95 percent confidence limit are 5.958 ± 0.645 ton/ha·yr.

In the only example of litter quantity measurement for tropical montane forest, Edwards and Grubb (1977) placed 64 traps, 16 in each of four plots, in montane forest at around 2,400–2,500 m in New Guinea with an aboveground biomass of 310 ton/ha and a maximum tree height of 37 m, and found the average annual litter quantity to be 7.55 ton/ha·yr. This is considerably higher than the value for Mount Pangrango, even though the latter plot was at a far lower altitude.

The composition of the litter collected from Mount Pangrango is shown in table 3-8. The leaves comprise all of the vegetation types represented in plot 1, including trees, shrubs, ferns, grasses, epiphytes, and climbers. The barks, lichens and mosses, and so on are those that were in the trap independently rather than attached to fallen branches.

## Table 3-8
### Litter Production, by Component

| COMPONENT | PRODUCTION (ton/ha·yr) | PERCENT |
|---|---|---|
| Leaves | 4.504 | 75.60 |
| Branches | 0.942 | 15.80 |
| Bark | 0.030 | 0.50 |
| Lichens and mosses | 0.046 | 0.76 |
| Flowers | 0.189 | 3.18 |
| Fruits and seeds | 0.201 | 3.38 |
| Feces | 0.011 | 0.19 |
| Animal bodies | 0.001 | 0.02 |
| Others | 0.034 | 0.56 |
| Total | 5.958 | 100.00 |

Flowers in some cases consisted of petals only, and in others the peduncle was attached. Fruits and seeds likewise consisted of fruits only in some cases and seeds only in others, but these were classified together. Insect feces can be differentiated largely by their shape and color, but they were not identified as to type of insect. Insect remains were not classified according to species. The majority of the materials included in "others" were markedly broken down, and the category was restricted to those whose original form could not be ascertained. The materials further classified according to tree species were leaves, flowers, and fruits and seeds. Materials for which this was not possible were treated as unclassifiable within their respective elements.

Quantitatively, leaves were the largest category, accounting for 75.6 percent of the total. Omura and Ando (1970) reported that leaf litter accounted for 74.1 percent of total litter fall in an evergreen forest in Kyushu. In a tropical rain forest in Pasoh, Malaysia, the proportion of leaf litter was said to be 60 percent (Ogawa, 1974). In a planted forest of *Chamaecyparis obtusa,* Saito (1972) found leaves to account for 75.0–81.1 percent of litter. In all cases, leaves made up the bulk of litter.

Fallen branches accounted for 15.8 percent of the litter fall in plot 1, compared with 11.3 percent in the evergreen forest in Kyushu, 26 percent in Pasoh, and 11.0–12.9 percent in the planted *C. obtusa* forest.

The quantity of flowers was slightly less than for fruits and seeds, and together they accounted for 6.5 percent of the total. In the evergreen forest in Kyushu, fruits and seeds exceeded flowers more than threefold, and together they accounted for a moderately high proportion of 14.6 percent of the total. In Pasoh and in the *C. obtusa* forest, these proportions were smaller, at 4.2 percent and 3.4 percent, respectively.

Insect feces, barks, and other elements each accounted for less than 1 percent of the total litter fall.

### Seasonal Variations in Litter Fall

Figure 3-15 shows the seasonal variations in the total quantity of litter fall and in the quantities of each element over the year. The total volume (fig. 3-15a) fluctuates markedly, with a peak in August-September, a major peak in late December, and another peak in February. However, no clear trend is apparent.

The seasonal changes for each of the constituent elements reveal that each element has its own characteristics. The changes in leaf fall, quantitatively the largest element, have a major peak in August-September (fig. 3-15b). Leaf fall occurs to some extent in all months, however, and fluctuates considerably.

Seasonality is even more apparent in the volume of fallen flowers. Figure 3-15c reveals two sharp peaks, in July and December-January, with appreciably less fall in other months. These peaks in flower fall are followed by the fall of fruits and seeds, with a major peak in August-September and a second peak in March (fig. 3-15d).

The fall of branches (fig. 3-15e) shows a different pattern to the seasonality of flowers, fruits, and seeds. The major peak in September and the minor peaks in November, December, February, March, and April are all considerably narrow. From May to August, no peaks appear.

Of the other litter elements not displayed on the figure, lichens and mosses and others showed similar patterns to branches. Insect remains and feces showed slight increases in the dry season.

To observe the scattering in the total litter quantity, coefficients of variation were calculated from the total quantities collected in the 37 traps at weekly intervals. Figure 3-16a shows the coefficients of variation over the one-year period. From May to mid-December, the coefficients fluctuate around the value of about 0.5, of which those from early July to mid-November show constantly low values of below 0.5. From late December, the coefficient rises to 1.25, and thereafter it shows high values above 1.0 on five occasions.

Each of the constituent elements of litter shows a higher coefficient of variation than the total litter fall (Saito 1972). Here, the weekly fluctuations of the coefficients of variation for the flowers and fruits of identified species in the 37 traps are plotted in figure 3-16b. Compared with the total litter fall, the coefficients of variation are much higher, with values of 6.6 for flowers on 8 October and 6.2 for fruits on 1 April. The overall pattern for flowers and fruits also differs from total litter fall in showing more frequent peaks and troughs.

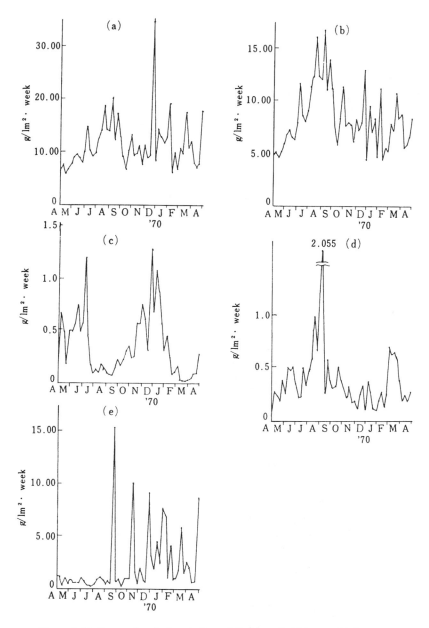

Figure 3-15. Seasonal variations in litter fall: (a) total; (b) leaves; (c) flowers; (d) fruit and seed; (e) branches.

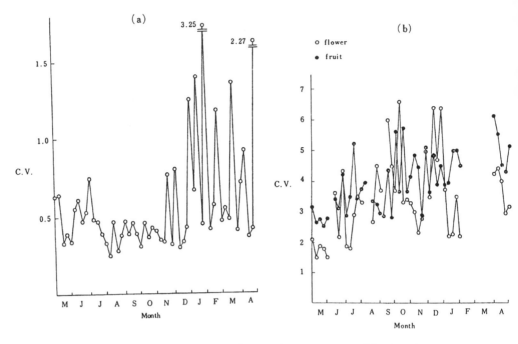

Figure 3-16. Fluctuations in coefficient of variance in litter fall: (a) total mass;
(b) flowers and fruit.

For total litter fall, the fairly localized fall of branches and other elements can be considered the major factor in the high values of coefficient of variance. For flowers, the coefficient of variance tends to be greater at times of low fall and to decrease with increasing quantity of flower fall. No such tendency is apparent for fruits.

Here, let us examine seasonal changes in the distribution of rainfall during the period of litter collection. Since the rain gauge at Cibodas Botanic Garden was not operative at that time, I adopted data from the Pacet meteorologic station, located about 6 kilometers east of Cibodas.

Figure 3-17 shows the monthly rainfall volumes and numbers of rainy days for the two years 1969 and 1970. It is clear from the figure that there is considerable variation in rainfall throughout the year. In 1969, rainfall was high from January to April and from September to December, while from May to August it decreased considerably. In 1970, rainfall was high until May and decreased from July to October. In 1969, monthly rainfall was below 100 mm from June to August. In 1970, while it did not drop below 100 mm, it decreased in the same period. As stated earlier, rainfall is high in the period predominated by the westerly monsoon, and the highest rainfall recorded in the two-year period was a

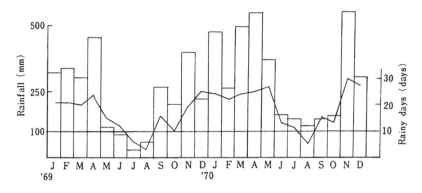

Figure 3-17. Monthly rainfall, 1969–1970, at Pachet meteorological station.

value of more than 500 mm in April 1970. From these tendencies, the period of litter collection can be described as comprising a dry season from May to August 1969 and a wet season from September 1969 to April 1970.

If we look at the year in terms of a dry season and a wet season, the tendencies in the constituent elements of litter fall are as follows. For leaves, the peak occurred from the dry season into the early rainy season; for flowers, peaks occurred in each of the dry and wet seasons; and for fruits, there was a major peak from the dry season into the early wet season and a minor peak at the end of the wet season. In contrast, the fall of branches was centered on the rainy season. The fact that mosses and lichens also center on the rainy season can be seen as demonstrating that, like branches, their fall is mainly the result of rain and accompanying strong winds.

The fall of leaves, flowers, and fruits is influenced not only by physical forces like wind and rain but also by inherent characters of the plants themselves that accompany climatic changes. Such a physioecologic reaction is thought to be species-specific, and litter fall will be analyzed by species in the following.

## Seasonal Variations in Leaf Fall by Species

As we have seen, leaf fall accounts for 75.6 percent of the total litter fall. Having been classified according to species, all identifiable materials were grouped into large trees (51 species), small trees (14 species), herbaceous plants (6 species), ferns (26 species), woody epiphytes (23 species), and climbers (24 species). Fifty-one tree species were represented in the litter, almost 90 percent of the 57 species identified in plot 1. For the other groups, except for the herbs, representation in the litter collected was of a similar or higher proportion. Not unexpectedly, trees accounted for the greatest proportion of the leaf litter. The analysis so far has centered on the large and small trees, and I shall continue to focus on these in the

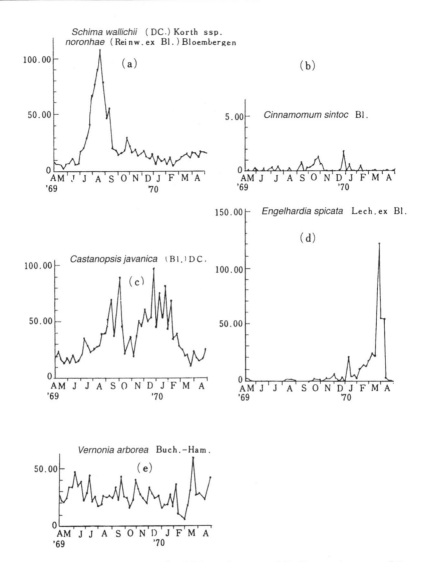

Figure 3-18. Seasonal types of leaf fall: (a) dry season fall; (b) pre-rainy season fall; (c) mid-rainy season fall; (d) late rainy season fall; (e) nonseasonal (unit: g/37 m² · week).

following. Species with a small quantity of leaf fall and in which seasonality could not be ascertained are excluded, leaving 38 species of large and small trees with a discernible seasonality. The seasonal variations observed in these species can be classified into the following five types.

***Dry season type.*** This type shows a peak of leaf fall in the dry season, as typified by *Schima wallichii,* which is illustrated in figure 3-18a. Leaf fall shows a very distinct peak in the dry season, with only a small quantity at other times of the year. Seven species (18 percent) showed this pattern.

*Early rainy season type.* As shown for *Cinnamomum sintoc* in figure 3-18b, this type has a peak in the first part of the rainy season, in September, October, and November. This is the most common type, represented by 13 species (34 percent).

*Middle rainy season type.* As can be seen for *Castanopsis javanica* in figure 3-18c, leaf fall peaks in January and February, in the middle of the rainy season. Five species (13 percent) show this pattern.

*Late rainy season type.* As typified by *Engelhardtia spicata* in figure 3-18d, leaf fall peaks in the late rainy season, in March, April, and May. Eight species are represented (21 percent).

*Nonseasonal type.* As can be seen for *Vernonia arborea* in figure 3-18e, there is no great variation in leaf fall throughout the year and no apparent seasonality. Five species (13 percent) show this type.

As mentioned, the canopy in plot 1 at 1,600 m where litter fall was measured is stratified into layers $I_0$ to III, and the canopy positions of the species for which the leaf fall type could be determined are also known. Those species distributed over several layers were classified as belonging to the layer in which their crown volume is maximal. Thus the relation between the leaf fall type and canopy position of each species can be depicted as shown in table 3-9.

The representative species of the five leaf fall types all occupy the overstory, and their seasonalities were distinct. Of the understory species, 11 species were recognized here. *Castanopsis argentea* was present as a small tree in plot 1, but since it can attain the proportions of a tall tree in this region, I shall exclude it here and classify the remaining 10 species as essentially understory species. These 10 species show far less marked seasonal variation than the overstory species. Three of them show fairly clear seasonal variation, *Strobilanthes cernua* as an early rainy season type, and *Saurauia blumiana* and *Decaspermum fruticosum* as late rainy season types, but the other 7 species are far from clear. Figure 3-19 compares the cumulative amounts of leaf fall for 6 of these 7 species and for the 5 representative overstory species. The understory species clearly show similar variations to the nonseasonal overstory type, *Vernonia arborea*.

## Leaf Fall Seasonality and Growth Type

Examination of the amount of leaf fall during the year allowed species to be grouped into five seasonal types (see table 3-9). Most of the overstory species

Table 3-9
Leaf Fall Type by Layer

| | TREES AND SHRUBS | | | |
|---|---|---|---|---|
| | $I_0$ LAYER | I LAYER | II LAYER | III LAYER |
| Dry season | *Schima wallichii*<br>*Lithocarpus rotundatus* | *Glochidion macrocarpum*<br>*Litsea mappacea* | *Polyosma ilicifolia* | *Saurauia reinwardtiana* |
| Early rainy season | *Platea latifolia* | *Cinnamomum sintoc* | *Litsea resinosa*<br>*Meliosma nervosa*<br>*Persea rimosa*<br>*Polyosma integrifolia*<br>*Laplacea integerrima*<br>*Syzygium antisepticum* | *Symplocos fasciculata*<br>*Villebrunea rubescens*<br>*Ardisia fuliginosa*<br>*Strobilanthes cernua*<br>*Antidesma tetrandrum* |
| Middle rainy season | *Lithocarpus indutus* | *Castanopsis javanica*<br>*Prunus arborea*<br>*Lithocarpus pseudomoluccus* | | *Castanopsis argentea* |
| Late rainy season | *Engelhardia spicata* | *Lithocarpus elegans* | *Flacourtia rukam*<br>*Tarenna fragrans* | *Decaspermum fruticosum*<br>*Saurauia blumiana*<br>*Turpinia sphaerocarpa* |
| Nonseasonal | | *Vernonia arborea* | *Astronia spectabilis*<br>*Macropanax dispermus*<br>*Syzygium rostratum* | *Saurauia pendula* |

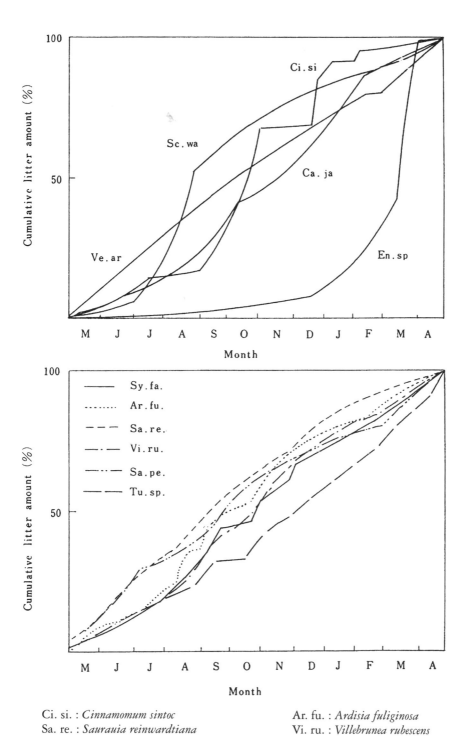

Ci. si. : *Cinnamomum sintoc*          Ar. fu. : *Ardisia fuliginosa*
Sa. re. : *Saurauia reinwardtiana*      Vi. ru. : *Villebrunea rubescens*

Figure 3-19. Cumulative litter amounts for tree species showing deciduous fall.
NOTE: The abbreviations are the same as in figure 3-4.

showed a clear seasonality, and only *Vernonia arborea* showed a nearly nonseasonal pattern. The understory species, on the other hand, were virtually nonseasonal. Such differences can be thought to be influenced by environmental differences between the overstory and the understory.

Climatic conditions differ greatly between the canopy and the understory (Richards, 1952). In the understory, the range of temperature variation is narrower, air movements are calmer, and changes in relative humidity are smaller than in the overstory; thus, climatic conditions are buffered and stabilized. In the tropics, temperature variations are small, but temperature may drop temporarily, after rainfall, for example, and Koriba (1947) states that this may promote leaf fall. We have also seen that an appreciable number of overstory species tend to shed their leaves with the change from the dry to the wet season. Climatic changes are buffered by the overstory layers, however, and reach the forest floor in milder form. The species occupying the overstory receive the full brunt of climatic fluctuations, while the understory species grow throughout the year in relatively unchanging conditions. This difference in growth environment might conceivably produce changes in leaf fall type.

Richards (1952) cites several examples of stratification and seasonality in various tropical regions. In British Guyana, for example, the higher strata A and B have two main flowering seasons, while in the lower stratum C, flowering occurs throughout the year; in the Belgian Congo, the proportion of deciduous species increases from the center of the forest toward the periphery, where the dry season is longer; in wet evergreen forest in Nigeria, most species in stratum A shed leaves in the dry season, whereas most stratum B species do not; and in Trinidad, the proportion of deciduous trees is overwhelmingly greater in the overstory. These examples also substantiate a difference in seasonality between overstory and understory.

Koriba (1947) made detailed observations of the vegetation in the Singapore Botanic Garden and in the primary forest remaining on the island, recording the seasonality of branch growth and leaf fall, and proposed four types of periodicity for tropical trees: evergrowing, manifold, intermittent, and deciduous. In the evergrowing type, leaves are produced continuously. In the manifold type, the growth period differs from branch to branch on the same tree, and alternation between branches is irregular, although overall the tree is growing continuously. The intermittent type produces a certain quantity of leaves in a certain period then rests, and repeats this intermittently. Koriba states further that growth can basically be divided into two types, evergrowing and intermittent, since manifold growth can be regarded as an intermediate stage between these two.

Having examined the relation between growth type and leaf renewal, Koriba states that while the evergrowing type produces new leaves continuously, the intermittent type shows three patterns of leaf production: seasonal, synchronized, and aseasonal. The seasonal type produces flushes at set seasons, one to three times annually; the synchronized type produces flushes in phase, three or four times annually in response to climatic change; and the aseasonal type produces new leaves without regard to season. In addition, the completion of the abscission layer and the shedding of old leaves is not sequentially related to the production of new leaves—the two are separate processes. Koriba states that if the completion of the abscission layer is advanced and the development of the new leaves is retarded, a deciduous tree results; and the opposite relation results in an evergreen.

While Koriba notes that the evergrowing type produces new leaves continuously, he does not relate this to leaf fall. Although it does not necessarily follow that continuous leaf production will be associated with continuous abscission, such behavior would seem most likely, since it is difficult to imagine that a branch continuously growing and continuously producing new leaves would shed its old leaves seasonally or periodically. An intermittent habit of abscission in the evergrowing type probably should be treated as a special case. In the absence of available data on the relation between leaf production and leaf fall on branches of the evergrowing type, a survey will be necessary to confirm or reject this speculation.

Neither does Koriba refer to the relation between leaf production and leaf fall in the intermittent growth type. It might be expected, however, that three patterns of leaf fall can be discerned, corresponding to the three patterns of leaf production.

Continuous growth can thus be considered to be associated with continuous, aseasonal leaf fall; and intermittent growth with intermittent, seasonal leaf fall. If these relations hold, then a tree showing aseasonal leaf fall can be regarded as evergrowing, and one showing seasonal leaf fall can be regarded as the intermittent growth type.

Application of this argument to the data on leaf fall in plot 1, classifying those species showing a peak in leaf fall in either the dry or wet season as intermittent growth type, and those species showing no seasonality as the evergrowing type, reveals that most of the species listed in table 3-9 are of the intermittent growth type. Those of the evergrowing type are 1 species (*Vernonia arborea*) in layers $I_0$ and I, 3 species (*Astronia spectabilis, Macropanax dispermus,* and *Syzygium rostratum*) in layer II, and 1 species (*Saurauia pendula*) in layer III. The species in layer III showing seasonality, including *Turpinia sphaerocarpa, Saurauia reinwardtiana, Ardisia fuliginosa, Villebrunea rubescens, Symplocos fasciculata,* and *Antidesma tetrandrum,* should probably be classified as evergrowing, since they show variations that are extremely close to being aseasonal.

As to stratification, the majority of species whose distribution centers in the overstory are of the intermittent growth type. Although this mountain is located in the tropics, its climate is seasonal, with a period of low rainfall from May to August, and a rainy season centered on November to March brought by the westerly monsoon. This seasonality can be thought to influence the habits of tree species.

*Vernonia arborea* is the only evergrowing overstory species. It has a wide range of distribution, being a fast-growing species that invades gaps where trees have fallen. Among the examples of evergrowing species cited by Koriba, a species group can be found, including *Macaranga, Mallotus,* and *Duabanga,* that is typical of secondary forest and predominates as pioneer species on logged land, in gaps, and similar places. Like *Vernonia arborea*, all of these species have the characteristics of rapid growth and preference for open land. For such a species group, the evergrowing habit is of major significance.

For the understory species, as described earlier, external environmental changes are buffered by the presence of the canopy, and the resulting stable, aseasonal conditions are a factor promoting the evergrowing type. Koriba states that shrubs are generally evergrowing. Among the species surveyed and classified according to growth type by Koriba (1958), one species also found on Mount Pangrango, *Symplocos fasciculata,* Koriba classifies as evergrowing. In the present survey, the leaf fall of this species was provisionally classified as the early rainy season type, but as I have stated repeatedly, the fluctuation was extremely close to being aseasonal. Consequently, this species can be classified as evergrowing. In the same way, all understory species that show the same kind of fluctuation can probably be classified as evergrowing.

To summarize, 38 species of large and small trees were classified according to seasonality of leaf fall, of which 27 species (71 percent) show the intermittent growth type and 11 species (29 percent) the evergrowing type. In stratification, the majority of overstory species show a seasonal type of leaf fall and thus an intermittent growth habit, with only one early maturing species with extremely rapid growth being of the evergrowing type. In the understory, because climatic changes do not have a direct influence, species appear showing aseasonal leaf fall, which were inferred to have the evergrowing growth habit.

Koriba (1958) estimates that for native Malay trees, the evergrowing and manifold habits together account for 20 percent of all species, the deciduous habit for 5 percent, and the intermittent habit for the remaining 75 percent. In tropical rain forest, the proportion of evergrowing species likewise is not very high. In temperate zones, the fact that not only the deciduous species but also the evergreens show seasonality of leaf fall attests to the fact that all are of the intermittent type. In growth habit, the transition from tropical rain forest to temperate forest corresponds to the replacement of the 20 percent of evergrowing species in tropical rain forest by species showing the intermittent growth type in temperate forest. The fact that a higher proportion of the evergrowing type was found on Mount Pangrango derives in part from the fact that Koriba did not include small trees and shrubs as the main object of his survey.

## 4. The Forest Carbon Cycle on Mount Pangrango

### *Estimation of Biomass*

In plot 1 (1,600 m), measurements were conducted in 1969 of tree height, height to first branch, diameter at breast height, litter quantity, soil respiration, and accumulated organic matter in the $A_0$ horizon. The positions of DBH measurements were marked with white paint. In 1976, DBH was again measured, and from the results of the two surveys, I could estimate the diameter growth and the number and diameter of trees that had died. Biomass was estimated indirectly by the following method.

The relation between diameter (D) and height (H) in February 1969 (fig. 3-20) is expressed by the following equation:

$$1 / H = (0.6923 / D) + 0.0192$$

It has been observed in many forests both in temperate and tropical climates that a linear double logarithmic relation obtains between timber volume and the product of the square of the diameter ($D^2$) and height (H). Thus, if D and H values are measured by census for trees of known timber volume, the relationship between them can be applied to the whole forest. And if the specific density of the timber is known, the volume can be converted to weight.

Table 3-10 reproduces a table from *Tectona*, volume 28, published by the Bogor Forest Research Institute, showing timber volumes for several species that had been surveyed in the same area as the present survey site (Wolff von Wülfing, 1935). It is based on measurements of from 51 to 132 well-formed specimens of such species as *Quercus, Castanopsis,* and *Engelhardtia* at altitudes of 1,700–2,100 m in Priangan, West Java (between Bogor and Bandung), and includes data such as diameters with and without bark, circumference, tree height, height of first branch, and timber volume to first branch (commercial bole).

A double logarithmic plot of $D^2H$ against the timber volume to first branch (V) appears in figure 3-21, and despite some differences between species, the relation can be tolerably approximated by the equation

$$\log V = 0.9753 \log D^2H - 4.5266$$

Using this relation, I determined the timber volume to the first branch from the DBH and height measurements of individual trees in the survey plot. The

# Table 3-10
## Timber Volumes

| CIRCUMFERENCE (o) AND DIAMETER ($_o$d) WITH BARK AT THE HEIGHT ABOVE THE GROUND OF: | | | | | | | | VOLUME | | HEIGHT OF TREE (h) |
|---|---|---|---|---|---|---|---|---|---|---|
| 1.50 m | | 2.00 m | | 3.00 m | | 4.00 m | | COMMERCIAL BOLE WITHOUT BARK | SQUARE TIMBER OBTAINED | |
| o cm | $_o$d cm | o cm | $_o$d cm | o cm | $_o$d cm | o cm | $_o$d cm | m³ | m³ | m |
| 1 | 2 | 3 | 4 | 5 | 6 | 7 | 8 | 9 | 10 | 11 |
| 80 | 25.5 | 78.5 | 25.0 | — | — | — | — | 0.29 | 0.17 | 18.7 |
| 90 | 28.6 | 88.0 | 28.0 | — | — | — | — | 0.39 | 0.23 | 20.8 |
| 100 | 31.8 | 97.7 | 31.1 | 93.6 | 29.8 | — | — | 0.53 | 0.29 | 22.1 |
| 110 | 35.0 | 107.4 | 34.2 | 103.0 | 32.8 | — | — | 0.66 | 0.36 | 22.9 |
| 120 | 38.2 | 117.5 | 37.4 | 112.8 | 35.9 | 109.6 | 34.9 | 0.82 | 0.43 | 23.6 |
| 130 | 41.4 | 127.2 | 40.5 | 122.5 | 39.0 | 118.8 | 37.8 | 0.98 | 0.50 | 24.2 |
| 140 | 44.6 | 137.3 | 43.7 | 131.9 | 42.0 | 127.9 | 40.7 | 1.16 | 0.58 | 24.8 |
| 150 | 47.7 | 147.0 | 46.8 | 141.4 | 45.0 | 136.7 | 43.5 | 1.35 | 0.66 | 25.2 |
| 160 | 50.9 | 156.8 | 49.9 | 151.1 | 48.1 | 145.8 | 46.4 | 1.56 | 0.73 | 25.7 |
| 170 | 54.1 | 166.5 | 53.0 | 160.5 | 51.1 | 155.2 | 49.4 | 1.78 | 0.82 | 26.2 |
| 180 | 57.3 | 176.6 | 56.2 | 170.3 | 54.2 | 164.6 | 52.4 | 2.02 | 0.92 | 26.7 |
| 190 | 60.5 | 186.6 | 59.4 | 180.0 | 57.3 | 174.0 | 55.4 | 2.26 | 1.02 | 27.2 |
| 200 | 63.7 | 196.4 | 62.5 | 189.4 | 60.3 | 183.2 | 58.3 | 2.53 | 1.13 | 27.6 |
| 210 | 66.8 | 205.8 | 65.5 | 198.9 | 63.3 | 192.3 | 61.2 | 2.80 | 1.24 | 28.1 |
| 220 | 70.0 | 215.5 | 68.6 | 208.3 | 66.3 | 201.7 | 64.2 | 3.10 | 1.35 | 28.5 |
| 230 | 73.2 | 225.6 | 71.8 | 218.0 | 69.4 | 211.1 | 67.2 | 3.43 | 1.48 | 29.0 |
| 240 | 76.4 | 235.3 | 74.9 | 227.8 | 72.5 | 220.5 | 70.2 | 3.77 | 1.61 | 29.4 |
| 250 | 79.6 | 245.4 | 78.1 | 237.2 | 75.5 | 230.0 | 73.2 | 4.11 | 1.76 | 29.9 |
| 260 | 82.8 | 255.1 | 81.2 | 246.9 | 78.6 | 239.4 | 76.2 | 4.48 | 1.91 | 30.3 |
| 270 | 85.9 | 264.8 | 84.3 | 256.0 | 81.5 | 248.5 | 79.1 | 4.85 | 2.06 | 30.8 |
| 280 | 89.1 | 274.9 | 87.5 | 265.8 | 84.6 | 257.9 | 82.1 | 5.25 | 2.21 | 31.2 |
| 290 | 92.3 | 284.6 | 90.6 | 275.5 | 87.7 | 267.3 | 85.1 | 5.68 | 2.38 | 31.6 |
| 300 | 95.5 | 294.7 | 93.8 | 284.9 | 90.7 | 276.8 | 88.1 | 6.08 | 2.54 | 32.0 |
| 310 | 98.7 | 304.4 | 96.9 | 294.7 | 93.8 | 286.2 | 91.1 | 6.53 | 2.73 | 32.5 |
| 320 | 101.9 | 314.2 | 100.0 | 304.1 | 96.8 | 295.3 | 94.0 | 7.01 | 2.93 | 32.9 |
| 330 | 105.0 | 323.6 | 103.0 | 313.5 | 99.8 | 304.7 | 97.0 | 7.50 | 3.13 | 33.2 |
| 340 | 108.2 | 333.6 | 106.2 | 323.0 | 102.8 | 314.2 | 100.0 | 8.02 | 3.35 | 33.7 |
| 350 | 111.4 | 343.4 | 109.3 | 332.4 | 105.8 | 323.3 | 102.9 | 8.54 | 3.57 | 34.1 |

| HEIGHT OF STUMP | HEIGHT OF FIRST BRANCH | | COMMERCIAL BOLE WITHOUT BARK | | | | | TIMBER PERCENTAGE | DIAMETER AND CIRCUMFERENCE WITH BARK AT HEIGHT OF 1.50 m | |
|---|---|---|---|---|---|---|---|---|---|---|
| | (ka) | (ka%) | LENGTH | AT SMALL END | | AT MIDDLE | | | | |
| | | | | circum. $(o')$ | diam. $(_od')$ | circum. $(o')$ | diam. $(_od')$ | % | | |
| m | in m | in m | m | cm | cm | cm | cm | | $_od$ cm | o cm |
| 12 | 13 | 14 | 15 | 16 | 17 | 18 | 19 | 20 | 21 | 22 |
| 0.6 | 9.4 | 50.5 | 7.6 | 59.4 | 18.9 | 67.5 | 21.5 | 61.5 | 25.5 | 80 |
| 0.7 | 10.5 | 50.3 | 8.5 | 66.0 | 21.0 | 75.1 | 23.9 | 58.8 | 28.6 | 90 |
| 0.7 | 11.1 | 50.1 | 9.2 | 73.5 | 23.4 | 82.9 | 26.4 | 56.3 | 31.8 | 100 |
| 0.7 | 11.4 | 49.9 | 9.6 | 77.9 | 24.8 | 91.7 | 29.2 | 54.0 | 35.0 | 110 |
| 0.7 | 11.7 | 49.6 | 10.0 | 88.6 | 28.2 | 99.6 | 31.7 | 52.2 | 38.2 | 120 |
| 0.7 | 11.9 | 49.3 | 10.3 | 96.1 | 30.6 | 108.1 | 34.4 | 50.7 | 41.4 | 130 |
| 0.7 | 12.2 | 49.0 | 10.5 | 104.0 | 33.1 | 116.2 | 37.0 | 49.4 | 44.6 | 140 |
| 0.7 | 12.2 | 48.6 | 10.6 | 111.5 | 35.5 | 124.4 | 39.6 | 48.3 | 47.7 | 150 |
| 0.7 | 12.4 | 48.1 | 10.7 | 119.4 | 38.0 | 132.9 | 42.3 | 47.3 | 50.9 | 160 |
| 0.7 | 12.5 | 47.6 | 10.8 | 127.2 | 40.5 | 141.4 | 45.0 | 46.4 | 54.1 | 170 |
| 0.7 | 12.5 | 46.9 | 10.9 | 134.8 | 42.9 | 149.5 | 47.6 | 45.7 | 57.3 | 180 |
| 0.7 | 12.6 | 46.2 | 11.0 | 142.6 | 45.4 | 158.0 | 50.3 | 45.0 | 60.5 | 190 |
| 0.7 | 12.6 | 45.5 | 11.1 | 150.2 | 47.8 | 165.9 | 52.8 | 44.6 | 63.7 | 200 |
| 0.7 | 12.6 | 44.9 | 11.2 | 157.4 | 50.1 | 174.0 | 55.4 | 44.1 | 66.8 | 210 |
| 0.7 | 12.6 | 44.3 | 11.3 | 165.2 | 52.6 | 182.2 | 58.0 | 43.7 | 70.0 | 220 |
| 0.7 | 12.7 | 43.8 | 11.4 | 172.8 | 55.0 | 190.7 | 60.7 | 43.3 | 73.2 | 230 |
| 0.7 | 12.7 | 43.3 | 11.5 | 180.6 | 57.5 | 198.5 | 63.2 | 43.0 | 76.4 | 240 |
| 0.7 | 12.8 | 42.7 | 11.6 | 188.5 | 60.0 | 206.7 | 65.8 | 42.8 | 79.6 | 250 |
| 0.8 | 12.8 | 42.3 | 11.7 | 196.0 | 62.4 | 214.6 | 68.3 | 42.5 | 82.8 | 260 |
| 0.8 | 12.9 | 41.9 | 11.8 | 203.3 | 64.7 | 222.7 | 70.9 | 42.4 | 85.9 | 270 |
| 0.8 | 13.0 | 41.7 | 11.9 | 210.8 | 67.1 | 230.6 | 73.4 | 42.2 | 89.1 | 280 |
| 0.8 | 13.1 | 41.5 | 12.0 | 218.0 | 69.4 | 238.8 | 76.0 | 42.1 | 92.3 | 290 |
| 0.8 | 13.2 | 41.3 | 12.0 | 225.9 | 71.9 | 247.2 | 78.7 | 42.0 | 95.5 | 300 |
| 0.8 | 13.4 | 41.2 | 12.1 | 233.7 | 74.4 | 255.4 | 81.3 | 41.9 | 98.7 | 310 |
| 0.8 | 13.5 | 41.1 | 12.2 | 241.0 | 76.7 | 263.6 | 83.9 | 41.9 | 101.9 | 320 |
| 0.8 | 13.6 | 41.1 | 12.3 | 248.2 | 79.0 | 271.4 | 86.4 | 41.8 | 105.0 | 330 |
| 0.8 | 13.8 | 41.0 | 12.4 | 255.7 | 81.4 | 279.6 | 89.0 | 41.8 | 108.2 | 340 |
| 0.8 | 13.9 | 40.9 | 12.5 | 263.3 | 83.8 | 287.8 | 91.6 | 41.7 | 111.4 | 350 |

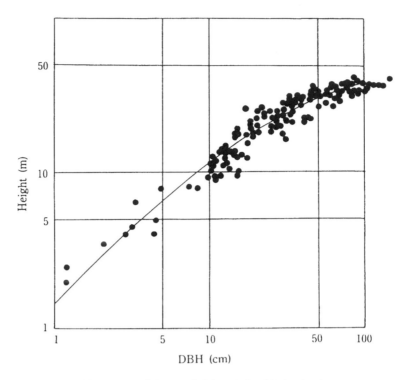

Figure 3-20. Diameter-height relationship in plot 1.

values in the table are only for circumferences from 80 cm to 350 cm, but the same relation was also applied outside this range.

For portions above the first branch, volumes were estimated from $D_B$ (diameter at the height of the first branch) and $H - H_B$ (crown depth, expressed as the difference between the total height and the height to the first branch). Plots of $D_B$ against DBH values from the timber volume table revealed a clear linear relationship for each species (fig. 3-22).

While the relationships for each species differ, the slopes of the plots are virtually the same. The upper and lower lines were taken as the upper and lower limits of $D_B$ in this forest. Although DBH was measured at 1.5 m aboveground in the timber volume table but at 1.3 m in the present survey, the above relation was applied to the results of the present survey without adjustment.

The plot in figure 3-23 of $H - H_B$ against DBH for the census shows considerable scattering of points. The figure also shows the curve for average $H - H_B$ against DBH. If the section of trunk within the crown is regarded as cylindrical, its volume is given as $\pi / 4D_B^2(H - H_B)$; but because it is probably more appro-

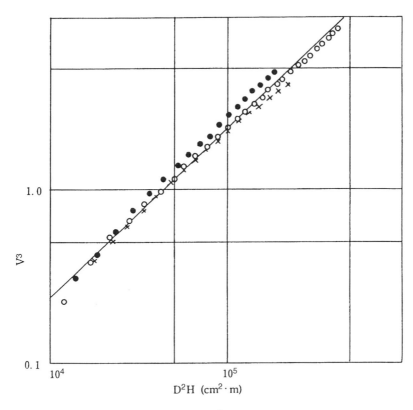

Figure 3-21. Plot of D²H against V.

priate to think of it as conical, I used the formula $\pi / 12 \cdot D_B^2(H - H_B)$. I took the value of $D_B$ to approximate the middle line in figure 3-22.

Concerning specific gravity, the Bogor Forestry Research Station has published the maximum, minimum, and average values of specific gravity after drying for individual species (Oey Djoen Seng, 1964). The list includes many of the species found in plot 1. The specific gravities of different species span a fairly wide range from 0.36 to 1.09; and even within the same species, the difference between maximum and minimum values is fairly large. As an average for all the species present, I took a value of 0.62. The specific gravity after air-drying is measured for a moisture content of 15 percent, and this must be taken into account in converting values to stem dry weight. From the sum of timber volumes below and above the first branch and the above specific gravity (15 percent water content), a value was obtained for the dry weight of the forest of 229.3 ton/ha.

Based on measured values of biomass in forests of Southeast Asia (Ogawa,

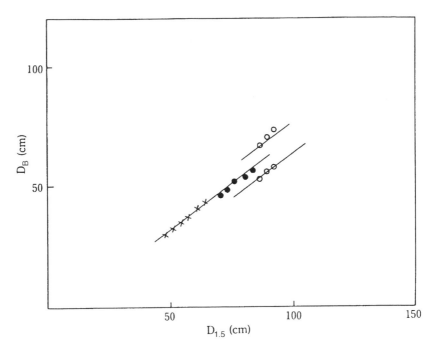

Figure 3-22. Relation between diameter at breast height and diameter at first branch.

1974), branch weight was taken as 40 percent of the dry weight, which gave a value of 119.7 ton/ha. For leaf weight, the figure of 1.3 times the leaf litter weight reported for Khao Chong in South Thailand was adopted, which gives a value for plot 1 of 5.9 ton/ha.

These values of 299.3 ton/ha for trunks, 119.7 ton/ha for branches, 5.9 ton/ha for leaves give a total aboveground biomass of 424.9 ton/ha for this forest. If the top-to-root (T/R) ratio is taken to be 4, the belowground biomass amounts to 106.2 ton/ha.

### Growth and Net Production

Diameter growth delta-D was calculated from the DBH measurements of February 1969 and October 1976. Figure 3-24 shows a plot of delta-D against D for all standing trees.

Despite the fairly long period of 7.6 years between measurements, some negative growth was seen. Ogino et al. (1967) found in a forest in Thailand that for trees of larger diameter, diameter growth tends to become negative while the leaf volume approaches its upper limit; and if the quantity of assimilated matter tends to level off, less matter will be accumulated in the lower trunk. Nevertheless, the consumption of matter as a result of respiration and other activities can

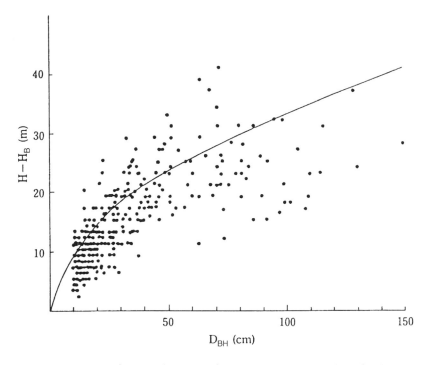

Figure 3-23. Relation between diameter at breast height and height above first branch.

be expected to increase, and this can result in the shrinkage of DBH. In the present survey, negative diameter growth was not limited to trees of large diameter but also was seen in individuals of 20-30 cm DBH. Because mosses, epiphytes, and climbers flourish on the trunks, the trunk is extremely uneven in the vicinity of breast height. Even though the position of measurement was clearly marked and the later measurements were taken in the same positions as the earlier ones, considerable error probably resulted from the unevenness of the surface.

In estimating biomass from the measurements of October 1976, the timber volumes of the lower trunk were obtained by the same method as in 1969, but because tree heights were not measured in 1976, they were estimated directly from the D-H relation of February 1969 (fig. 3-20). The timber volume above the first branch was estimated by assuming that it increased in the same proportion as the timber volume below the first branch. These figures were combined to give the total trunk timber volume for 1976, and the difference between this and the 1969 trunk timber volume was regarded as the quantity of growth in this period. Converted to weight, this is equivalent to 0.64 ton/ha·yr. For branches, a value of 40 percent of the trunk volume was again adopted, giving an

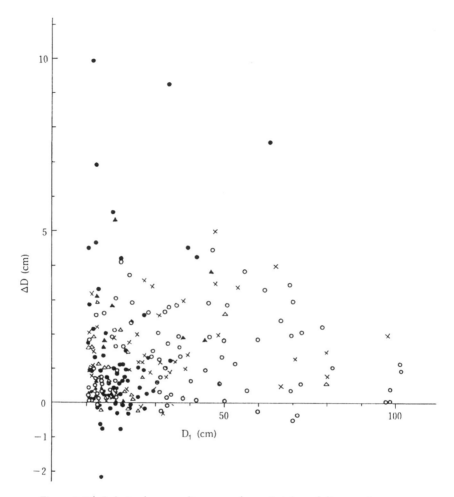

Figure 3-24. Relation between diameter at breast height and diameter increment.

increment of 0.261 ton/ha·yr. For leaves, no estimate was made in 1976, but to judge from estimates appearing in the literature, the increment can be regarded as negligibly small. For roots, I follow Kira (1978) in estimating this at one-quarter of the total for trunks and branches, namely, 0.23 ton/ha·yr. Trees which had newly attained diameters of more than 10 cm in this period were calculated to account for 0.27 ton/ha·yr. The total of the above gives an increase in the biomass in this forest of 1.4 ton/ha·yr.

The volume of fallen branches, as stated earlier, was 0.942 ton/ha·yr, that of fallen leaves 4.504 ton/ha·yr, and that of other litter 0.512 ton/ha·yr, giving a total 5.958 ton/ha·yr. The biomass of trees that died during the survey period

Table 3-11
Gross Production of Plot 1

|  | | TRUNKS | BRANCHES | ROOTS | LEAVES | OTHERS | TOTAL |
|---|---|---|---|---|---|---|---|
| Biomass (1969) (ton/ha) | | 299.3 | 119.7 | 106.2 | 5.885 | — | 531.085 |
| Biomass (1976) (ton/ha) | | 304.2 | 121.7 | — | — | 0.27 | — |
| Biomass increment (ton/ha·yr) | | 0.639 | 0.261 | 0.255 | — | 0.27 | 1.425 |
| Small litter | (ton/ha·yr) | — | 0.942 | 1.489 | 4.504 | 0.512 | 7.447 |
| Large litter | (ton/ha·yr) | 2.413 | — | — | — | — | 2.413 |
| Total litter | (ton/ha·yr) | 2.413 | 0.942 | 1.489 | 4.504 | 0.512 | 9.860 |
| Volume eaten | (ton/ha·yr) | — | — | — | — | 0.013 | 0.013 |
| Total (Biomass increment + Total litter + Volume eaten) | | | | | | | 11.298 |

was estimated from their D and H values while alive to be 2.41 ton/ha·yr. The biomass of dead roots is completely obscure, but if it is assumed to be equivalent to one-quarter of the total aboveground litter, a value of 1.5 ton/ha·yr is obtained. Thus the total quantity of dead matter for this forest is estimated to be 9.9 ton/ha·yr. The quantity of leaves consumed by insects has been reported to be 1.2-fold the quantity of insect feces (Yoda, 1971). The quantity of insect feces collected in the litter traps was 0.011 ton/ha·yr, which gives a quantity of leaves consumed of 0.013 ton/ha·yr.

The totals thus obtained for growth, death (including leaf and branch fall), and consumption give a grand total of 11.3 ton/ha·yr. These are listed in table 3-11 and represent the net production of the forest.

### Accumulation of Organic Matter in the $A_0$ Horizon

Square frames measuring 1 × 1 m were placed at random on the forest floor, and the quantity of organic matter in the $A_0$ horizon was measured. Fallen leaves and branches were weighed separately, and samples were taken away and weighed after drying for 48 hours at 80°C to give the dry weight.

Table 3-12 shows the quantities of organic matter in the $A_0$ horizon in plot 1 measured at monthly intervals. The variations in these values probably should be attributed to sampling error rather than seasonal fluctuation. These figures give an overall average accumulation of organic matter in the $A_0$ horizon of 6.8 ton/ha.

### Soil Respiration

Soil respiration in plot 1 was measured by the carbon dioxide absorption method in closed chambers. The lower part of a brick cylinder 20 cm in diameter and 25

## Table 3-12
### Accumulation in $A_0$ Horizon of Plot 1

|          | JULY  | AUGUST | SEPTEMBER | OCTOBER | NOVEMBER | DECEMBER |
|----------|-------|--------|-----------|---------|----------|----------|
| Leaves   | 1.886 | 2.335  | 3.330     | 3.148   | 5.491    | 5.190    |
| Branches | 6.156 | 4.415  | 2.237     | 0.628   | 3.815    | 1.990    |
| Total    | 8.042 | 6.750  | 5.567     | 3.776   | 9.306    | 7.180    |

Unit: ton/ha

## Table 3-13
### Soil Respiration in Plot 1

|                                   | AUGUST | SEPTEMBER | OCTOBER | NOVEMBER | DECEMBER |
|-----------------------------------|--------|-----------|---------|----------|----------|
| Soil respiration ($mg\ CO_2/m^2 \cdot yr$ | 168.5  | 135.1     | 197.2   | 238.1    | 277.5    |

cm high was buried in the forest floor and left for several days. A glass bottle containing 20 ml of 1 N KOH was placed in the cylinder, the cylinder was sealed for twenty-four hours, and then the bottle was removed and the contents were titrated. The maximum and minimum temperatures during the twenty-four-hour period were also measured. Brick cylinders were set at ten locations, and measurements were repeated every two to four days.

Table 3-13 shows the results. Measurements made during the five-month period from August to December show that soil respiration increases from the dry season into the rainy season. Such seasonal fluctuation in soil respiration is expected, but from these limited data it is not possible to discuss the nature of this fluctuation. To gain some idea of the quantity of organic matter that is equivalent to this volume of respiration, the minimum and maximum values were converted and indicated a range of 8.1–16.6 ton/ha·yr.

### Summary: Focusing on the Carbon Cycle

The results for plot 1 may be summarized as follows. The total aboveground biomass is 425 ton/ha, consisting of 299 ton/ha in trunks, 120 ton/ha in branches, and 6 ton/ha in leaves. The growth increment of trunks and branches is 1.17 ton/ha·yr. Through the shedding of leaves and branches and death, 8.4 ton/ha·yr of organic matter is deposited on the forest floor, equivalent to as much as six

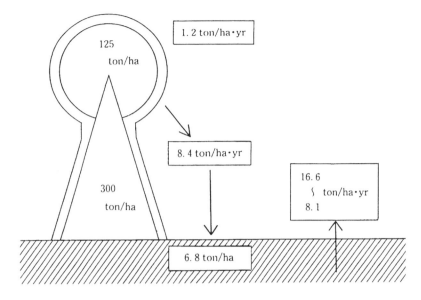

Figure 3-25. Recycling in plot 1.

times the growth increment. Thus, the greater part of the net production of 11.3 ton/ha·yr is returned to the forest floor. The 6.8 ton/ha·yr of organic matter accumulated in the $A_0$ horizon represents about 81 percent of the material deposited on the forest floor. Soil respiration is equivalent to 8.1–16.6 ton/ha·yr of organic matter. This is accounted for in part by the respiration of organisms living in the soil and thus is not exactly equivalent to the amount of decomposed organic matter, but nevertheless, the amount of matter falling to the forest floor is counter-balanced almost exactly by the amount disappearing from the surface.

These findings are summarized in figure 3-25. Considerable doubt remains concerning the accuracy of the figures shown, but they probably can be regarded as fairly accurate.

Several examples of biomass measurements made in tropical forests are summarized in table 3-14. The value obtained for the tropical montane forest of Mount Pangrango is smaller than for the rain forest in West Malaysia but larger than the evergreen seasonal forest in South Thailand. Table 3-15 shows estimates of net production. Mount Pangrango has the smallest figure. Kira and Shidei (1967) proposed the following explanation for the relation between forest age and productivity. Once a forest canopy becomes closed, the leaf weight of the forest does not increase above a certain value. Gross production and leaf respiration vary proportionally to leaf weight. In contrast, the biomass of nonphotosynthetic parts continues to increase, and their respiration, too, continues to

## Table 3-14
### Biomass of Tropical Forests

| | TRUNKS | BRANCHES | LEAVES | TOTAL ABOVE-GROUND | ROOTS | TOTAL | AREA SURVEYED |
|---|---|---|---|---|---|---|---|
| | ton/ha | ton/ha | ton/ha | ton/ha | ton/ha | ton/ha | ha |
| Tropical rain forest (Pasoh) | 528 | 126 | 8.8 | 663 | | | 0.01 |
| Tropical evergreen seasonal forest (Khao Chong) | 230 | 93 | 7.8 | 331 | 31 | 362 | 0.32 |
| Tropical evergreen seasonal forest (Cheko) | 211 | 98 | 6.4 | 315 | 60 | 375 | 0.50 |
| Tropical evergreen seasonal forest (l'Anguededou) | 240 | | 2.5 | 243 | | | 0.25 |
| Monsoon forest (Ping Kong) | 209 | 53 | 3.8 | 266 | | | 0.16 |
| Savanna forest (Ping Kong) | 55 | 11 | 2.1 | 68 | | | 0.16 |
| Tropical montane forest (Pangrango) | 299.3 | 119.7 | 5.9 | 424.9 | 106.2 | 531.1 | 1.00 |

SOURCE: Ogawa, 1974

## Table 3-15
### Estimated Net Production of Terrestrial Plant Communities in Various Localities

| Vegetation type | Locality | Age (yr) | Net Production (ton/ha·yr) |
|---|---|---|---|
| Tropical rain forest | Malaysia | Mature | 27.2 |
| Tropical seasonal forest | Africa | Mature | 13.4 |
| Laurel forest | Kyushu, Japan | Mature | 20.6 |
| Secondary *Castanopsis* forest | Kyushu, Japan | 14 | 22.7 |
| Secondary *Pinus densiflora* forest | Kanto, Japan | 15 | 15.8 |
| Secondary beech forest | Niigata, Japan | 30–70 | 15.3 |
| Tropical montane forest | Indonesia | Mature | 11.3 |

SOURCE: Kira, 1976

increase. As a result, net production reaches its maximum value at about the time the canopy closes, and thereafter it gradually declines.

Compared with the aboveground biomass, the growth increment is extremely small, while the supply of organic matter to the $A_0$ horizon (leaf and branch fall and death) is of the same order as the disappearance of matter from this horizon. For the carbon recycling system of the forest, this means that carbon is not being accumulated in any of the components of this system, and ecologically, the forest is in a state of stable equilibrium.

# 4.
# Conservation of Tropical Rain Forest Genetic Resources

## 1. Problems in Tropical Rain Forests

In recent years, the problems of the tropical rain forests have become an almost daily topic of conversation. Nevertheless, it seems that the substance of the problems involved is unclear. One reason for this is that the tropical rain forests are still a far-away world. It is the task of those of us who work in the tropical rain forests to bring this distant world closer to home and thereby, in a very important sense, open up a better future for the world.

What exactly are the problems of the rain forests, and what sort of approaches are desirable? First, I will give my own opinions, then discuss developments relating to the conservation of genetic resources.

The tropical rain forests that are presently the focus of concern occupy the tropical rain forest climatic zones centered on the equatorial regions of Southeast Asia, America, and Africa. Richards (1952) identified the common features of the forests in these three tropical zones and succeeded for the first time in generalizing tropical forest research. Since then, research has progressed to a stage where it is no longer easy to treat these three regions as one. Specialists in each of the three regions have conducted more detailed research and delineated the characteristics of each region. Even so, it still seems to me to be possible to cite common problems. I am familiar only with Southeast Asia, and my knowledge of South America and Africa is limited to the literature, the accounts of people who have been there to conduct surveys, and the compilations of information from many research students each year in the various countries. Were I to visit the places in question, my views might change; but here I would like to consider common problems of tropical forests based on these sources of information and my own experience in Southeast Asia. As a first step, the causes of these problems can be divided into three broad categories, human, natural, and biological, as listed in table 4-1.

## Table 4-1
### Problems in Tropical Rain Forests

| HUMAN FACTORS | NATURAL FACTORS | BIOLOGICAL FACTORS |
|---|---|---|
| 1. Information | 1. Climate | 1. Diversity |
| 2. Infrastructure | 2. Rainfall | 2. Distribution |
| 3. Purpose | 3. Soil | 3. Variation |
| 4. Philosophy | 4. Landform | 4. Breeding systems |
| 5. Technology | 5. Fire | 5. Phenology |
| 6. Capital | 6. Grazing | 6. Morphology |
| 7. Human resources | | 7. Life span |

## Human Factors

The first human factor that can be cited is lack of information. If, for example, you want to inspect a typical *Chamaecyparis obtusa* forest of eighty years' standing in the Chubu region of Japan, then by following the proper channels, you can quickly receive a copy of detailed information kept at the local forestry office, including a location map and data on the timber volume in such a forest, and the status of the surrounding forests. In the tropics, such information is available for very few forests and for only a limited number of useful tree species.

A further problem is access to the forest. In Japan, a comprehensive network of forest roads exists, and detailed information on a locality, including seasonal road conditions, is readily obtainable by telephone and other methods of communication. This also means that a coherent system is in place for yield surveys and felling and extraction for the market. Few such systems exist for the tropical forests. Countries where topographic maps on a scale of 50,000:1, soil maps, vegetation maps, and the like have been prepared are few, and countries where these are available for inspection are fewer still.

For a foreigner, the road to a forest of interest is particularly long. Having completed immigration formalities, you must obtain travel permission from the interior ministry and the police, from the governor, police, and army chiefs of the province concerned, and before entering the village you are aiming for, further permission is needed at lower administrative levels. Having reached the village, you require permission of the village headman, and must rely on the accounts of local elders and find a guide familiar with the mountains, then on occasion organize a caravan for a trek of several days into the mountains. In a small country like Brunei, the information network and infrastructure are comparatively good along the main routes, but one step away from these the situation is again difficult. In larger countries like Indonesia, it takes at least two weeks to reach the mountains in the outer regions. Even though such a trip may be a pleasure rather than a hardship, conditions are undeniably adverse in terms of efficiency.

The same difficulties apply to the problem at hand, the conservation of genetic resources. Experience in various countries has shown that a long search may be required to find a desired species or population. When I was in Indonesia in 1969, two Australians and a young worker from a herbarium met with an accident in Sulawesi in which two of them died. They had been searching for seeds in a *Eucalyptus deglupta* forest, though without much success, and on their long journey home their truck left the road and overturned. One of those killed, Dr. Prijanto, a young scholar specializing in pollen analysis, was a close friend of mine, and thus I was deeply affected. The forest roads, where they existed, were in a poor state of repair and lacked roadside ditches. After heavy rain, even the main roads flooded, making it impossible to tell where the road ended and the adjacent paddy fields began. I have also on many occasions run into trouble on treacherously slippery roads after rain. This underdevelopment of the infrastructure will continue to be a major obstacle to forestry, an industry that requires a comprehensive infrastructure.

A further problem is people's attitude toward the forest, which is expressed in a lack of purpose and philosophy. With few exceptions, the mountains and forests of the tropics are seen either as something to be plundered or something sacred. From either viewpoint, the maintenance and regeneration that are the essence of forestry have been practiced only for teak. In Japan, forestry took shape as far back as six hundred years ago in Kyoto's Kitayama mountains, producing the timbers needed for construction of Kyoto's temples. Through this, forestry gained widespread recognition, to the extent that famous forestry regions developed in various parts of the country, and in each region systems of specific skills were developed that passed from generation to generation and spread to surrounding areas.

Except for teak in the monsoon zone of continental Southeast Asia and in East Java, such historically important forestry systems and philosophies did not develop in the tropics. Until very recently, forests were regarded the same as oil, coal, or mineral resources: as a natural product to be exploited unilaterally, without concern for regeneration. It is no exaggeration to say that regeneration was not considered until the effects of exploitation on future national production became apparent, and that the management policy, still found in some countries, was to continue felling as long as the forest remained. Even in Malaysia, the Philippines, and other countries where policies were formulated for management of virgin forest, these have not always been successful. Basically, forests have the capacity to regenerate, and the central purpose of forestry should be to manage this capacity efficiently. Forestry in the tropics, to a greater or lesser extent, has been characterized by a lack of the basic planning required to meet this purpose. The same has been true of virgin forest management in the temperate zones.

In the absence of an ideological background, technology does not develop. In the *Cryptomeria japonica* forest of Kitayama, the technique of propagating superior phenotypes through cutting was established long before modern breeding techniques, as was the breeding of trees for barrel making at Yoshino. Notable forestry technology of the tropics is limited to the *taungya* and stump methods developed for teak.

Capital and human resources also contribute to technological development. Funding is still far from satisfactory even in temperate zone countries, and the situation relating to human resources raises further concern. Talented scientists and technicians with overseas training work energetically on returning to their homelands, but salary structures often entice them soon to transfer to administrative positions without putting their hard-gained technical skills to use. Various countries in the temperate zones are helping greatly to ease the problems of technology, funding, and human resources in the tropics, but ultimately the recipient countries must be willing to operate their own forest technology systems based on a long-term perspective, involving their own technicians and technology and depending on their own funding. A long road still lies ahead.

## Natural Factors

In this category I have cited six factors, of which the most influential is climate. A climate with high temperature and heavy rainfall is harsh for human habitation compared with temperate and cold climates, but it makes for a generally favorable environment for plant life. This climate, however, can also produce adverse effects, as in the recent droughts caused by the El Niño phenomenon.

In tropical climates, rainfall causes greater problems than temperature. Temperature does not vary greatly in the tropics, and except in the small areas of desert, it does not have a large effect on plant growth. The volume of rainfall, on the other hand, has a major impact. The equatorial tropical rain forest climate zone receives a certain amount of rain throughout the year. A slight decrease in rainfall volume brings about flowering and fruiting, and prolonged drought causes forest fire and is a major hindrance to afforestation works. Sudden massive downpours wash away large volumes of topsoil, removing nutrients that generally already are scarce in tropical soils, causing the appearance in some regions of abnormal soil morphology that renders afforestation impossible. For example, in Pantabangan in the Philippines, where afforestation work is being carried out with Japanese cooperation, it is necessary in some areas to use an iron bar to make holes for planting. Just as rainfall volume influences the agricultural calendar in temperate zones, it must be taken into consideration in planning forestry operations, from nursery preparation to transplanting. Such operational calendars have been drawn up for the afforestation works that have recently been expanding in the tropics, but these plans often have been spoiled by a late onset of the rainy season.

Few substantial research projects have been conducted on the soils of tropical forests, but it is well known that their nutrient contents are generally lower than those in temperate zones. The productivity of tropical forests depends on the constant supply of dead plant material, its rapid conversion to inorganic matter, and its reuse by living plants. Once the supply of plant material is interrupted, the soil rapidly becomes impoverished, and heavy rainfall and strong sunlight cause further deterioration of its physical properties. In addition, many areas have peculiar soils, from the coastal brackish water zones to the peat zones, and zones of podzols, limestone, and so on, on which little research has been conducted. Landforms also vary widely, from totally flat swamps to steeply sloping terraces, limestone formations, and mountains. These landforms, soils, and vegetation require further analysis, and more data need to be accumulated.

While fire and grazing are manmade rather than natural occurrences, their deleterious impact on tropical forest is significant. On a journey through the length of the Philippines, I frequently observed brush fires on the hills between the lower mountain slopes and the plains in central and northern Luzon. In many instances these were fires that had been set in swiddens or pastureland and had not been extinguished once the intended area had burned, but had continued to burn up the middle slopes to the ridges before dying out. With pastureland, peasants claim that old leaves of *alang-alang* injure the mouths of cattle and must therefore be burned to produce soft new shoots.

Through this combination of fire and grazing, the forest land deteriorates even further, and the tracks trodden by cattle criss-crossing the slopes greatly hinder regeneration. In Southeast Asia, such land is a common sight in Luzon, the Khorat plateau, the hills of North Thailand, and elsewhere. In Africa, more extreme examples can be seen in the drier savanna regions of East Africa and the Sahel region of North Africa. In South America, burning for pasturage on the land parceled out along the trans-Amazon highway is obvious on Landsat images. Burning and grazing represent the invasion of stock farming into the forest, which on an appropriate scale presents no problem. However, serious problems arise when fires run out of control and when unthinking expansion greatly exceeds the capacity of land for regeneration.

These natural conditions are not independent of each other but are mutually interrelated. Apart from the individual problems, attempts to understand their interactions are also required, but few such comprehensive studies have been made.

## Biological Factors

The first of the seven biological factors in forest-related problems listed in table 4-1 is the often-cited issue of diversity. Here, I am thinking not only of diversity of species, although it goes without saying that this is overwhelmingly larger than in other climatic regions. For example, more than 170 species of trees exceeding 10 cm in DBH have been counted in one hectare of mixed dipterocarp forest in Malaysia, with no species represented by more than a handful of individuals. Forests where the majority of trees are of one particular species, as in Japan's temperate forests, are exceptional in the tropics. In addition, the range of distribution is clearly known for only a few species. The *Flora Malesiana* is a taxonomic compilation of the vegetation of the Malesia region begun in 1950. This is now complete for the smaller families, but for the larger families such as Lauraceae, Sapotaceae, and Anonaceae, it is not known how long it will take to reach completion. For example, *Schima wallichii* is a species widely distributed from the Himalayas to Malaysia, and it has become clear that what were formerly classified as several species are in fact members of the same species. Much remains unknown about such intraspecific variants.

Breeding systems are also diverse, and it is in the tropics that the greatest diversity of agents involved in cross-fertilization have developed, including wind, birds, bees, moths, insects, and small animals. These systems, are understood only for small plants, however, or for a small fraction of useful species, while for the vast majority of forestry trees they are unknown. Flowering and fruiting in the tropics, as has been described frequently in the literature and as we have seen in earlier chapters in this volume, do not follow the rhythm of spring flowering and autumn fruiting that is seen in temperate regions but differ from one species to another, from one individual to another of the same species, and even from branch to branch of the same individual.

Concerning morphology, recent research into branching modes and architecture has revealed a great diversity of forms. These studies have focused mainly on the smaller species like orchids, herbs, and forest floor vegetation, and much remains unknown about the giant species, whose lifespans can exceed one hundred years.

While these biological features of tropical forests have gradually been clarified, much work remains. Recent progress in production ecology is elucidating the mechanisms of productivity. Rapid progress is also under way in the fields of population ecology, reproductive biology, and species biology, but again much remains unknown. The large sizes and long growth periods of forest trees, even in the temperate zones, mean that progress inevitably tends to be slower than in other fields of biology, particularly in the tropics.

## *Future Directions*
### Global Scale

As typified by the major impact of chlorofluorocarbons on the earth's atmosphere, the present age is one when events in one country alone can have global repercussions. Apart from nuclear power generation and toxic substances, the smoke of shifting cultivation in the tropics appears large on Landsat images, and debates on such topics as its climatic impact and the problem of carbon dioxide concentration have become a daily occurrence. This demonstrates the need for constant awareness of the global situation in dealing with domestic problems.

Environmental problems were first taken up as matters of major importance at the 1972 United Nations environmental conference in Stockholm, which spurred the many debates just mentioned. Not one problem, however, has yet been resolved. The reason is that such global holism is not coordinated with the work that is going on in the field.

### From Region to Locality

Discussion in general becomes more interesting the larger its scope, but it also becomes further removed from reality. Nobody knows less about the situation on the ground than the person who talks on a global scale. People who have long lived by shifting cultivation will not listen if they simply are told not to set fires. Describing the global effect of their actions will not strike home to people at one small point on the earth. Between these two extremes are administrators and educators at the national and local levels, and at present their actions fail to evince an understanding of both standpoints. While this situation persists the problems of the tropical forests will continue to increase in severity.

What is now required of forestry in tropical countries is action that above all is grounded in a knowledge of the actual situation in the forests and at the same time maintains a global perspective.

### Long-Term and Practicable Plans

In view of the lengthy nature of forestry operations, plans must be long-term, spanning several decades. Those that terminate after a few years are futile. Plans must also be practicable. Too often in the past, totally impracticable plans have been pushed through. These two elements seem to be the most deficient in forestry planning in many tropical countries.

### Ecological and Genetic Viewpoints

Finally, I wish to stress the importance both of ecology and genetics in the theoretical background of forest studies. In the past, emphasis has been placed on ecology at the expense of research in the fields of genetics and physiology. A genetic perspective is particularly important, and in dealing with a collectivity of so many diverse species, a basis in population genetics is important.

## 2. Developments Toward Genetic Resource Conservation

### Global Developments

At the international level, the Food and Agriculture Organization of the United Nations is at the center of the movement toward conservation of forest genetic resources. The FAO Panel of Experts on Forest Gene Resources publishes reports of its sessions, and more general information appears annually in *Forest Genetic Resources Information*. Because of the international nature of these activities, many of the reports cover a single topic from a particular country and compare the situation there with a general theory rather than presenting comprehensive results. Seed collection, for example, is a major task currently in hand, and it is readily obvious from the reports which countries are actively engaged in this. In Southeast Asia, Australia is making notable efforts, leading the field with regard to the genetic resources of *Eucalyptus* and *Acacia*. Several of the earlier reports discuss the difficulty of seed collection, while recently all countries have become more organized and systematic, and work is progressing steadily.

In Japan, both the Ministry of Agriculture and the Science and Technology Agency are independently carrying out conservation work related to agriculture, and for forest trees, work centers on the Kanto Forest Tree Breeding Institute. Having consolidated its work on domestic genetic resources, the institute plans to extend operations overseas, but with a small budget and a large volume of administrative work, those in charge are faced with an awesome task. This work is far beyond the scope of a university to accomplish and requires the organizational capacity of a government research establishment.

From around the time of the 1972 United Nations environmental conference in Stockholm, which brought the problem of genetic resource conservation to the fore, several works were published that touched on the principles of genetic resources.

Frankel and Bennett (1970), in the introduction to their book *Genetic Resources in Plants,* published by the International Biological Programme (IBP), summarize the tasks of conservation as follows:

1. Determination of the location and nature of genetic resources in the field
2. Survey of material in existing collections
3. Utilization of genetic resources

4. Conservation of genetic resources

5. Documentation

6. International coordination, guidance, and administrative backing

Each of these tasks is discussed from various standpoints in this book; and the FAO's *Forest Genetic Resources Information* also generally discusses regional problems in relation to these tasks. In developing countries in particular, various problems are pointed out and the difficulty of conservation of genetic resources is stressed.

With regard to forest trees, Callaham (1970) has summarized the general principles of geographic variation as follows:

1. A diverse environment throughout the range of a species leads to genetically variable species. Widespread species tend to be more variable than restricted species.

2. Patterns of inherent variation parallel patterns of environmental variation. Discontinuities in patterns of inherent variation are related to breaks in the distribution of species or rapid changes in environmental factors.

3. Races of a species growing in different climatic regions may differ in genetic adaptation to environmental factors. In one region a certain factor of the environment may be critical; in another region this factor may be less important than some other critical factor.

4. Sympatric species will be similar but not identical in inherent adaptations to the same environment. Limiting factors generally are not always the same for cohabiting species.

5. Two or more successive seed source trials will be necessary to determine an optimum seed source. Most species and environments are too variable to be analyzed completely in one experiment.

6. Seed source studies of native species undisturbed by humans generally show the local seed source to be the best adapted but not necessarily the most productive. Exotic populations usually do not equal local populations in adaptation to the unique combination of factors in the local environment.

7. The local seed source is safest if little is known about variation in a native species. If local seed is not available, seed sources from environments comparable to the planting site should be used.

8. Strong genotype-environment interactions can be expected in all seed source studies. Widely separated provenances of a species should be tried in several diverse environments to find the best source for each environment.

9. Performance will be unpredictable for a species grown for a long time under cultivation or disturbance by man or for species transferred to radically different environments, as in the case of species of *Eucalyptus* and *Cedrus*.

10. Exploration should not be confined to locating and cataloguing of

forest species. Rather it should be concerned with exposing variability that exists between and within species, particularly in relation to environmental changes.

11. Exploration routes should follow major environmental gradients, which must be understood before exploration starts. Exploration routes should pass from warm to cold regions, from long-day zones at high latitudes to short-day zones at low latitudes, and from moist to dry regions across mountain ranges.

12. Explorations should seek out the margins of distribution of each species. Outlying populations should be found and sampled to detect possible significant evolutionary changes that have occurred in these populations.

These points precisely express the principles for search and collection of forest tree genetic resources. Putting these principles into practice, however, is not a simple matter.

Zobel (1970), for example, describes the actualities of field sampling in connection with collection of genetic resources of Mexican pines.

First, it is necessary to decide the species or species complex to be studied and to make notes on the environmental conditions of the collecting sites. Next, in sampling seeds of a given population, he states that samples of up to 25 trees are preferable, though sometimes a sample of as few as 5 trees was used. Sound seeds from ripe cones were collected, and the samples and original trees were permanently labeled. By this method, approximately 42,000 cones were collected from 128 trees representing 18 species. Two 10-mm wood samples were taken from each tree, together with herbarium specimens, needle samples, and black-and-white and color photographs.

The 18 species of pines thus collected were planted in other parts of the world, but with the exception of *Pinus patula,* the trials were not successful. This was in part because geographic and ecoclinal effects were ignored. Careful matching of seed source with planting site is required, but even so, misjudgment about climatic factors and lack of proper mycorrhizal infection can lead to failure.

Much has been published about pines, particularly those of Central and South America. The United Kingdom has been central in conducting provenance trials, and results are emerging from various parts of the world and being reported at conferences of the International Union of Forestry Research Organizations (IUFRO).

Ranking with the tropical pines, the eucalypts, of which great numbers have been planted in the tropics, are the subject of a large body of literature. Larsen and Cromer (1970) have summarized the exploration, evaluation, utilization, and conservation of eucalypt gene resources. The genus *Eucalyptus* is one of the largest and most complex genera of woody plants in the world. More than 1,000 species and varieties have been recorded, of which 675 are considered to be valid.

There are also many hybrids. Some 100 eucalypt species are now commonly planted in more than fifty countries. The most important species are *Eucalyptus alba* in Brazil; *E. camaldulensis* in arid zones of the Mediterranean and Pakistan; *E. citriodora, E. globulus,* and *E. grandis/saligna* in Africa; and *E. robusta* and *E. tereticornis* in India. In more restricted regions, the following six species are also important: *E. deglupta, E. gomphocephala, E. maidenii, E. microtheca, E. occidentalis, E. regnans,* and *E. viminalis.*

Eucalypt seeds are very small, and a single tree may yield 5 pounds of seed, which, for a species such as *E. grandis,* would equal about 3 million viable seeds, enough to plant about 2,000 acres of forest. Most eucalypts produce seed at a very early age. Since 1964, systematic trial plantations of eucalypts have been established in twenty-five countries, but it is still too early to assess the results.

Research on the conservation of these species parallels that described by Zobel for pines. Work has included the collection of seed, botanic specimens and wood samples, with a sampling intensity of 10 trees of a selected species or variety in an area of from 5 to several hundred acres in its natural range; permanent marking of parent trees, or preservation of these by grafting; establishment of seedlings of each parent; study of variation of parent material and seedlings; and distribution of seed to other countries.

The main problems encountered in this work have been the determination of sampling intensity and the preservation of the seed parents. It is not possible to root cuttings from mature eucalypts; grafting, too, is difficult, the main problems centering on getting the scion material to the grafting site in satisfactory condition, and a high mortality after planting-out, possibly because of stock-scion incompatibility.

Eucalypts grow naturally from 43° S in Tasmania to 7° N in the Philippines, from the humid tropics to the arid desert in Central Australia, and in the alpine environment of the Snowy Mountains. Tree forms range from the small to some of the tallest in the world, and the range of distribution varies from a few square miles for some species to some 2 million square miles for *E. camaldulensis.* Future tasks, according to Larsen and Cromer, include the study of intraspecific variation, testing of seed sources, and conservation; these efforts require international coordination.

In an overview of genetic conservation, Frankel (1970) argues that the strategy for conservation is determined by the nature of the material, namely, the length of the life cycle, the mode of reproduction, the size of individuals, and their ecological status—wild or domesticated; and by the objective—research, introduction, breeding, and so on. Gene pool conservation has much in common with nature conservation in general, but differs in one important point. While nature conservation aims to protect areas representing habitats and communities that have been identified, gene pool conservation goes further in being concerned with genetic differences that often can be only surmised but not identified. It is concerned with population samples, possibly along latitudinal or altitudinal transects, and often over extensive areas. Thus a "genetic reserve" should include a spectrum of ecological variability to provide a spectrum of genetic variability. It may therefore have to be extensive or scattered, the latter being difficult to manage. He concludes by stressing that the conservation of genetic resources for future generations should be the joint responsibility of all nations, and coordination should be vested in a United Nations agency.

Bouvarel (1970) deals with the conservation of gene resources of forest trees, setting forth principles for practical conservation that derive from three essential characteristics of forest trees: their size is large, therefore collections will cover large areas and each type cannot be represented by a great number of individuals; they are long-lived; and they are wild species, predominantly allogamous, which grow over large natural regions and possess great genetic variability at the individual level and at the intraspecific level.

Five years after the publication of the works just cited, a volume edited by Frankel and Hawkes (1975) was published as part of a series resulting from the International Biological Programme (IBP). This volume expounds the need for concrete action. Bradshaw (1975) explains the principles of population structure and concludes by supporting the two-stage collecting process proposed by Bennett (1970), involving a combination of collection on a wide basis followed by detailed collection in limited areas of interest. Marshall and Brown (1975) discuss the theoretical background to optimization of sampling strategies within the given limitations. Frankel (1975) reports on the situation of exploration and collection in various parts of the world; then various authors give more detailed accounts of the situation in particular areas.

The methodology of conservation of endangered plants is detailed in the proceedings of an international conference edited by Synge (1980). Preservation of seed has been treated by Norton (1986) and Frankel and Soule (1981); Ford-Lloyd and Jackson (1986) discuss the conservation of genetic resources for various crops, and in a separate section on forest tree resources, they outline the problems of conservation of genetic resources of forest trees from a global perspective.

On a global scale, the problem of genetic resources has been taken up by the FAO, and as a result, various countries, particularly developing countries, are embarking on genetic resource works.

## Regional Developments in Southeast Asia

While general discussions of a comprehensive nature or of basic principles are necessary to establish guidelines, these do not always have a clear grasp of regional characteristics, and in relation to tree species, large intraspecific variation is involved. Japan is fortunate in having outstanding forest trees like *Cryptomeria japonica* and *Chamaecyparis obtusa,* which have been the subject of ongoing improvement and the development of forestry techniques for several centuries; but most tropical countries have lacked the idea of afforestation and have simply plundered the forests for their products, with the result that regeneration and the concomitant treatment of genetic resources, including genetic improvement, is an awesome task.

Bogor, Indonesia, has a long tradition as a center of tropical botany, and it was here, in 1975, that an international symposium was held on Southeast Asian plant genetic resources (Williams et al. 1975). Together with detailed expositions covering fruits, tubers, vegetables, plantation crops, forest products, and so on, the situation in various countries was described.

The Bogor Botanic Garden, its branch at Cibodas and its herbarium, and the zoologic museum, together with the Treub Laboratory, are a vast treasure-house of zoologic and botanic specimens dating from the Dutch period. These materials, including the living specimens, represent valuable genetic resources, and continued efforts are required to expand these materials. Because of the vast area of Indonesia, exploration and collection will require a long time and huge funding. This will call for international cooperation.

Malaysia is one of the more active countries in Southeast Asia on forest research, and genetic resource works are in progress centered on the Forest Research Institute, Kepong.

In the Philippines, the destruction of forest by swidden farming is considerable, with 172,000 ha being lost annually; and consideration is being given to conservation of species in danger of extinction. Together with this, a forest conservation treaty was enacted in 1963 that designated afforestation and forest conservation areas, preserved species, and so on. Consideration was also given to building a botanic garden at Makiling, with basic research to be conducted centered on the University of the Philippines, Los Baños.

Thailand traditionally has supplied teak, and the breeding of this and tropical pine has been carried out there in cooperation with Denmark, the former at the TIC (Thai-Danish Teak Improvement Center) established in Lampang in 1965, the latter at the PIC (Thai-Danish Pine Improvement Center) established in Chieng Mai in 1969. At a conference on the Southeast Asian Tree Improvement and Seed Procurement Cooperative Programme, held in Chieng Mai in February 1979, Thanom Premrasmi (1980) presented a proposal for the establishment of such a program and argued for designation of seed zones with descriptions of their ecological and climatic conditions, and selection of standard seed sources within each zone, initiation of a seed certification scheme, improvement of research facilities, and establishment of in-situ conservation stands of threatened provenances of *Pinus merkusii* and *P. kesiya.*

Keiding (1980) discussed the constraints on international cooperation in seed procurement as one keenly aware of the problems, having long been involved in forest tree breeding in Thailand. These include restrictions on the import and export of seed, the need for exploration, collection and distribution of seed for internationally coordinated provenance trials and gene resource conservation, communication difficulties in exchange of information, differences in the development of seed procurement organization, and a lack of common systems for registration, seed certification, and standardized methods for assessment of experiments.

Nikles (1980a) proposed twelve major functions and a possible structure for the Southeast Asian cooperative seed procurement and tree improvement program, which would strengthen and coordinate local projects at the regional level, and through this be linked to the global network. He noted that the species currently receiving greatest attention in the region included *Pinus merkusii, Albizia, Eucalyptus, Pinus caribaea* var. *hondurensis, Tectona, Shorea, Pinus kesiya, Araucaria, Agathis, Acacia mangium, Gmelina,* and *Swietenia.* He also estimated the budgetary requirements for the coordinating center of a regional cooperative program.

In a separate article, Nikles (1980b) stresses the need to retain expert local staff for six to eight years and to develop their expertise in technical work to a high level. He summarized as follows a survey of coniferous tree breeding programs in the tropics in 1977–1978:

1. Breeding programs operate in many countries, but often without communication between countries.

2. Large variations exist in resources available for breeding programs, the stage reached, breeder expertise, and the type of material needed to develop high-yielding cultivars.

3. Some breeding programs have reached an advanced stage, and samples of their products could be of benefit in new or less-developed programs.

4. Even advanced programs could benefit from testing and use of their material elsewhere.

5. There are common problems: securing gene resources through their natural ranges; documentation of seed sources, including detailed information on genetic composition; devising efficient breeding strategy; developing technology for breeding procedures and seed production; obtaining and handling data efficiently; researching problems such as family-environmental interactions and maximizing seed production in orchards; undertaking studies of wood quality; and training personnel.

6. Seed of a chosen species sometimes cannot be produced locally and is often unavailable in large quantities internationally.

7. The failure of tree breeders to achieve more than localized successes in genetic improvement of the major plantation species is not because of a lack of technology but because of a lack of its application and a failure to make widespread use of the improved material that has been developed.

Roche goes on to stress the importance of cooperation in tree breeding, citing such examples from the United States as the North Carolina State University–Industry Cooperative Tree Improvement Program, which concentrates on *Pinus taeda* and has been in operation for more than twenty years; the Florida Cooperative, which works mostly on *Pinus elliottii;* and the Western Gulf Forest Tree Improvement Program, mainly for loblolly pine.

In the Association of Southeast Asian Nations (ASEAN), Canada has subsequently instituted a cooperative program centered on seed, and New Zealand and Australia have also been involved in very close cooperation. Japanese cooperation, on the other hand, has centered on afforestation, and research has recently been expanded, but it is yet to come to grips with the more fundamental question of seed.

At present, Japanese funding for cooperation related to tropical forests is said to exceed that of the FAO's Forestry Department. In the fields of afforestation and ecology, Japanese work has been acclaimed internationally, and since the 1990s international cooperative projects in the field of forest breeding have begun in Indonesia, Uruguay, China, and elsewhere, and it is pleasing to see that breeding operations are beginning to run smoothly.

## 3. In-Situ Conservation

### *Difficulty of In-Situ Conservation*

In 1975, with support from the United Nations Environment Program (UNEP), FAO launched a project on Conservation of Forest Genetic Resources, encompassing both in-situ and ex-situ conservation. For the latter, progress was made in seed collection for conservation and the establishment of ex-situ conservation stands in several countries in Africa and Asia. Progress in in-situ conservation, however, lagged noticeably behind, which led FAO to organize an Expert Consultation on In Situ Conservation of Forest Genetic Resources, which convened in December 1980 (FAO 1981). Its purpose was to draft guidelines for selection and management of in-situ genetic conservation areas.

The Consultation pointed to the need for more detailed guidelines for in-situ conservation, and these were drawn up by Roche and Dourojeanni (1984), who gave a detailed discussion of the principles and practice of gene resource conservation. Published at about the same time were a treatment of the scientific and technical base of in-situ conservation (Ingram 1984) and a status review and action plan for in-situ conservation (FAO 1984). More recently, an overview of in-situ and ex-situ conservation has appeared (Palmberg 1987).

The principles outlined in these papers can be applied to all countries and are well regarded; but local conditions include human elements that are often problematic. For example, it is necessary to post guards to patrol the in-situ conservation areas, but because they are poorly paid they may neglect their duties and even take bribes to allow entry to the forest. To anyone familiar with local conditions, such examples are commonplace. Palmberg (1987) points out that conservation policies for forest genetic resources that have been too hastily conceived and are at variance with local conditions have caused unnecessary friction. He holds that conservation policies will not succeed unless local residents perceive such policies to be beneficial to the region. He states that first priority should be given to tree species used for timber production or those with high economic value, centering on endangered species. For necessary population size, he sets targets of 50 trees in the short term and 500 in the long term. He also states that in selection of populations within a species, it is necessary to search for intraspecific genetic variations arising from variations in habitat. Further, because genetic resources are a common heritage of humankind, there is a need for (1) base collection of plant genetic resources, (2) status survey of in-situ conservation, (3) an international information system, and (4) training in the fields of plant genetic resources, plant breeding, and seed production.

A report by FAO (1975) summarizes the methodology of conservation of forest genetic resources. It neatly arranges basic information, much of which is subsequently duplicated in Roche and Dourojeanni (1984). From a technical standpoint, it cites the following principles:

1. Genetic resource conservation should be seen comprehensively as one form of forest management.

2. Because of large local variations in tree species, and also in administrative systems, conservation should accord with local conditions, with natural ecosystems conserved in situ, artificial stands ex situ, and seed in seed banks.

3. Basic maps should be used.

4. In-situ conservation of entire ecosystems is most desirable.

5. Forest gene resources should be conserved in association with the conservation of wildlife, watersheds, and national parks.

6. Strict forest reserves should be established, surrounded by a buffer zone in which forest exploitation and tourism are permissible.

7. Consideration must be given to intraspecific variation and minimum numbers of breeding individuals for a viable gene pool, and a judgment made whether to establish one large reserve or several separate small reserves.

8. In areas to be cleared for agriculture or other use, seed from endangered species should be collected and deposited in seed banks or conserved ex situ in forest plantations in safe locations.

From the administrative and financial standpoints, the following recommendations were made: (1) priority in research and international cooperation should be given to the tropical and subtropical zones and the Mediterranean arid zone; and (2) because of a shortage of funds in many developing countries, international funding programs and cooperation are necessary.

## *Area and Population Size Requirements*

Opinion is divided on the area necessary for genetic resource conservation. Richards (1971) states that 100 ha is necessary for conservation of typical tree species of Malaysia; van Steenis (1971) proposes 500 ha; Hedberg and Hedberg (1968) cite 1,000 ha for a forest reserve in Africa; Shanklin (1951) gives 400 ha for one type of conservation in North America; the Forest Service Natural Area in America has a minimum of 120 ha; and Franklin and Trappe (1967) give 200 ha. The above figures, while having no scientific basis, range from 100 to 1,000 ha. Ashton (1976), on the other hand, from species number–area curves for Sarawak, set a standard of 2,000 ha containing 200 trees, which is probably an appropriate figure. Of course, for species with wide latitudinal and altitudinal ranges, one location is inadequate, and several sites should be obtained, which take into account ecological variation.

The next question is population size, and on this the works of Stern and Roche (1974), Koski (1974), and others are instructive.

Intraspecific variation has been extensively investigated by use of isozymes, but nothing definite has emerged concerning effective population size. As a rule, the northern conifers are anemophilous and xenogamous, and therefore the effective population in a continuous forest is large, Toda (1965) giving a figure of 10,000. Southern trees, on the other hand, are largely pollinated by insects, birds, bats, and the like, while species represented by one or two individuals per ha show various breeding systems and self-pollination is general, which means that effective breeding populations are smaller than in temperate forest. In practice, however, few concrete figures for effective populations are available. From the reports of Franklin (1980), Frankel and Soule (1981), Burley and Namkoong (1980), Marshall and Brown (1975), and others, figures emerge of 50 trees in the short term and 500 in the long term.

Further, because of so-called edge effects, it is proposed that a buffer zone be established around a conservation area.

Roche (1975a) cites the following ten areas for future investigation:

1. Seed storage and testing
2. Intraspecific variation and its causes
3. Breeding systems
4. Phenology of fruiting and flowering
5. Effective population size
6. Size of strict natural reserves to ensure integrity of the ecosystem
7. Size of buffer zone surrounding such reserves
8. Sexual and vegetative propagation
9. Feasibility of seed orchard establishment for particular species
10. Management of natural forest ecosystems in the tropics

While such themes represent the broader picture, I have made the following specific proposals concerning the problems of tropical forest conservation (Yamada 1982). These take into account ecological conditions and can be considered the minimum requirement for a data base.

1. Ecosystem survey: Subdivision of conservation areas on the basis of vegetation and soil surveys, and establishment of permanent plots in each subdivision. Ongoing surveys of productivity and material cycling within the permanent plots. Surveys to track the process of seedling growth, and regeneration studies on logging sites and naturally open land.

2. Superior genetic resources survey: identification of plus trees and long-term recording of such habits as flowering and fruiting, leaf fall, pollination, and growth by periodic growth survey and detailed phenological observation. Research into distribution of species. For species forming pure stands, ecological and phenological surveys of the community and analysis of family differences by means of isozymes and the like.

3. Preservation and trial of seed and seedlings: collection and preservation of bumper-year seeds, germination trials, collection of natural seedlings, and growth experiments on density and illumination.

4. Seed forests, seed orchards, and scion gardens: for a stable seed supply, designation of in-situ seed forests for species forming pure stands. Establishment of ex-situ seed orchards and scion gardens. Either method requires research into breeding systems.

5. Preservation of materials and data: organization and preservation of seed, herbarium specimens and wood samples, and data on growth environment of parent materials and various experimental data. For ex-situ conservation, seed

stores, facilities for preservation of plant material, clonal orchards, arboreta, botanic gardens, and so on.

6. Information exchange: together with exchange of information between conservation areas, establishment of an information center for Southeast Asia at Bogor for collection of all information and linkage with similar centers and networks in South America and Africa. Besides information, materials, too, could be distributed.

## Strict Conservation Areas

In tropical forest, the number of trees of each species is small, but the existence of small clumps encourages cross-pollination. Also, dioecious species, namely, absolutely xenogamous species, are common in normal tropical forest ecosystems. In Sarawak, 26 percent of 711 species were found to be dioecious, a high proportion compared with 2 percent for Great Britain and 5 percent for seed plants worldwide. These dioecious species are found in the shrub layer, and the vectors are mainly birds, insects, bats, and small animals, rarely wind. The interdependence of the fauna and flora of the rain forest is often the result of coadaptation. For example, each of the 40 species of *Ficus* in Central America has its own unique vector. Introduced *Ficus* does not produce seed in the absence of a vector. This means that if the ecosystem is severely disturbed, there will cease to be a sufficient number of vectors to ensure fertilization, and species will become extinct.

In the tropics, wind dispersal of seed is rare, though there are such exceptions as *Bombax buonopozense* and *Ceiba pentandra*. Fruits high in food materials are predominant, and dispersal by animals is common.

Like teak, eucalypts, and others, the *Khaya grandifolia* of Nigeria has a wide distribution and shows intraspecific variation, which militates against seed preservation independent of the ecosystem.

In such instances, the criteria for selection of strict conservation areas come into question. Examples from Nigeria and Kenya are instructive.

Nigeria
1. The areas must contain adequate samples of typical major ecological formations, which exhibit major disturbance.

2. The areas must support species of plants of outstanding interest or great rarity.

3. The areas must contain endangered species or species undergoing genetic impoverishment.

4. The areas must be accessible but not too close to highways, plantations, or settlements.

5. The areas must be sufficiently large to prevent the vegetation type from being disrupted by change in the vegetation surrounding it.

Kenya

1. Potential areas are demarcated on aerial photographs.

2. A reconnaissance survey is carried out in the demarcated area to locate the particular area suited for constitution as a Strict Nature Reserve.

3. Approval is sought from the Chief Conservator of Forests of the state concerned to constitute the selected portion of the forest reserve as a Strict Nature Reserve.

4. When approval is given, the area is surveyed and its boundaries demarcated by cement pillars.

5. The vegetation is described, and a detailed list of the flora of the reserve is made. A profile of the forest structure is drawn, and the soil type is determined.

6. A file is opened for the reserve, and all subsequent records are maintained in this file.


These methods will be more or less the same for any country. In practice, however, the strict conservation areas are in name only, since their boundaries are being ignored and the forests are being felled. Let me cite examples I have seen in Indonesia.

An extreme instance is the Gumai Pasmah conservation area in South Sumatra. This is a protected forest of montane vegetation located on a medium-sized mountain at about 1,500 m in the center of South Sumatra. Following a year of frost damage to coffee in Brazil, the area around Lahat in South Sumatra became a major coffee-producing region, and production gradually expanded to higher ground. During a survey in 1978, I found that coffee had extended to an altitude of 1,400 m, there was no sign of boundary markers and no awareness among local residents that this was a forest reserve, and the natural forest had been felled for coffee cultivation. The Nature Conservation Agency official who accompanied me commented that the forest in this region would probably disappear within a few years.

The Gede-Pangrango Strict Nature Reserve in West Java was the first area of its kind in Indonesia, and it has been strictly protected from the Cibodas Botanic Garden at 1,400 m up to the summit at 3,000 m. Management has passed from the National Biological Institute to the Ministry of Forestry, fences have been erected in areas of frequent intrusion, and mountain climbers are allowed entry only through a single guarded gate. On both sides of the reserve, steep cliffs drop down to rivers, forming natural barriers that make intrusion difficult. People nevertheless often come to trap birds, and the sound of logging can be heard. Watchmen constantly patrol the forest and attempt to dissuade local residents from entry. The fact that the forest remains on the Gede-Pangrango mountains even though the surrounding mountains have been totally cleared, and in the worst places swiddens have been made at up to 2,400 m, is due to these management systems and, it should not be overlooked, the frequent presence of researchers. The Bogor Botanic Garden is a center for tropical biology and is visited by many researchers, a proportion of whom also visit Cibodas and inspect the montane forest. Anything untoward would always be reported to the relevant authorities. A report in the international *Flora Malesiana Bulletin* would be particularly effective.

Another example is Ujung Kelong, at the tip of West Java. This is the last remaining habitat of the Java rhinoceros and is frequently visited by World Wildlife Fund observers. Tourism, too, has increased recently. It is also significant that there is not one settlement in this area of 40,000 ha. A single village formerly located on the coast was swept away following the eruption of Krakatoa. Access from the closest village takes half a day by sea or two days by land, and every strategic point has a guard post. All of the houses in the reserve are for research purposes. It has recently been announced that the Java rhinoceros, once close to extinction, is increasing in numbers. Nevertheless, poachers seeking rhinoceros horn are sometimes caught.

There appears to be little future in establishing strict nature reserves simply by marking them on a map. It is difficult in the short term to change the attitudes of local residents; and, indeed, a balance needs to be drawn in "protecting" natural forest from the impoverished local inhabitants who have long been collecting forest products on a daily basis.

Probably the best protection of a forest reserve is afforded by regular visits by researchers and conservation workers. But for researchers to visit a forest reserve, it must contain sufficient materials of scientific interest and be worthy of regular survey over the long term. There are no longer many such places. In the vast tropics, only small areas of such worthwhile natural forest remain. In this situation, areas of in-situ conservation of genetic resources must overlap with ecosystem conservation areas.


# 4. Ex-Situ Conservation

## *Viewpoints of Ex-Situ Conservation*

It is clear that ex-situ conservation should progress more effectively than in-situ conservation. Nevertheless, it is not easy in the tropics to carry out effective ex-situ conservation.

Guldager (1975) presents two arguments for ex-situ conservation. One is that the security of seed supply of promising exotic species is imperative for tree breeding and plantation programs. The second is that, because of expanding exploitation of natural resources, the genetic potential of many species is in danger of being eroded. He presents basic genetic considerations in determining strategies in ex-situ conservation: (1) to establish and maintain conservation stands characterized as far as possible by the same genotype frequencies as the original population—static conservation of genotypes, by vegetative propagation; (2) to establish and maintain conservation stands characterized as far as possible by the same gene frequencies—static conservation of gene pools, in this instance with seed collection; (3) to establish conservation stands where gene frequencies are allowed to change freely according to natural selective forces—evolutionary conservation; and (4) to establish conservation stands in which gene frequencies are deliberately changed by man—selective conservation. To maintain genetic integrity in ex-situ conservation, it is necessary to consider (1) the sampling of genotypes (seed) in the original population; (2) the survival and growth of sampled genotypes ex situ; and (3) the mating between sampled genotypes ex situ.

The first question focuses on population sampling. For a given stand of conifers, 15–25 unstressed trees are selected at random and 10–50 liters of cones are collected. In this instance, even if collection is done with a grid system along an environmental gradient, it is not known whether the collection site represents the region where the population grows. Also, the sample size is too small. And the fact that only genotypes that have fruited are collected is also problematic. Nevertheless, in practice, sample size is probably determined by local conditions, labor, and funds. And priority is given to species of high economic value, those from which reprocurement of seed is difficult, and endangered species. Desirable sites for ex-situ conservation stands should afford high seed production, freedom from pollen contamination, and safety from damage by humans, animals, fire, erosion, flooding, and so on, but have easy access, a ready supply of labor, and be suited for mechanical maintenance. These criteria match exactly those employed in Japan in, for example, establishing a seed orchard. Ideally, if a plant or forest is facing extinction in a certain larger region, a conservation stand should be sited within the same climatic zone, in an area with similar topography, soil, and other conditions, that is fairly close to human habitation but is not highly urbanized and has considerable room to spare. Further, to spread risk, the same material should not be in one place only but scattered in several locations.

### Methods of Ex-Situ Conservation

Methods of ex-situ conservation include arboreta, botanic gardens, clone orchards, seed orchards, scion gardens, seed stores, and tissue stores. The former three center on conservation rather than reproduction, and the former two in particular combine educational purposes with conservation. In the past, botanic gardens often performed functions derived from the concept of the medicinal herb garden, and today they play an important role in the ex-situ conservation of valuable genetic resources, with frequent exchanges of seed between them. The clone orchards often preserve particularly outstanding phenotypes by grafting. In Japan, ten grafts are made for each plus tree selected, which are subsequently thinned to leave five. Nevertheless, this places pressure on available land, and it will be necessary to secure a considerable area in advance to avoid this problem.

Seed orchards and scion gardens have the purpose of propagation, and the clones employed should show outstanding phenotypes representative of their respective regions. With seed orchards, consideration must be given to the placement of clones so that random natural crossing can occur. Pruning and topping require intensive labor, making mechanization indispensable, which in turn requires an adequate planting interval. In Japan, this is 5 × 5 m for *Cryptomeria japonica* and 7 × 7 m for pine; in Sweden it is 5 × 5 m for conifers; and in America, 5 × 5 m for slash and taeda pines, sometimes 10 × 10 m, and the maximum seed yield per hectare is said to be obtained with 8 × 8 m.

As many clones as possible should be planted together, and an effort should be made to ensure that the characters of the seeds produced are as uniform as possible. The following three conditions, as far as feasible, should be met:

1. The chances of crossing between different clones should be equal.
2. The chances of crossing within the same clone should be minimized.
3. The above conditions should remain unchanged after thinning.

While the chances of crossing are governed by such factors as the landform and relief of the seed orchard, wind direction, and differences in the flowering conditions and flowering periods of each clone, basically, the distances between clones should be equal and the same clones should not be adjacent to each other.

A seed orchard should be located close to a considerable labor force, on land that is not steeply sloping and has high soil fertility, and as far as possible from forest containing the same species. In Sweden, conditions are set at a minimum 100 m from inferior trees and at least 1,000 m from an inferior forest. I believe a separation of 500 to 1,000 m is desirable, and if this is not possible, at least 100 m. Dispersal distances of pollen, both of internal and external origin, come into question, and in the tropics there are few instances in which these have been measured. In the temperate and cool temperate zones, maximum distances of 300 to 500 km have been recorded for pines, while species with distances of around 50 km include spruce, birch, and beech. In experiments in Japan with a variety of *Cryptomeria japonica,* the effective range of pollination was surprisingly small, around 10–20 m. This has bearing on the planting distance between clones, and also on whether machinery can be introduced.

The area of a seed orchard ideally should match the volume of seed required, but at least 1 ha is necessary. Smaller areas are more susceptible to external influences and are inconvenient; larger areas allow more seedlings to be planted and increase the probability of the even mixing of pollens. A circular or square shape is best. The number of clones should be from 9 to 25. If possible, orchards should be established in several separate locations to distribute risk.

For scion gardens, species with superior rooting ability are desirable, like *Cryptomeria japonica* in Japan. In the tropics, however, the establishment of scion gardens must await progress in development of the basic technology for vegetative propagation. In particular, it is necessary to investigate the extent to which tropical trees with their outstanding growth characteristics can tolerate topping and pruning. If results are favorable, scion gardens will probably be useful for tropical broad-leaved trees with short seed-storage periods.

The same is true for clone orchards. The first requirement here is to develop the technology for propagation by grafting. The number of trees preserved will depend on available space, but a minimum of 5 should be left. At the Kanto Forest Tree Breeding Institute, 10 trees are planted and later thinned to leave 5. In temperate zones, the problem of space is great, but in the tropics a considerable area can often be secured. If plans are made in advance, it should be possible to establish adequate conservation stands.

Arboreta and botanic gardens have a long history, and there are many outstanding examples such as the Bogor Botanic Garden in Indonesia. These do not simply arrange clones in rows but, with Bogor, display together specimens of the same line group from around the world. The locations for conservation are chosen to suit the climatic zones of growth, with tropical species in Bogor, slightly cool temperate zone species in Cibodas, and monsoon zone species in Malang. Today, with air-conditioned greenhouses, tropical and savanna vegetation is displayed even in temperate zones.

Many botanic gardens in Europe began as gardens for medicinal herbs, and like Kew Botanic Gardens in England, they served mainly for the introduction of tree species providing useful products, such as rubber and quinine, which contributed to the national economy. Bogor Botanic Garden still has the parent trees of the oil palm that spread across Southeast Asia. Botanic gardens also play important roles in education and tourism. Recently, with the question of genetic resources looming large, the significance of botanic gardens and arboreta is being reconfirmed as a basis for advanced bioengineering. Many botanic gardens exchange seed and are providing adjunct facilities for seed storage. In the developed countries, this is being taken further with the provision of comprehensive systems from recording to conservation and distribution, making use of computers and the latest air-conditioning equipment.

## *Directions of Basic Research*

A vast number of problems remain concerning the ex-situ conservation of tropical forest resources. In the thirty-year history of forest tree breeding in Japan, a
huge amount of basic survey was necessary for the selection of plus trees, the
creation of seed orchards, and the establishment of progeny test plantations,
even for a small number of species, which include *Cryptomeria japonica, Chamaecyparis obsusa, Pinus densiflora, Larix leptolepsis,* and *Abies sachaliensis.* Even so, many
projects had to be started without gathering all the necessary information. With
such a situation even in the temperate zones, the number of surveys required in
the tropics, where species numbers and varieties are far more diverse, is somewhat daunting, but problems nevertheless must be tackled and solved one by
one. To this end, the following basic viewpoint is necessary.

### Choice of Species

In the tropics, it will be necessary to limit work to species that are taxonomically
confirmed. Basically, these will include species facing extinction, species of high
economic potential, and those already being commercially exploited. In addition, I would like to include tree species with a high regenerative capability. At
present, much work has been done on the tropical pines, eucalypts, *Acacia,* teak,
and others, but little on other local species. In particular, apart from the work by
the Forest Research Institute of Malaysia, virtually nothing of note has been done
on the Southeast Asian dipterocarps.

### Urgent Research Themes

For each species, research is required into geographic distribution, phenology,
breeding system, seed storage, vegetative propagation, and shade tolerance.

*Geographic distribution.* Researchers need to examine centers of distribution and
isolated areas. Detailed surveys are also required to determine the ecological environments of distribution along latitudinal and altitudinal gradients and the relation between the mode of distribution within a community and other species in
the community, as well as analysis of races by means of isozymes and other means.

*Phenology.* Unlike the simple phenology of temperate zones, in the tropics, the
seasonal variation depending on species, individual, and part of the individual is
great. It is necessary to examine geographic variation not only in characteristics
of species but also in quantitative seed production, and to consider the linkage
with climatic variation. As discussed earlier, reports have appeared on many tree
species in Southeast Asia, and in general the overstory tends to show a clear

seasonality, while the understory does not. Phenology has applications in breeding and is related basically to the field of reproductive biology, and as such it will continue to be an important field of study.

*Breeding system.* Forest breeding in temperate areas is considered to center mainly on pollination depending on wind dispersal, but in the tropics a variety of pollination vectors operate, including wind, insects, small animals, and water. Research into this topic has been conducted only in the past two decades in Central and South America by reproductive biologists of the American school, while in Southeast Asia very little has been done. Research in this field will be required in the future.

*Seed storage.* While the seeds of tropical pines, *Acacia,* eucalypts, and others are said to be storable for more than fifteen years, those of the dipterocarps and other broad-leaved trees have a short viability of only a few weeks. Wang (1975) has discussed the storage of seed and pollen, while Sasaki (1976) and Ng (1977b) have described the problems of Malaysian trees. At that time, the maximum survival period of dipterocarp seeds was ten weeks, and subsequent trials at around 15°C have extended this to no more than six months. Further work on this problem is required.

*Vegetative propagation.* Since forestry practice premised on regeneration is a recent innovation, propagative techniques by means other than seed are particularly underdeveloped. It is necessary to develop basic technology for cutting, grafting, and air layering. Tissue culture also seems likely to become an important theme in the future. In Japan, propagation using embryos has been attempted for *Quercus acutissima,* which is difficult to propagate vegetatively, and only acclimation remains a problem. In the future it appears necessary to conduct such tissue culture in the laboratory in parallel with field techniques.

*Shade tolerance.* Although tropical silviculture is widespread, shade tolerance and density trials in seedling beds are disproportionately lacking. The former are particularly important for species groups, like the Dipterocarpaceae, which change their habit depending on the stage of growth, and experiments with different degrees of illumination due to shading are strongly recommended.

Even for ex-situ conservation, then, the problems are monumental. To tackle such tasks, the appropriate facilities and personnel, support organizations, and funds are required; and at present most developing countries rely in large measure on overseas aid, which is seldom adequate. While these basic problems remain unresolved, many species are being lost daily.

# 5. Examples of Ex-Situ Conservation

## *Eucalyptus deglupta*

Davidson (1977) has reported in detail on the genetic conservation of *Eucalyptus deglupta*. In general, the strategy leading to full use of existing genetic resources involves five phases: exploration, collection, evaluation, conservation, and utilization. Exploration is further divided into botanic exploration, which focuses on taxonomic identification and range of distribution, and genecological exploration, which studies ecological and phenotypic variation within the natural range.

Let us look first at botanic exploration. *Eucalyptus deglupta* was first described by Rumphius (1743) based on material from Seram collected in 1668, which he called *Arbor versicolor*. Subsequently, it was described validly by Blume (1849) under the name *Populus deglubata,* and six further synonyms have been published. This early confusion arose from the inadequacy and variability of the materials collected and a lack of communication between botanists and herbaria. The natural range of *E. deglupta,* which was ascertained in detail in the 1960s, extends from Mindanao in the Philippines, through Sulawesi, Seram, and Irian Jaya in Indonesia, to New Britain in Papua New Guinea. Trials of this species have been conducted in Australia, British Honduras, Ecuador, Fiji, Gold Coast, Hawaiʻi, India, Indonesia, Kenya, Malaysia, Madagascar, Nigeria, Papua New Guinea, the Philippines, the Solomon Islands, South Africa, Surinam, Tanzania, Thailand, Trinidad, and Uganda.

Genecological exploration in the 1980s included studies of variations in morphology and growth within the natural range. Surveys covered tree height and shape, bark, and dimensions and shape of buds, fruit, and leaves, with remarkable variation found in leaves. Although the variations in leaves did not warrant the establishment of subspecies, two varieties can be recognized: *E. deglupta* var. *schlechteri,* having short, broad leaves, and *E. deglupta* var. *deglupta,* having long, narrow leaves with pronounced drip-tips.

Collection was combined with or followed on from genecological exploration. In 1969, rangewide collections of relatively small samples of seed were made from nine sites in Mindanao, Sulawesi, and Papua New Guinea. These were distributed from Australia and Papua New Guinea for evaluation in provenance trials in various countries. For reforestation of wet tropical lowlands, seeds from New Britain, Sulawesi, and Mindanao gave good results.

Conservation measures for this species are required for four reasons:

1. It has great potential for industrial fiber production on deforested tropical lowland. Its rapid growth rate and excellent wood properties for high-quality timber have attracted various forest agencies in the tropics.

2. *E. deglupta* is intolerant. Older stands are invaded by rain forest, preventing its regeneration.

3. Mature stands are being heavily cut over for building, fence and canoe timbers and firewood.

4. This species occurs on fertile river flats and levee banks, which are rapidly being cleared for swiddening and other agricultural uses.

Methods of conservation include in-situ and ex-situ conservation and storage of germ plasm, of which in-situ conservation is the most desirable. The conditions for in-situ conservation include the following:

1. Stands should be relatively pure and extensive to minimize encroachment by rain forest species.

2. They should be fully protected from natural and manmade hazards.

3. The genetic resource should be available for collection and use both within and outside the country of origin.

4. Necessary actions should be taken immediately, including legislation, marking of boundaries, public relations, and education.

In the Philippines, the government has declared an area of *E. deglupta* as a forest reserve, and commercial interests have reserved areas for future seed collection.

Logically, the area surrounding a forest reserve should also be conserved, although little is known about the minimum area needed to form a viable unit for conservation. Because *E. deglupta* is a widespread species possessing great diversity within its natural range, a series of reserves representing different provenances is preferable to a single large reserve.

Ex-situ conservation can be considered where populations are endangered but in-situ conservation is unlikely. Two provenances from Mindanao have been established in blocks of up to 10 ha each at three sites in Papua New Guinea.

Storage of seed is difficult for *E. deglupta,* and even under refrigeration its viability is largely lost after two years.

Concerning utilization, information is being gathered about what provenances of seed are most suitable for what areas. Seeds have been supplied by governments and commercial operators. It is recommended that compact blocks of at least 15 ha of the best adapted provenances be established in importing countries to serve as a seed source and a basis for local breeding. For efficient selection of the first generation of plus trees, it is necessary to have about 500 ha of plantations of good provenance. Programs for tree improvement are well advanced in Papua New Guinea, but in other countries they have scarcely started, because of lack of expertise and shortage of finance.

## Pinus kesiya and Pinus merkusii

Turnbull (1972) has reported on *Pinus kesiya* in the Philippines, including its distribution, habits, and seed sources. This pine occurs on Luzon Island from 15°30' N to 18°15' N at elevations of 450–2,450 meters. The main areas of distribution are approximately 300,000 hectares in the Cordillera Central Mountains of northern Luzon, and small stands in the Caraballo Mountains (10,000 hectares) and Zambales Mountains. These areas have a monsoon climate with a dry season of from five to seven months. The dry season is longer at lower elevations, from December to April, while at high elevations it is milder, lasting only for the two months of February and March.

*Pinus kesiya* forms extensive pure stands, which may be dense at higher elevations and more open at lower elevations. The most vigorous stands occur in the cool, moist areas at 1,500–2,000 m, where the tree can reach a height of 45 m. This pine can colonize clearings in broad-leaved forest caused by swiddening, fire, erosion, and other factors, but at low altitudes its regeneration is influenced by overfelling, fire, and grazing. It was first used as mining timber, and also for house construction and as fuel. Racial variations of this pine can probably be found in the isolated stands of the Zambales and Caraballo Mountains, while in the Cordillera Central Mountains variations in such characters as growth and drought resistance can be found between lower and higher elevations. The presence also of phenotypic variants within a stand indicates that selection would be advantageous. At present, seeds are collected without regulation from the logging area near Baguio.

*P. kesiya* is a complex of three-needled Southeast Asian pines and is thought to include *P. khasya* Royle, which is found in Assam, Tibet, Burma, Laos, Yunnan, Vietnam, and elsewhere, *P. insularis* Endlicher from the Philippines, *P. langbianensis* A. Chev from southern Vietnam, and probably *P. yunnanensis* from China. In the Philippines it is commonly called Benguet pine, occurring mainly in Benguet Province. It occurs on a variety of soil types, generally yellow clay or clay-loam, and in various depths of soil. It originally occurred as scattered stands in areas of evergreen broad-leaved forest that, because of swiddening and other disturbance by humans, has been replaced by pure pine forest.

This tree is large and well formed, with neither buttresses nor taper. It grows rapidly, reaching more than 30 m high with mean diameters in the range of 40–60 cm. The largest trees occur in the Boboc region at elevations over 1,800 m, where the largest specimen recorded was 45.7 m tall and 1.6 m in DBH. Growth is rapid for the first twenty-five years, slowing thereafter. Considerable variation is seen in its branching habit, and because this is under genetic control, improvement is possible.

Mature cones are borne toward the ends of branches, singly, in twos or threes.

They are normally ovate and symmetrical, 5–10 cm long, 2.5–4.0 cm wide, and light brown when ripe. Variations in size are considerable. Cones from Kapaya in Nueva Vizcaya Province and Langangilang in Abra Province are large, while those from Zambales Province are small. Cones ripen from October to March, most seed being shed in January and February. In Boboc, trees first produced cones at twelve to fifteen years, in Zambales at fifteen to eighteen years, showing regional and individual variations. Seed weight has been estimated variously, at 52,900 seeds/kg (Cooling 1968), and 59,000–76,300 seeds/kg (Turnbull 1972). Germination rate is high, about 80 percent (60–90 percent), with most seed germinating in seven to ten days under suitable conditions, and is unaffected by refrigeration before germination.

In logging areas, parent trees are left at a rate of 15 trees/ha. To facilitate seed dispersal these are left along mountain ridges, and the outcome has been successful. Planting in the forest also has been carried out. Seeds are replanted into bamboo tubes after broadcasting in seed beds, or sown directly in bamboo tubes. After four to six months in the shade, they are transplanted into the forest.

Breeding operations have only just begun, and plus trees are beginning to be chosen. In the mountains of northern Luzon, these pines are readily apparent, being as conspicuous as the *Pinus densiflora* of Japan. In-situ conservation will be difficult because of swiddening in the area. Also, because this is a secondary forest species, it is conceivable that in-situ conservation stands will be replaced in the long term by climax species. So, while in-situ conservation is necessary for forests of exceptional quality, this will be difficult under the given local conditions. Ex-situ conservation appears to be more effective, with as many variants as possible chosen along latitudinal and altitudinal gradients in each region. A review of this species has appeared from Oxford University, giving detailed descriptions including the state of planting in various parts of the world (Armitage and Burley 1980).

Another Southeast Asian pine is *Pinus merkusii* Jungh. and de Vriese, which has been studied by Cooling (1972). *Pinus merkusii* is a two-needled pine distributed from northeastern India through Burma, Thailand, Laos, Cambodia, Vietnam, Sumatra, and Hainan to the Philippines. It was discovered by Junghuhn in the Batak region of Sumatra, where it was known as *toesam*. He provisionally named it *Pinus sumatrana*. He sent herbarium material to Governor Merkus, who passed it on to de Vriese, who applied the name *Pinus merkusii* Junghuhn and de Vriese when he published the first botanical description in 1845.

*P. merkusii* is a variable species, and recently the name *Pinus merkusiana* sp. nov. has been proposed for the mainland populations (Cooling and Gaussen 1970). In Sumatra, morphological variations between Aceh and Tapanuli populations are readily apparent; and the Philippine *P. merkusii,* while said to resemble the mainland provenances, requires further taxonomic study. This variation will prove extremely important in future trials and improvements. In particular, the grass stage of *P. merkusiana* requires suitable seed beds and grafting techniques, and the length of the period of weeding and other factors are also problematic. In Zambia, *P. merkusiana* has proven difficult to establish in the subhumid zone, but once height growth has begun, it proceeds rapidly. In Java, *P. merkusii* grows extremely well in moist areas, far outpacing the mainland provenances. Morphological differences between the mainland and Sumatran provenances in the early stages of growth are also widely recognized in the results of international provenance trials (Burley and Cooling 1972).

Concerning seed collection from these two types of pines, Keiding (1972) gives detailed opinions based on his long experience in cooperative works between Thailand and Denmark. Demand for both species is high. *P. kesiya* favors higher altitudes of 800–2,000 m, while *P. merkusii* favors lower altitudes up to about 1,000 m. The fact that the latter species in particular can be grown at low altitude has drawn attention to the types without a grass stage from Mindoro in the Philippines, the Kirirom plateau in Cambodia, the Srisaket area of eastern Thailand, and Sumatra. Sampling is necessary from various altitudes and from relatively limited areas for the relation between such factors as different soil types and mixed forests. The problems involved in such sampling can be described as follows:

*Accessibility.* War has prevented entry into Vietnam, Laos, and Cambodia, and entry into Burma has not been permitted. Most mountains are inaccessible to motor vehicles, and the alternatives are several days on foot or horseback. Where there are roads, roadside sampling alone is insufficient, while in the interior there are swiddens and logging areas.

*Temporal factors.* Seed collection takes time, sometimes spanning several seasons, necessitating arrangements whereby seed can be collected continuously from accessible locations.

*Information.* On *P. merkusii,* Cooling's (1968) monograph is good; for *P. kesiya,* preparations are being made at Oxford.

*Regional cooperation.* Without the cooperation of the regions involved, efficiency will not increase in Southeast Asia. It is necessary to clarify direct and indirect benefits and to communicate objectives clearly to each country. Although every country has its forestry program, their seed sources are not specified. It is important to promote local forestry breeding programs and research. In Thailand, such programs were instituted in 1965 for teak and in 1969 for pine.

Seed is more readily obtainable in large quantity for *P. kesiya,* and its germination rate is high (70–80 percent). For *P. merkusii*, results show considerable scatter, and the germination rate is sometimes as low as 10–30 percent. For the former, the times of flowering, fruiting, and cone ripening are regular and seasonal; but for the latter, much remains unknown about cone development.

According to Bryndum (1974), the cones of *P. merkusii* generally ripen from mid-February to May, but this may vary from one region to another and also from one tree to another. In germination trials, the highest rate was obtained when the cones were brown and the scales just beginning to separate. Collection of young cones reduced germination rates. Cones take three to four months to ripen, and to ensure that collections can be made at the optimum time, it is best to chose a site where repeated visits are possible.

*P. kesiya,* on the other hand, gives good germination even with green cones. The season for collection is from January to February in Thailand, and from November to February in the Philippines. A climber is employed for collection, and simple climbing equipment can be used. In Sumatra, the season of cone maturation is irregular, being sparsest in December, followed by May, October, January, and February. In the Takegon area, seed was collected in the dry season in July and August.

The tropical pines, even those of the same species, thus show irregularity of flowering season from one individual to another, in the same way as the Dipterocarpaceae. It is particularly interesting that this tendency is strong in *P. merkusii*, which favors the humid tropics. Examination of a large sample, however, possibly would reveal that flowering and fruiting tend predominantly to concentrate at the end of the dry season. A program to designate parent trees and conduct regular surveys of their phenology would be beneficial.

## *Central American Pines*

A similar situation is found concerning the pines of Central America (Kemp 1972). The area of natural pine forest in Central America exceeds 45,000 km², consisting of 27,000 km² in Honduras, 14,000 km² in Guatemala, 4,000 km² in Nicaragua, and smaller areas in El Salvador and British Honduras. These natural forests are not dense, pure stands; more than half of the total consists of open, savanna-like pine forest, mixed pine forest, or mixed forest with broad-leaved trees, and the accessible dense forest has already been felled. The natural distribution is continuous from the northwest Atlantic coast of Nicaragua at 12°13' N to the Mexican border at 18° N. In Mexico, the number of species decreases progressively southward, and in Nicaragua only three species are found (*P. caribaea* Morelet, *P. oocarpa* Schide, and *P. pseudostrobus* Lindl). *P. caribaea* and *P. oocarpa* grow over a wide range of environments, and are found particularly in areas influenced by humans. Morphological variations are considerable, but detailed research has not been conducted.

*P. caribaea* occurs exclusively as var. *hondurensis* Barr. et Golf, and it appears in two distinct ecological zones. One is represented by coastal lowlands on the Atlantic side at 12°13' N. This area is flat alluvium with a high groundwater table, subject to seasonal flooding and permanently marshy, and about 1 km inland. Annual rainfall is 4,000 mm, monthly rainfall is below 76 mm in only one month, and in no month does the relative humidity drop below 70 percent. Pines here are small, scattered among grassland, gramineous wetland, and small forests of broad-leaf trees. Such pine forests can be found intermittently along the coast to the Mexican border at 18° N. Annual rainfall in this range is from less than 2,000 mm to 3,200 mm. Some forests are also found away from the coastal alluvia, inland on well-drained, gravel-like soils at an elevation of about 100 m.

Inland forests, on the other hand, occur under a much drier climate, typically standing on the well-drained lower slopes of the spinal mountain ranges, and also in montane valleys. *P. caribaea* is found at elevations below 800 m, and *P. oocarpa* at higher elevations. Average annual rainfall in these areas is 1,600 mm or less, and the dry season is far more severe than on the Atlantic coast. *P. caribaea* grows in areas of annual rainfall up to 900 mm and can withstand a long dry season.

Thus, *P. caribaea* occurs in two markedly different locations, a wet zone and a dry zone.

*P. oocarpa* is extremely widely distributed in Central America. It grows in areas with annual rainfall of 2,000 mm or less, at elevations of 300–2,000 m, on shallow, stony, acidic soils that dry out seasonally. The southern limit of its natural distribution is at 12°45' N. It is distributed continuously northward from Sierra de Diepilto in Honduras over an area measuring 160 km north-south and 300 km east-west, and because it grows on a plateau with fairly steep slopes that impede access, it has escaped logging. A particularly fine forest stretches northward from the Nicaraguan border to Baja Verapaz in Guatemala.

This pine is characterized by its appearance in very dry areas, growing even where annual rainfall is less than 900 mm and the dry season is long and harsh, with six consecutive months of less than 100 mm and only 25 mm in four of those months. On the other hand, it also grows in areas with 2,000 mm of rainfall annually and no dry season. Its morphological variations are great, but detailed research is lacking.

Finally, *Pinus pseudostrobus* is distributed in Honduras and Nicaragua. It appears on high land at above 1,200 m with annual rainfall of more than 1,500 mm, while in northwest Honduras it also appears at 850 m. Its southward distribution is similar to that of *P. oocarpa,* and at Volcan de Yali (13°15' N), extremely large, well-formed trees grow on volcanic soils at 1,200 m.

We turn now to seed collection from these three types of pines. For the lowland *P. caribaea,* the extreme north-south orientation of its distribution means that it will suffice to collect from ten or twelve locations at intervals of 60 km or 0.5 degrees of latitude, using boat or aircraft. For the inland species, again ten to twelve provenances will be necessary, with particular attention paid to extremely dry locations. *P. oocarpa* should be selected by latitude, with consideration given to provenances at the southern limit of distribution, extremely dry areas, and different altitudes. Further investigation is needed in the extensive area of distribution in central Honduras, and more intensive sampling will be required. Collection here should accord with the size of the population. Finally, *P. pseudostrobus* is the most scattered, but five or six provenances should suffice, with particular attention paid to low elevations.

For the former two of these three species, demand exceeds supply domestically and internationally, and prospects are good. Small-scale conservation is practiced in each country, and it can be considered that, in terms of international conservation of genetic resources, endangered populations growing at the limits of the range will be conserved.

## 6. Future Prospects

Kemp et al. (1976) note that while international efforts directed toward genetic resource conservation are increasing, in practice, food production takes precedence in the poorer tropical countries, with the result that social and economic problems locally pose a greater obstacle to in-situ conservation than the technical problems of conserving species. They point out that, in practice, therefore, the problems of genetic resources should be tackled in conjunction with the objectives of food production, water resource development, national parks, tourism, education, and others; that even for ex-situ conservation, the number of tree species is extremely limited because of shortages of funds and personnel; and that international cooperation is needed that combines funds and personnel. The same is true of Southeast Asia, and there is no task so easily stated yet so difficult in practice as the conservation of genetic resources. The gap between ideal and reality can successfully be bridged only by the accumulation of modest endeavors over a long period. These should be carried out, moreover, not by individuals or small groups but as national or international operations. Traditionally, in this field, it has been the Oxford Forestry Institute in England that has achieved results over the long term. Here, based on the stockpile from the colonial era, information is gathered from around the world, and through annual summer courses and acceptance of overseas students, new information is continually being added. The project begun in the 1960s with the exploration and collection of Central and South American pines has now reached the stage where results are beginning to emerge from provenance trials of seeds that were distributed worldwide.

I visited this institute in September 1989 and met with people who had been active in the field. Immediate work on the Central and South American pines had been completed, and it remained only to enter the results of worldwide provenance trials into the data base. Data already accumulated had been fed into a data base for each species, and a system was in place to allow primary selection of species suited to given environmental conditions.

The Oxford Forestry Institute has the most voluminous collection of information worldwide, with 200,000 volumes of books, 1,800 periodical and 4,000 nonperiodical magazine titles, and 3,500 reels of microfilm. Based on the information in this library, the institute prepares abstracts in cooperation with the Commonwealth Agricultural Bureau and independently publishes basic monographs on tropical forests. In preparing a monograph, the staff read an average of three hundred to four hundred articles over a period of four months and themselves prepare abstracts, which are first collected together. Based on these, monographs are then prepared for each species containing detailed descriptions from taxonomy to use. These publications are highly regarded as essential reading on the world's tropical gene resources.

The institute is now beginning work on the African *Acacia* group, focusing on a limited number of species and surveying their ranges of distribution. The work encompasses not only gene resources but also ecology, conservation, and use, and the systematic collection of leaf, wood, and other specimens. At the same time, several researchers spoke to me of the many problems they encountered in the field, particularly in exploration, collection, provenance trials, and other aspects of the work on gene resources. Because provenance trials are entrusted to the countries involved, the results are often questionable, and many trials have ended in failure. Nevertheless, because of the large number of trials conducted, the project is still fully viable.

Besides its history of more than twenty-five years, the institute draws an unassailable strength in this field from Britain's past colonial connections with many countries. Beyond this, the institute's constant efforts to absorb new ideas and invite new personnel from across the world is a matter for serious consideration by us Japanese. Personal connections are of paramount importance, as all who have worked in the tropics would concur. Overseas students and visitors for short study courses always maintain contact with Oxford after returning to their own countries to continue their research. These relationships forge links between research establishments and between nations, and embrace regional communities of several countries. This attitude toward research activities is particularly noteworthy in Britain.

Besides Britain, Denmark, as mentioned earlier, is cooperating with Thailand in setting up forestry breeding stations for teak and pine, which the Thai government is operating smoothly. Canada is active in the ASEAN seed center. Australia is distributing seeds and seedlings of eucalypts and *Acacia* worldwide. In addition, many large commercial enterprises are known to be collecting superior genetic resources worldwide in what is called the "seed war."

Japan, unfortunately, lags far behind in this field. A superbly equipped center for agriculture has been established at Tsukuba, but for forestry, prospects seem distant. Even if only for a limited area or species, it appears high time for Japan to develop some strength.

As an ideal for ex-situ conservation, I envision reserving a natural forest that would later become a gene conservation forest, with a facility for ex-situ conservation located near its buffer zone, modeled on the Hokkaido Forest Tree Breeding Institute and the Cibodas Botanic Garden with suitable improvement. Because trees are long-lived and large, a large area should be set aside in advance, preferably in at least two separate locations. Roche (1975b) recommends about 10–30 ha per provenance in at least two locations. Of Japan's forest breeding stations, the Kanto station is slightly less than 40 ha and seems cramped, although the Tohoku station has 70 ha, which is probably sufficient for conservation of several species, including seed orchards, clone orchards, nurseries, and various facilities. If we assume that 500 provenances are to be preserved with 10 trees per provenance, this means sufficient area will be needed for planting 5,000 trees. Planting density will differ from species to species, but broad-leaved trees require more space than conifers. Even with 500 trees per hectare, this will mean that 10 ha is necessary for a clone orchard for just one species. An additional 4 ha is needed for seedling beds, 4 ha for associated facilities, and 4 ha for seed orchards, and thus a total of 70 ha will accommodate only a handful of representative species. Fortunately, securing the necessary land is often less difficult in developing countries than developed countries, and large areas can probably be reserved. Operation, including planting maintenance and management, will require fixed annual amounts of labor and funds and should therefore be undertaken on a national scale.

My idea is to establish ex-situ conservation facilities in several locations for each forest type. Tropical forest types are various, from coastal mangrove, through peat swamp, lowland Dipterocarp forest, to montane forest, subalpine forest, *kerangas* forest, and so on. Ex-situ conservation facilities would be located in the vicinity of representative forests, from which plus trees of representative species would be chosen for breeding. As to climate, Southeast Asia can be divided into a monsoon zone and a tropical rain forest zone, represented by the following areas:

*Tropical rain forest zone*
 Mangrove: South Sumatra, New Guinea
 Peat swamp forest: Sarawak, Brunei, South Sumatra, New Guinea

Lowland Dipterocarp forest: Kalimantan, Malaysia, Mindanao, Central
  Sumatra
Montane forest: Kinabalu, Cibodas, New Guinea
Kerangas forest: Sarawak

*Monsoon zone*

Mangrove: South Thailand
Freshwater swamp forest: Rangoon
Lowland forest: Khao Yai
Montane forest: Doi Inthanon
Pine: North Sumatra, Chieng Mai
Teak: Chieng Mai

Such facilities would most conveniently be run at the national level, but because international cooperation would be needed, an information center should be established at or near Bogor, as mentioned, to exchange information with similar centers in Central and South America and Africa. These facilities would also, of course, maintain close contact with international bodies and conduct international joint ventures.

This concept is idealistic, and in practice various problems are likely to arise. Nevertheless, research centers have been established at strategic locations and, although small, they have slowly but surely expanded their activities. It is most desirable to make use of these existing research facilities and further expand their activities.

Of the experiment and research organizations concerned with forestry, it is the breeding stations that have materials immediately available and are always engaged in experimentation. Ideally, university experimental plantations and breeding stations should combine and have experimental facilities on a par with forestry experiment stations. This is difficult in the developed countries that already have established facilities, but in the developing countries the possibilities for the future are still great, and it is clear from the present developments in gene resource conservation that the need for such facilities will continue to grow. These should be named "forest gene resource research centers," prefixed by the name of the forest type.

The biggest problem is personnel. Experience in various places has shown that it is more difficult to attract personnel if the facilities are located too far from an urban area. Consequently, facilities are best located within commuting distance of the major city in the area. It is also good to attract not only local staff but also overseas researchers on a regular basis. The Japan International Cooperation Agency has more than a decade of achievements behind it, and in the field of forest research, it has cooperated in research in Thailand, Indonesia, Brunei, and Peru, and it has newly established a cooperative project with Papua New Guinea.

In afforestation projects, it has worked steadily in Thailand, the Philippines, Indonesia, Malaysia, Paraguay, Kenya, and other countries around the world. The fact that the results of this work have not been as highly evaluated as might be expected from the sums invested is because of shortcomings in the cooperative structures and the lack of a single rational axis. In comparison, countries such as Canada, New Zealand, Australia, the United Kingdom, and Holland, while operating on a small scale, have a firm grip on the most basic point—the supply of seed. Requests to Japan for cooperation will undoubtedly continue to increase, and the costs involved will grow year by year. What proportion of this budget is allocated to problems of gene resource conservation, the world's strongest demand at present, will influence the world's evaluation of Japanese cooperation in the future.

Let us now turn to domestic problems, reputedly the most difficult part of international cooperation. Many problems derive from the inflexibility of hierarchic administration and the difficulty of personnel transfer, but the greatest obstacle is a lack of trained personnel. Very few people are willing to spend a lifetime working on the tropical forests, and many give up after their first visit. Even those who wish to return may not easily find the opportunity. Returnees from the Japan Overseas Cooperation Volunteers often cannot find suitable posts. Universities offer few opportunities for overseas research. With regard to accepting foreign students and conducting overseas surveys, there is much waste and inefficiency. In short, from top to bottom, the framework in Japan is not one that allows people to work overseas. Promotion ceases for those who go overseas, and when they return, there often is not a suitable post waiting for them. Even as demands from overseas grow, these problems will not be resolved without major changes at home.

The basic period of residence overseas is currently two years, which is too short. Other countries, particularly in the area of forestry, think in terms of five to eight years, with short periods of leave to the home country allowed as a matter of course. The important point is for the same person to maintain responsibility for a project until it has taken shape to a certain degree. In Japan, longer residence is said to lead to attachments to the locality, and for this reason it is usual for people to be replaced at short intervals, but this argument has no international currency.

In this situation, people who reside overseas for a long period again face problems when they return home. One way to solve this problem is to create an agency that pools personnel required for work overseas. This means not sending back to the regional forestry offices the people who have done outstanding work overseas, as at present, but allowing those who wish to work overseas again to concentrate lifelong on overseas research. At the Tropical Agriculture Research Center, the organization concerned with forestry is too small for the actual situation. This organization should be brought to full strength, perhaps under the new name Tropical Forestry Research Center.

Forestry in tropical countries is the object of cooperation by such countries as the United Kingdom, the United States, West Germany, Holland, Australia, New Zealand, and Canada, each operating their own systems. FAO, UNESCO, and other international bodies are also actively involved. Close cooperation between these organizations will be necessary in the future.

In the training of personnel, universities not only need to expand their facilities for science and research related to the tropics, they should also have overseas campuses and experimental plantations. Overseas students could more effectively receive all their training in the tropics, rather than traveling all the way to Japan to learn their work in an environment that does not match the conditions in their homelands. The United States and the United Kingdom already have overseas bases in various localities, both tropical and temperate. In this, Japan lags far behind. Just as many Japanese industries have factories overseas, it seems to me necessary for Japanese universities, particularly those university institutes that aim at area study, to move toward establishing firm bases overseas. America's Smithsonian Institution has had a base in Panama for more than forty years, and the Organization for Tropical Studies (OTS) has had a base in Costa Rica for more than twenty-five years. These have been engaged in tropical research and education, and have built the strong foundations on which tropical research in the United States today rests.

*Appendix 1*

*Seasonal Changes in Litter Mass of Trees in a 1-ha Plot on Mount Pangrango*

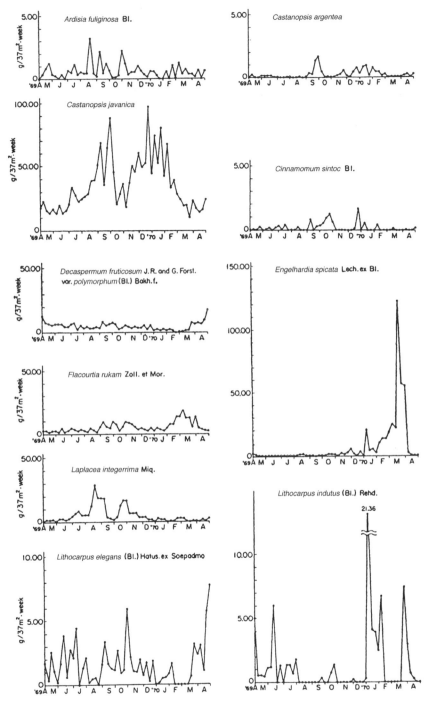

Figure A-1 (1)–(5). Seasonal changes in litter mass of trees in a 1-ha plot on Mount Pangrango.

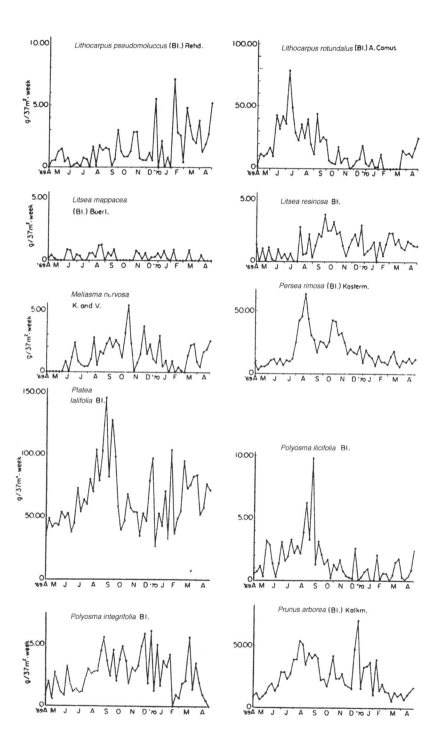

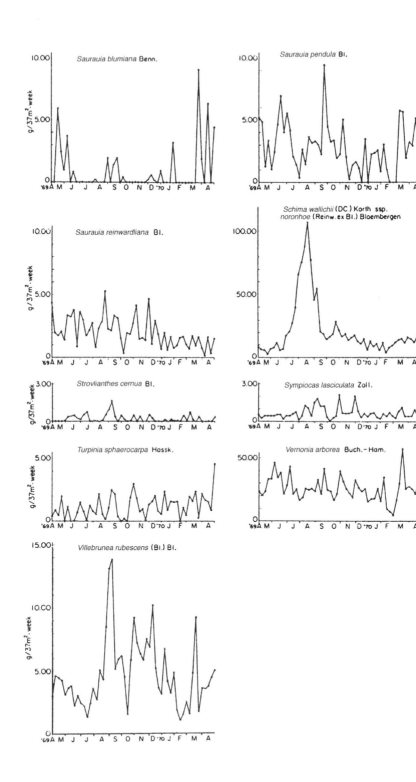

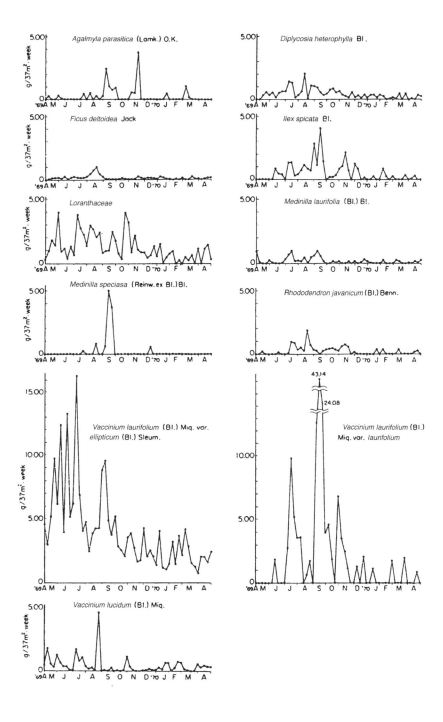

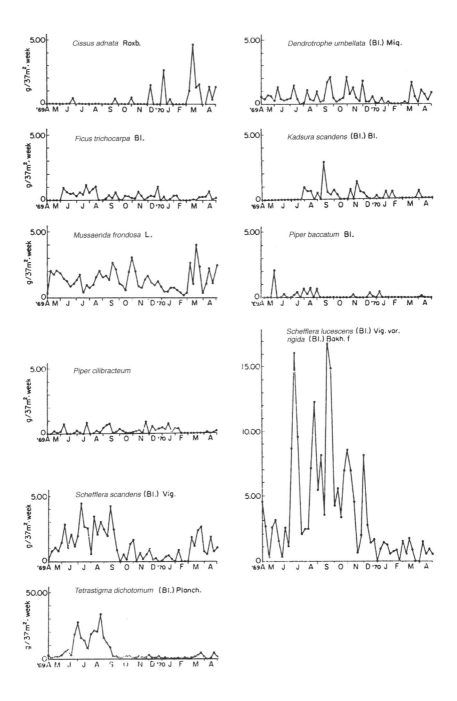

*Appendix 2*
# Survey Results for Species Composition and Structure Study of Nine Plots on Mount Pangrango, West Java

Table A-1

Number (N) and Relative Dominance (R.D.; Relative Basal Area) of All Species
with DBH of 10 cm and above in Mount Pangrango Plot 1

| SPECIES | 1st LAYER | | 2nd LAYER | | 3rd LAYER | | BROKEN TREES | | TOTAL | |
|---|---|---|---|---|---|---|---|---|---|---|
| | N | R.D. (%) | N | R.D. (%) | N | R.D. (%) | N | R.D. (%) | N | R.D. (%) |
| Schima wallichii spp. noronhae (Reinw. ex Bl.) Bloemb. | 35 | 30.57 | 9 | 8.76 | 3 | 1.34 | | | 47 | 25.12 |
| Saurauia pendula Bl. | | | | | 44 | 35.67 | 2 | 1.73 | 46 | 2.24 |
| Castanopsis javanica (Bl.) DC. | 20 | 25.02 | 11 | 10.15 | 6 | 2.54 | | | 37 | 21.01 |
| Persea rimosa (Bl.) Kosterm. | 11 | 6.81 | 13 | 9.59 | 4 | 1.45 | 2 | 29.72 | 30 | 7.61 |
| Turpinia sphaerocarpa Hassk. | | | 5 | 3.27 | 14 | 10.37 | 5 | 5.88 | 24 | 1.23 |
| Litsocarpus pseudomoluccus (Bl.) Rehd. | 9 | 5.29 | 6 | 4.13 | 7 | 3.08 | | | 22 | 4.83 |
| Decaspermum fruticosum var. polymorphum (Bl.) Bakh. f. | | | 13 | 4.86 | 7 | 2.97 | | | 20 | 0.76 |
| Vernonia arborea Buch.-Ham. | 6 | 4.68 | 9 | 7.34 | 4 | 1.63 | | | 19 | 4.64 |
| Symplocos fasciculata Zoll. | | | 2 | 0.78 | 13 | 5.83 | 1 | 1.18 | 16 | 0.49 |
| Polyosma integrifolia Bl. | 3 | 0.61 | 9 | 6.83 | 4 | 1.36 | | | 16 | 1.37 |
| Polyosma ilicifolia Bl. | | | 11 | 9.39 | 4 | 1.47 | | | 15 | 1.20 |
| Ficus ribes Reinw. ex Bl. | | | | | 10 | 4.76 | | | 10 | 0.29 |
| Macropanax dispermus (Bl.) O.K. | 1 | 0.48 | 2 | 4.90 | 4 | 4.44 | | | 7 | 1.23 |
| Flacourtia rukam Zoll. et Mor. | | | 5 | 2.92 | 1 | 0.41 | | | 6 | 0.37 |
| Syzygium antisepticum (Bl.) Merry & Perry | 4 | 2.73 | 2 | 1.83 | | | | | 6 | 2.36 |

(continued on next page)

Table A-1   (continued)

| SPECIES | 1st LAYER | | 2nd LAYER | | 3rd LAYER | | BROKEN TREES | | TOTAL | |
|---|---|---|---|---|---|---|---|---|---|---|
| | N | R.D. (%) | N | R.D. (%) | N | R.D. (%) | N | R.D. (%) | N | R.D. (%) |
| *Astronia spectabilis* Bl. | | | 3 | 1.42 | 2 | 0.75 | | | 5 | 0.21 |
| *Castanopsis argentea* (Bl.) DC. | | | | | 3 | 1.10 | 2 | 1.27 | 5 | 0.11 |
| *Lithocarpus rotundatus* (Bl.) A. Camus | 4 | 5.14 | | | | | 1 | 49.22 | 5 | 5.75 |
| *Mischocarpus fuscescens* Bl. | | | 3 | 2.50 | 2 | 0.99 | | | 5 | 0.36 |
| *Saurauia blumiana* Benn. | | | | | 5 | 4.12 | | | 5 | 0.25 |
| *Villebrunea rubescens* (Bl.) Bl. | | | | | 5 | 2.27 | | | 5 | 0.14 |
| *Antidesma tetrandrum* Bl. | | | | | 4 | 1.79 | | | 4 | 0.11 |
| *Lithocarpus indutus* (Bl.) Rehd. | 3 | 1.78 | 1 | 0.50 | | | | | 4 | 1.45 |
| *Pygeum parviflorum* Teysm. et Benn. | 4 | 1.50 | | | | | | | 4 | 1.17 |
| *Syzygium rostratum* (Bl.) DC. | | | 3 | 2.35 | 1 | 0.35 | | | 4 | 0.30 |
| *Viburnum sambucinum* Bl. | | | | | 3 | 3.36 | 1 | 1.14 | 4 | 0.25 |
| *Acronychia laurifolia* Bl. | | | 3 | 4.94 | | | | | 3 | 0.58 |
| *Casearia tuberculata* Bl. | 1 | 0.33 | 1 | 1.65 | 1 | 0.74 | | | 3 | 0.50 |
| *Eurya acuminata* DC. | | | 1 | 0.29 | 2 | 1.04 | | | 3 | 0.10 |
| *Lithocarpus elegans* (Bl.) Hatus. ex Soepadmo | 1 | 0.96 | | | 2 | 0.83 | | | 3 | 0.81 |
| *Litsea resinosa* Bl. | 3 | 0.86 | | | | | | | 3 | 0.68 |
| *Manglietia glauca* Bl. | | | 2 | 1.11 | 1 | 0.47 | | | 3 | 0.16 |
| *Michelia montana* Bl. | 1 | 0.32 | 2 | 3.78 | | | | | 3 | 0.70 |

(continued on next page)

Table A-1   (continued)

| SPECIES | 1st LAYER | | 2nd LAYER | | 3rd LAYER | | BROKEN TREES | | TOTAL | |
|---|---|---|---|---|---|---|---|---|---|---|
| | N | R.D. (%) | N | R.D. (%) | N | R.D. (%) | N | R.D. (%) | N | R.D. (%) |
| Platea latifolia Bl. | 3 | 4.61 | | | | | | | 3 | 3.62 |
| Acronodia punctata Bl. | | | 2 | 1.86 | | | | | 2 | 0.22 |
| Dysoxylum alliaceum Bl. | | | 1 | 0.95 | 1 | 0.45 | | | 2 | 0.14 |
| Elaeocarpus stipularis Bl. | | | 1 | 0.25 | 1 | 0.27 | | | 2 | 0.05 |
| Engelhardia spicata Lech. ex Bl. | 2 | 2.49 | | | | | | | 2 | 1.96 |
| Hypobathrum frutescens Bl. | | | | | 2 | 0.70 | | | 2 | 0.04 |
| Litsea mappacea (Bl.) Boerl. | | | 1 | 0.91 | 1 | 0.29 | | | 2 | 0.13 |
| Pithecellobium clypearia (Jack) Bth. | | | | | | | 2 | 9.33 | 2 | 0.33 |
| Pyrenaria serrata Bl. | | | | | 2 | 0.54 | | | 2 | 0.03 |
| Saurauia reinwardtiana Bl. | | | | | 1 | 0.25 | 1 | 0.52 | 2 | 0.03 |
| Cinnamomum parthenoxylon (Jack) Meissn. | 1 | 2.05 | | | | | | | 1 | 1.61 |
| Cinnamomum sintoc Bl. | 1 | 0.40 | | | | | | | 1 | 0.31 |
| Glochidion macrocarpum Bl. | | | 1 | 2.17 | | | | | 1 | 0.26 |
| Gordonia excelsa (Bl.) Bl. | 1 | 2.28 | | | | | | | 1 | 1.79 |
| Ilex cymosa Bl. | | | 1 | 0.59 | | | | | 1 | 0.07 |
| Lithocarpus tijsmannii (Bl.) Rehd. | 1 | 0.62 | | | | | | | 1 | 0.48 |
| Macropanax undulatus (Wall. ex G. Don) Seem | | | | | 1 | 0.26 | | | 1 | 0.02 |

(continued on next page)

Table A-1  *(continued)*

| SPECIES | 1st LAYER | | 2nd LAYER | | 3rd LAYER | | BROKEN TREES | | TOTAL | |
|---|---|---|---|---|---|---|---|---|---|---|
| | N | R.D. (%) | N | R.D. (%) | N | R.D. (%) | N | R.D. (%) | N | R.D. (%) |
| *Meliosma ferruginea* Bl. | | | | | 1 | 0.25 | | | 1 | 0.02 |
| *Podocarpus imbricatus* Bl. | 1 | 0.48 | | | | | | | 1 | 0.38 |
| *Prunus arborea* (Bl.) Kalkman | | | | | 1 | 0.27 | | | 1 | 0.02 |
| *Sauvauia nudiflora* DC. | | | | | 1 | 0.53 | | | 1 | 0.03 |
| *Tarenna laxiflora* (Bl.) K. & V. | | | | | 1 | 0.41 | | | 1 | 0.03 |
| *Viburnum coriaceum* Bl. | | | | | 1 | 0.27 | | | 1 | 0.02 |
| *Viburnum lutescens* Bl. | | | | | 1 | 0.38 | | | 1 | 0.02 |
| Total | 116 | 100.00 | 123 | 100.00 | 171 | 100.00 | 17 | 100.00 | 427 | 100.00 |

## Table A-2

### All Species with DBH less than 10 cm and height more than 1.3 m found in Ten Quadrants 10 × 10 m along a Diagonal in Mount Pangrango Plot 1

| Species | No. of Individuals | Frequency | Basal Area (cm²) | Mean Height (m) | SDR % (N'T' BA') | SDR % (N'F' BA'H') |
|---|---|---|---|---|---|---|
| Trees | | | | | | |
| *Castanopsis argentea* (Bl.) DC. | 5 | 3 | 35.86 | 3.9 | 13.9 | 22.7 |
| *Hypobathrum frutescens* Bl. | 4 | 4 | 44.64 | 4.0 | 17.8 | 25.9 |
| *Symplocos fasciculata* Zoll. | 4 | 2 | 29.73 | 4.0 | 9.7 | 19.8 |
| *Decaspermum fruticosum* J. R. & Forst. var. *polymorphum* (Bl.) Bakh. f. | 3 | 2 | 98.66 | 3.8 | 12.8 | 21.5 |
| *Saurauia pendula* Bl. | 3 | 3 | 39.97 | 3.6 | 13.7 | 21.5 |
| *Turpinia sphaerocarpa* Hassk. | 3 | 2 | 82.16 | 5.2 | 13.7 | 25.3 |
| *Apodytes cambodiana* Pierre | 2 | 2 | 28.31 | 5.8 | 9.2 | 25.0 |
| *Astronia spectabilis* Bl. | 2 | 2 | 9.06 | 4.0 | 8.3 | 18.7 |
| *Casearia tuberculata* Bl. | 2 | 2 | 23.83 | 3.5 | 9.0 | 17.7 |
| *Persea rimosa* (Bl.) Kosterm. | 2 | 2 | 65.83 | 7.3 | 11.0 | 31.1 |
| *Veronia arborea* Buch.-Ham. | 2 | 1 | 47.38 | 3.5 | 6.4 | 15.8 |
| *Antidesma tetrandrum* Bl. | 1 | 1 | 69.40 | 8.0 | 7.2 | 30.4 |
| *Eurya acuminata* DC. | 1 | 1 | 51.53 | 7.0 | 6.4 | 26.7 |
| *Ficus ribes* Reinw. ex Bl. | 1 | 1 | 49.02 | 6.0 | 6.3 | 23.5 |
| *Glochidion cyrtostylum* Miq. | 1 | 1 | 10.75 | 4.5 | 4.4 | 17.4 |

*(continued on next page)*

Table A-2   (continued)

| Species | No. of Individuals | Frequency | Basal Area (cm²) | Mean Height (m) | SDR % (N'T' BA') | SDR % (N'F' BA'H') |
|---|---|---|---|---|---|---|
| Glochidion macrocarpum Bl. | 1 | 1 | 2.84 | 4.0 | 4.1 | 15.6 |
| Macropanax dispermus (Bl.) O.K. | 1 | 1 | 14.52 | 6.0 | 4.6 | 22.2 |
| Polyosma integrifolia Bl. | 1 | 1 | 3.14 | 3.5 | 4.1 | 14.0 |
| Litbocarpus pseudomoluccus (Bl.) Rehd. | 1 | 1 | 43.01 | 8.0 | 6.0 | 29.5 |
| Litbocarpus tijsmannii (Bl.) Rehd. | 1 | 1 | 1.13 | 2.0 | 4.0 | 9.3 |
| Schima wallichii (DC.) Korth. ssp. noronhae (Reinw. ex Bl.) Bloembergen | 1 | 1 | 18.86 | 8.0 | 4.8 | 28.6 |
| Shrubs | | | | | | |
| Strobilanthes cernua Bl. | 145 | 9 | 180.90 | 2.3 | 75.2 | 63.6 |
| Ardisia fuliginosa Bl. | 63 | 9 | 205.85 | 3.0 | 57.5 | 52.5 |
| Talauma candollii Bl. | 8 | 4 | 33.34 | 4.0 | 18.4 | 26.3 |
| Saurauia reinwardtiana Bl. | 4 | 2 | 86.10 | 5.4 | 12.4 | 26.2 |
| Viburnum lutescens Bl. | 3 | 2 | 82.00 | 6.0 | 12.0 | 27.7 |
| Claoxylon glabrifolium Miq. | 1 | 1 | 0.95 | 2.0 | 4.0 | 9.2 |
| Lasianthus sp. | 1 | 1 | 3.46 | 3.0 | 4.1 | 12.5 |
| Saurauia blumiana Benn. | 1 | 1 | 5.73 | 3.5 | 4.2 | 14.1 |
| Palms | | | | | | |
| Pinanga coronata (Bl. ex Mart.) Bl. | 31 | 5 | 244.67 | 3.9 | 37.2 | 40.1 |

(continued on next page)

Table A-2  (continued)

| Species | No. of Individuals | Frequency | Basal Area (cm²) | Mean Height (m) | SDR % (N'T' BA') | SDR % (N'F' BA'H') |
|---------|--------------------|-----------| ----------------|------------------|------------------|---------------------|
| Tree ferns | | | | | | |
| Cyathea raciborskii Copel. | 37 | 9 | 705.17 | 3.1 | 75.2 | 66.1 |
| Cyathea contaminans (Wall. ex Hook.) Copel. | 1 | 1 | 86.59 | 7.5 | 8.0 | 29.5 |
| Herbs | | | | | | |
| Nicolaia solaris (Bl.) Horan. | 24 | 1 | 117.84 | 3.0 | 14.8 | 20.5 |
| Cyrtandra sp. | 5 | 1 | 3.95 | 1.5 | 5.0 | 8.5 |

## Table A-3

### All Plants less than 1.3 m high found in Ten Quadrants 1 × 1 m along a Diagonal in Mount Pangrango Plot 1

| SPECIES | FRE-QUENCY | NO. OF INDIVID-UALS | FRESH WEIGHT INCLUDING ROOT (g) | MEAN HEIGHT (cm) | SDR (F'N' H') |
|---|---|---|---|---|---|
| Trees | | | | | |
| *Saurauia pendula* Bl. | 1 | 2 | 16 | 11.0 | 7.7 |
| Shrubs | | | | | |
| *Strobilanthes cernua* Bl. | 8 | 35 | 1957 | 72.5 | 62.0 |
| *Psychotria divergens* Bl. | 4 | 5 | 251 | 42.8 | 29.8 |
| *Ardisia fuliginosa* Bl. | 3 | 5 | 334 | 63.3 | 31.3 |
| *Rubus moluccanus* L. | 2 | 3 | 45 | 30.0 | 17.4 |
| *Saurauia reinwardtiana* Bl. | 1 | 1 | 19 | 30.0 | 12.7 |
| *Talauma candollii* Bl. | 1 | 1 | 9 | 25.0 | 11.3 |
| Woody climbers | | | | | |
| *Tetrastigma* sp. | 3 | 3 | 14 | 4.3 | 14.4 |
| *Cissus adnata* Roxb. | 2 | 2 | 94 | 5.0 | 10.2 |
| *Ficus lanata* Bl. | 2 | 2 | 37 | 8.5 | 11.2 |
| *Ficus trichocarpa* Bl. | 1 | 1 | 47 | 20.0 | 10.0 |
| *Piper baccatum* Bl. | 1 | 1 | 12 | 13.0 | 8.0 |
| *Piper cilibracteum* DC. | 1 | 1 | 4 | 10.0 | 7.2 |
| *Tetrastigma papillosum* (Bl.) Planch. | 1 | 1 | 4 | 5.0 | 5.8 |
| Herbaceous climbers | | | | | |
| *Smilax* sp. | 1 | 1 | 10 | 40.0 | 15.5 |
| Woody creepers | | | | | |
| *Piper sulcatum* Bl. | 2 | 2 | 30 | 30.0 | 17.2 |
| *Piper* sp. | 1 | 1 | 14 | 15.0 | 8.6 |
| Non-woody creepers | | | | | |
| *Malaxis* sp. | 1 | 1 | 2 | 6.0 | 6.1 |
| *Scindapsus hederaceus* (Z. & M.) Miq. | 1 | 1 | 28 | 10.0 | 7.2 |
| Orchids | | | | | |
| A species of Orchidaceae | 2 | 2 | 11 | 12.5 | 12.3 |
| Palms | | | | | |
| *Pinanga coronata* (Bl. ex Mart.) Bl. | 2 | 2 | 676 | 85.0 | 32.4 |

*(continued on next page)*

Table A-3   *(continued)*

| SPECIES | FRE-QUENCY | No. OF INDIVID-UALS | FRESH WEIGHT INCLUDING ROOT (g) | MEAN HEIGHT (cm) | SDR (F'N' H') |
|---|---|---|---|---|---|
| **Ferns** | | | | | |
| *Cyathea raciborskii* Copel. | 3 | 4 | 1089 | 71.7 | 33.4 |
| *Thelypteris heterocarpa* (Bl.) Ching | 2 | 3 | 92 | 35.0 | 18.6 |
| *Nephrolepsis acumimata* (Houtt.) Kuhn | 2 | 3 | 15 | 15.0 | 13.2 |
| *Asplenium* sp. | 1 | 1 | 12 | 20.0 | 10.0 |
| *Diplazium* sp. | 1 | 3 | 30 | 20.0 | 10.5 |
| *Diplazium pallidum* (Bl.) Moore | 1 | 1 | 27 | 20.0 | 10.0 |
| *Dryopteris hirtipes* (Bl.) O. K. | 1 | 1 | 22 | 20.0 | 10.0 |
| *Dryopteris* sp. | 1 | 2 | 70 | 30.0 | 13.0 |
| *Egenolfia appendiculata* (Willd.) J. Sm. | 1 | 1 | 4 | 15.0 | 8.6 |
| *Trichomanes* sp. | 1 | 2 | 2 | 7.0 | 6.3 |
| *Diplazium dilatatum* Bl. | 1 | 1 | 620 | 110.0 | 35.0 |
| **Herbs** | | | | | |
| *Elatostema paludosum* (Bl.) Hassk. | 7 | 137 | 2285 | 20.7 | 68.2 |
| *Cyrtandra picta* Bl. | 6 | 18 | 507 | 30.0 | 37.7 |
| *Zingiber inflexum* Bl. | 3 | 7 | 35 | 17.7 | 19.1 |
| *Arisaema filiforme* Bl. | 2 | 3 | 26 | 25.0 | 16.0 |
| *Amomum hochreutineri* Val. | 1 | 1 | 32 | 30.0 | 12.7 |
| *Coleus galeatus* (Poir.) Bth. | 1 | 1 | 200 | 120.0 | 37.7 |
| *Cyrtandra* sp. | 1 | 6 | 540 | 60.0 | 22.3 |
| *Elatostema* sp. | 1 | 18 | 355 | 20.0 | 14.1 |
| *Forrestia marginata* (Bl.) Back. | 1 | 3 | 24 | 20.0 | 10.5 |
| *Forrestia* sp. | 1 | 1 | 20 | 35.0 | 14.1 |
| *Lycianthes laevis* (Dunal) Bitt. | 1 | 1 | 2 | 20.0 | 10.0 |
| *Musa acuminata* Colla | 1 | 2 | 2 | 4.0 | 5.8 |
| *Amomum pseudo-foetens* Val. | 1 | 1 | 65 | 35.0 | 14.1 |

## Table A-4

## Occurrence of Epiphytes in Relation to Diameter Class of Host Trees in 0.5 ha of Mount Pangrango Plot 1

| SPECIES NAME | AVERAGE NUMBER OF EPIPHYTIC SPECIES PER EPIPHYTE-BEARING TREE OF DIAMETER CLASS (cm) | | | | | | | | | | NUMBER OF EPIPHYTE-FREE TREES OF DIAMETER CLASS (cm) | | | | | | | | | |
|---|---|---|---|---|---|---|---|---|---|---|---|---|---|---|---|---|---|---|---|---|
| | 10 | 20 | 30 | 40 | 50 | 60 | 70 | 80 | 90 | 100 | 10 | 20 | 30 | 40 | 50 | 60 | 70 | 80 | 90 | 100 |
| *Schima wallichii* ssp. *noronhae* (Reinw. ex Bl.) Bloembergen | 1 | 1 | 3.5 | 3.2 | 6 | 4.8 | 4.3 | 7 | 4.5 | | 2 | 1 | | | | | | | | |
| *Saurauia pendula* Bl. | 2.5 | 1.7 | | | | | | | | | 7 | 2 | | | | | | | | |
| *Castanopsis javanica* (Bl.) DC. | 2 | 1.3 | 3.5 | 2 | | 3.5 | 5 | | 6 | 5 | 3 | 3 | 1 | 1 | | | | | | |
| *Persea rimosa* (Bl.) Kosterm. | 3 | | 3 | 2 | | 4 | | | | 8 | 2 | 1 | 1 | | | | | | | |
| *Turpinia sphaerocarpa* Hassk. | 3.5 | 5 | | | | | | | | | 4 | 1 | | | | | | | | |
| *Litocarpus pseudomoluccus* (Bl.) Rehd. | | 5 | 2.5 | 3.5 | 4 | 7 | 3 | 5 | | | | | | | | | | | | |
| *Decaspermum fruiticosum* var. *polymorphum* (Bl.) Bakh. f. | 2 | 1 | | | | | | | | | 4 | | | | | | | | | |
| *Vernonia arborea* Buch.-Ham. | | | 3 | | | | | | 3.5 | | | 1 | 1 | | | | | | | |
| *Symplocos fasciculata* Zoll. | | 3 | | | | | | | | | 11 | | | | | | | | | |
| *Polyosma integrifolia* Bl. | 1 | | 3 | | | | | | | | 2 | | 1 | | | | | | | |
| *Polyosma ilicifolia* Bl. | 2 | 3 | 3.7 | | | | | | | | 5 | | 1 | | | | | | | |
| *Ficus ribes* Reinw. ex Bl. | 1 | | | | | | | | | | 7 | | | | | | | | | |
| *Macropanax dispermus* (Bl.) O.K. | 1.5 | | 2 | | 5 | | | | | | 1 | | | | | | | | | |
| *Flacourtia rukam* Zoll. et Mor. | 1 | | | | | | | | | | 1 | 2 | | | | | | | | |
| *Syzygium antisepticum* (Bl.) Merry & Perry | | | | | | | | | | | | 1 | | | | | | | | |
| *Astronia spectabilis* Bl. | | 1 | | | | | | | | | 1 | | | | | | | | | |

*(continued on next page)*

Table A-4 *(continued)*

| SPECIES NAME | AVERAGE NUMBER OF EPIPHYTIC SPECIES PER EPIPHYTE-BEARING TREE OF DIAMETER CLASS (cm) | | | | | | | | | | NUMBER OF EPIPHYTE-FREE TREES OF DIAMETER CLASS (cm) | | | | | | | | | |
|---|---|---|---|---|---|---|---|---|---|---|---|---|---|---|---|---|---|---|---|---|
| | 10 | 20 | 30 | 40 | 50 | 60 | 70 | 80 | 90 | 100 | 10 | 20 | 30 | 40 | 50 | 60 | 70 | 80 | 90 | 100 |
| *Castanopsis argentea* (Bl.) DC. | | | | | | | | | | | 3 | | | | | | | | | |
| *Lithocarpus rotundatus* (Bl.) A. Camus | | | | | | 4 | | | | | | | | | | | | | | |
| *Mischocarpus fuscescens* Bl. | | 3 | | 6 | | | | | | | 2 | | | | | | | | | |
| *Villebrunea rubescens* (Bl.) Bl. | 1 | | | | | | | | | | 2 | | | | | | | | | |
| *Antidesma tetrandrum* Bl. | 1 | | | | | | | | | | 2 | | | | | | | | | |
| *Lithocarpus indutus* (Bl.) Rehd. | | | | | | | 6 | | | | 1 | 1 | | | | | | | | |
| *Pygeum parviflorum* Teysm. et Benn. | | | 2.5 | | | | | | | | | | | 1 | | | | | | |
| *Syzygium rostratum* (Bl.) DC. | | | | | | | | | | | 1 | | | | | | | | | |
| *Viburnum sambucinum* Bl. | 2 | | | | | | | | | | 2 | | | | | | | | | |
| *Acronychia laurifolia* Bl. | 2 | | 1 | 3 | | | | | | | | | | | | | | | | |
| *Casearia tuberculata* Bl. | 1 | | 3 | | | | | | | | | | | | | | | | | |
| *Eurya acuminata* DC. | 3 | | | | | | | | | | 1 | | | | | | | | | |
| *Lithocarpus elegans* (Bl.) Hatus. ex Soepadmo | | | | | | 6 | | | | | | | | | | | | | | |
| *Litsea resinosa* Bl. | | | 3 | | | | | | | | | 1 | 1 | | | | | | | |
| *Manglietia glauca* Bl. | 1 | | | | | | | | | | 1 | 1 | | | | | | | | |
| *Michelia montana* Bl. | | | | 3 | | | | | | | | | 1 | 1 | | | | | | |
| *Platea latifolia* Bl. | | | | 2 | | | | | | | | | | | | | | 2 | | |
| *Acronodia punctata* Bl. | 2.5 | | | | | | | | | | | | | | | | | | | |

*(continued on next page)*

Table A-4  (continued)

| SPECIES NAME | AVERAGE NUMBER OF EPIPHYTIC SPECIES PER EPIPHYTE-BEARING TREE OF DIAMETER CLASS (cm) | | | | | | | | | | NUMBER OF EPIPHYTE-FREE TREES OF DIAMETER CLASS (cm) | | | | | | | | | |
|---|---|---|---|---|---|---|---|---|---|---|---|---|---|---|---|---|---|---|---|---|
| | 10 | 20 | 30 | 40 | 50 | 60 | 70 | 80 | 90 | 100 | 10 | 20 | 30 | 40 | 50 | 60 | 70 | 80 | 90 | 100 |
| Elaeocarpus stipularis Bl. | 2 | | | | | | | | | | 1 | | | | | | | | | |
| Hypobathrum frutescens Bl. | | | | | | | | | | | 1 | | | | | | | | | |
| Litsea mappacea (Bl.) Boerl. | | 2 | | | | | | | | | 1 | | | | | | | | | |
| Pithecellobium clypearia (Jack) Bth. | | 1 | | | | | | | | | | | | | | | | | | |
| Pyrenaria serrata Bl. | 2 | | | | | | | | | | | | | | | | | | | |
| Saurauia reinwardtiana Bl. | 2 | | | | | | | | | | | | | | | | | | | |
| Cinnamomum parthenoxylon (Jack) Meissn. | | | | | | | | | | 5 | | | | | | | | | | |
| Cinnamomum sintoc Bl. | | | | 4 | | | | | | | | | | | | | | | | |
| Glochidion macrocarpum Bl. | | | | 1 | | | | | | | | | | | | | | | | |
| Gordonia excelsa (Bl.) Bl. | | | | | | | | | | 6 | | | | | | | | | | |
| Macropanax undulatus (Wall. ex G. Don) Seem | | | | | | | | | | | 1 | | | | | | | | | |
| Meliosma ferruginea Bl. | | | | | | | | | | | 1 | | | | | | | | | |
| Podocarpus imbricatus Bl. | | | | | | 3 | | | | | | | | | | | | | | |
| Pygeum arboreum (Bl.) Endl. ex F. v. M. | | | | | | | | | | | 1 | | | | | | | | | |
| Tarenna laxiflora (Bl.) K. & V. | 1 | | | | | | | | | | | | | | | | | | | |
| Viburnum coriaceum Bl. | 1 | | | | | | | | | | | | | | | | | | | |
| Viburnum lutescens Bl. | | | | | | | | | | | 1 | | | | | | | | | |

Table A-5

Terrestrial Small Trees, Shrubs, and Palms of Mount Pangrango

| Species | P-1 (1,600) | P-2 (1,700) | P-3 (1,900) | P-4 (2,100) | P-5 (2,300) | P-6 (2,400) | P-7 (2,600) | P-8 (2,800) | P-9 (3,000) |
|---|---|---|---|---|---|---|---|---|---|
| Acer laurinum Hassk. | | | | + | | | | | |
| Acronodia punctata Bl. | | | | | + | | | | |
| Ardisia fulginosa Bl. | + | | + | | | | | | |
| Ardisia javanica DC | | | | | | + | | | |
| Astronia spectabilis Bl. | | | | + | | | | | |
| Daphniphyllum glaucescens Bl. | | | | | | + | | | |
| Dichroa sylvatica (Reinw. ex. Bl.) Merr. | | | | | | + | | | |
| Eupatorium riparium Reg. | | | + | | | | | | |
| Eurya obovata (Bl.) Korth. | | | | | | + | | | |
| Hypobathrum frutescens Bl. | | + | | | | | | | |
| Lasianthus sp. | | | | + | | | | | |
| Litbocarpus pseudomoluccus (Bl.) Rehd. | | + | | | | | | | |
| Litsea diversifolia Bl. | | + | | | | | | | |
| Molineria capitulata (Lour.) Herb. | | + | + | + | | | | | |
| Mycetia cauliflora Reinw. | | + | | + | | | | | |
| Myrsine affinis DC. | | | | | | + | | + | + |
| Piper sulcatum Bl. | | | + | | | | | | |

(continued on next page)

Table A-5   *(continued)*

| SPECIES | P-1 (1,600) | P-2 (1,700) | P-3 (1,900) | P-4 (2,100) | P-5 (2,300) | P-6 (2,400) | P-7 (2,600) | P-8 (2,800) | P-9 (3,000) |
|---|---|---|---|---|---|---|---|---|---|
| *Podocarpus imbricatus* Bl. | | | | | + | + | + | | |
| *Polyosma ilicifolia* Bl. | | | | + | | + | | | |
| *Polyosma integrifolia* Bl. | | | | | | + | | | |
| *Psychotria divergens* Bl. | + | + | + | | | | | | |
| *Rubus alpestris* Bl. | | | | + | | | | + | |
| *Rubus moluccanus* L. | + | + | | | | | | | |
| *Saurauia pendula* Bl. | + | | + | | | | | | |
| *Saurauia reinwardtiana* Bl. | + | | | | | | | | |
| *Schima wallichii* ssp. *noronhae* (Reinw. ex Bl.) Bloemb. | | | | + | + | | | | |
| *Strobilanthes blumei* Bremek. | | | + | | | | | | |
| *Strobilanthes cernua* Bl. | + | + | + | | | | | | |
| *Symplocos sessilifolia* (Bl.) Gurke | | | | | + | | | + | |
| *Symplocos* sp. | | | | | | + | | | + |
| *Talauma candollii* Bl. | + | | | | | | | | |
| *Turpinia sphaerocarpa* Hassk. | | + | | | | | | | |
| *Vaccinium laurifolium* var. *ellipticum* (Bl.) Sleum. | | | | | | + | | | |
| *Vaccinium laurifolium* (Bl.) Miq. var. *laurifolium* | | | | | | + | | | |
| *Viburnum coriaceum* Bl. | | | | | | + | | | |

## Table A-6
### Terrestrial Climbers of Mount Pangrango

| SPECIES | P-1 (1,600) | P-2 (1,700) | P-3 (1,900) | P-4 (2,100) | P-5 (2,300) | P-6 (2,400) | P-7 (2,600) | P-8 (2,800) | P-9 (3,000) |
|---|---|---|---|---|---|---|---|---|---|
| *Cissus adnata* Roxb. | + | | | | | | | | |
| *Clematis smilacifolia* Wall. | | | + | | | | | | |
| *Ficus lanata* Bl. | + | | | | | | | | |
| *Ficus sagittata* Vahl. | | + | | | | | | | |
| *Ficus trichocarpa* Bl. | + | | | | | | | | |
| *Freycinetia insignis* Bl. | | + | | | | | | | |
| *Hoya* sp. | | | + | | | | | | |
| *Lonicera javanica* (Bl.) DC. | | | | | | + | | | + |
| *Malaxis* sp. | + | | | | | | | | |
| *Medinilla laurifolia* (Bl.) Bl. | | | + | + | | | | | |
| *Nertera granadensis* (Mutis ex L.f.) Druce | | | | | | + | | + | + |
| *Piper abbreviatum* Opiz | | | + | | | | | | |
| *Piper baccatum* Bl. | + | | | | | | | | |
| *Piper cilibracteum* DC. | + | | | | | | | | |
| *Piper sulcatum* Bl. | + | | | | | | | | |
| *Piper* sp. | + | | | | | | | | |
| *Psychotria sarmentosa* Bl. | | | + | | | | | | |

*(continued on next page)*

Table A-6   (continued)

| SPECIES | P-1 (1,500) | P-2 (1,700) | P-3 (1,900) | P-4 (2,100) | P-5 (2,300) | P-6 (2,400) | P-7 (2,600) | P-8 (2,800) | P-9 (3,000) |
|---|---|---|---|---|---|---|---|---|---|
| Rubus lineatus Reinw. ex Bl. | | | | | | | | | + |
| Scindapsus hederaceus (Z. & M.) Miq. | + | | | | | | | | |
| Smilax macrocarpa Bl. | | | + | | | | | | |
| Smilax odoratissima Bl. | | | + | | | | | | |
| Smilax sp. | + | | | | | | | | |
| Tetrastigma dichotomum (Bl.) Planch. | | + | | + | | | | | |
| Tetrastigma papillosum (Bl.) Planch. | + | | | | | | | | |
| Tetrastigma sp. | + | | | | | | | | |

## Table A-7
### Terrestrial Ferns of Mount Pangrango

| SPECIES | P-1 (1,600) | P-2 (1,700) | P-3 (1,900) | P-4 (2,100) | P-5 (2,300) | P-6 (2,400) | P-7 (2,600) | P-8 (2,800) | P-9 (3,000) |
|---|---|---|---|---|---|---|---|---|---|
| Asplenium caudatum Forst. | + | | | | | | | | |
| Asplenium sp. | + | | | | | | | | |
| Athyrium japonicum (Thunb.) Copel. | | + | | | | | | | |
| Athyrium sp. | | | | | | | | | + |
| Blechnum patersonii (R. Br.) Mett. | | | | | | + | | + | |
| Blechnum sp. | | | | | + | | | | |
| Coniogramme fraxinea (Don) Diels | | + | | | | | | | |
| Coniogramme sp. | | | + | + | | | | | |
| Cyathea junghuhniana (Kunze) Copel. | | + | | | | | | | |
| C. latebrosa (Wall. ex Hook.) Copel. | | | | + | + | | | | |
| C. orientalis (Kunze) Moore | | | | + | | | | | |
| C. raciborskii Copel. | + | | | | | | | | |
| C. spinulosa Wall. ex Hook. | | + | | | | | | | |
| Davallia sp. | | | | | | + | | | |
| Diplazium dilatatum Bl. | + | | | | | | | | |
| D. esculentum (Retz.) Sw. | | + | | | | | | | |
| D. pallidum (Bl.) Moore | + | + | | | | | | | |
| Diplazium sp. | + | | | | | | + | | |

(continued on next page)

Table A-7   (continued)

| SPECIES | P-1 (1,600) | P-2 (1,700) | P-3 (1,900) | P-4 (2,100) | P-5 (2,300) | P-6 (2,400) | P-7 (2,600) | P-8 (2,800) | P-9 (3,000) |
|---|---|---|---|---|---|---|---|---|---|
| *Dryopteris adnata* (Bl.) v.A.v.R. | | | | | | | | | + |
| *D. hirtipes* (Bl.) O. K. | | + | | | | | | | |
| *Dryopteris* sp. | + | | + | | | | | | |
| *Egenolfia appendiculata* (Willd.) J. Sm. | + | | | + | + | + | + | | |
| *Elaphoglossum callifolium* (Bl.) Moore | | | + | | | | | | + |
| *Lycopodium serratum* Thunb. | | | | | | + | | | |
| *Lycopodium* sp. | | | | + | | | | | |
| *Nephrolepis acuminata* (Houtt.) Kuhn | + | | | | | | | | |
| *Plagiogyria glauca* (Blume) Mett. | | | | + | + | + | | + | + |
| *Polypodium feei* (Bory) Mett. | | | | | | + | | | |
| *Polypodium* sp. | | | | | + | | | | |
| *Polystichum aculeatum* Schott | | + | | | | | | | |
| *Polystichum* sp. | | | | | + | | | | |
| *Pteris excelsa* Gaud. | | | + | + | | | | | |
| *Thelypteris callosa* (Bl.) K. Iwats. | | + | | | | | | | |
| *T. heterocarpa* (Bl.) Ching | + | + | | | | | | | |
| *T. opaca* (Don) Reed | | + | | | | | | | |
| *T. truncata* (Poir. in Lam.) K. Iwats. | | + | | | | | | | |
| *Trichomanes* sp. | + | | | | | | | | |
| *Woodardia* sp. | | | | | | | | + | |

Table A-8

Terrestrial Herbs of Mount Pangrango

| SPECIES | P-1 (1,600) | P-2 (1,700) | P-3 (1,900) | P-4 (2,100) | P-5 (2,300) | P-6 (2,400) | P-7 (2,600) | P-8 (2,800) | P-9 (3,000) |
|---|---|---|---|---|---|---|---|---|---|
| Amomum hochreutineri Val. | + | | | | | | | | |
| A. pseudo-foetens Val. | + | | | | | | | | |
| Arisaema filiforme Bl. | + | + | | | | | | | |
| Arisaema sp. | | | | + | | | | | |
| Balanophora elongata Bl. | | | | | | | | + | |
| Balanophora sp. | | | + | | | | | | |
| Begonia robusta Bl. | | | + | | | | | | |
| Calanthe flava (Bl.) Hassk. | | | + | | | | | | |
| Carex sp. | | | | | | | + | + | + |
| Coleus galeatus (Poir.) Bth. | + | | | | | | | | |
| Cyperus sp. | | | | | | + | | | |
| Cyrtandra picta Bl. | + | | | + | | | | | |
| Cyrtandra sp. | + | | | | | | | | |
| Disporum chinense (Ker-Gawl.) O. K. | | | | + | | | | | |
| Elatostema paludosum (Bl.) Hassk. | + | | | | | | | | |
| Elatostema sp. | + | + | | | | | + | + | |
| Forrestia marginata (Bl.) Back. | + | | | | | | | + | |
| Forrestia sp. | + | | | | | | | | |

(continued on next page)

Table A-8   (continued)

| SPECIES | P-1 (1,600) | P-2 (1,700) | P-3 (1,900) | P-4 (2,100) | P-5 (2,300) | P-6 (2,400) | P-7 (2,600) | P-8 (2,800) | P-9 (3,000) |
|---|---|---|---|---|---|---|---|---|---|
| Hedychium roxburghii Bl. | | + | | | | | | | |
| Lyciambes laevis (Dunal) Bitt. | + | + | | | | | | | |
| Musa acuminata Colla | + | | | | | | | | |
| Myriactis javanica (Bl.) DC. | | | | | | | + | | + |
| Pilea sp. | | | + | | | | | | |
| Primula prolifera Wall. | | | | | | | | + | |
| Ranunculus javanicus Bl. | | | | | | | | + | + |
| Sanicula elata Ham. ex D. Don | | | | | | | | + | |
| Swertia javanica Bl. | | | | | | | | | + |
| Viola pilosa Bl. | | | | | | + | | + | + |
| Zingiber inflexum Bl. | + | | | | | | | | |

Table A-9

Tree-Epiphytic Shrubs of Mount Pangrango

| SPECIES | P-1 (1,600) | P-2 (1,700) | P-3 (1,900) | P-4 (2,100) | P-5 (2,300) | P-6 (2,400) | P-7 (2,600) | P-8 (2,800) | P-9 (3,000) |
|---|---|---|---|---|---|---|---|---|---|
| *Aeschynanthus horsfieldii* R. Br. | + | | | | | | | | |
| *Aeschynanthus* sp. | | | | + | | | | | |
| *Agalmyla parasitica* (Lamk) O. K. | + | + | | | | | | | |
| *Diplycosia heterophylla* Bl. | + | | | | | | | | |
| *Fagraea ceilanica* Thunb. | + | | | | | | | | |
| *Fagraea* sp. | + | | | | | | | | |
| *Ficus deltoidea* Jack. | + | | | | | | | | |
| *Ficus sinuata* ssp. *cuspidata* (Reinw. ex Bl.) Corner | + | | | | | | | | |
| *Hedychium roxburghii* Bl. var. *roxburghii* | + | | | | | | | | |
| *Ilex spicata* Bl. | + | | | | | | | | |
| *Medinilla laurifolia* (Bl.) Bl. | + | + | | + | | | | | |
| *Medinilla speciosa* (Reinw. ex Bl.) Bl. | + | | | | | | | | |
| *M. verrucosa* (Bl.) Bl. | + | | | | | | | | |
| *Peperomia* sp. | | | + | | | | | | |
| *Polygonum chinense* L. | | + | | | | | | | |
| *Procris frutescens* Bl. | + | | | | | | | | |
| *Rhododendron javanicum* (Bl.) Benn. | + | | | | | | | | |

(continued on next page)

Table A-9   (continued)

| SPECIES | P-1 (1,600) | P-2 (1,700) | P-3 (1,900) | P-4 (2,100) | P-5 (2,300) | P-6 (2,400) | P-7 (2,600) | P-8 (2,800) | P-9 (3,000) |
|---|---|---|---|---|---|---|---|---|---|
| *Schefflera lucescens* var. *rigida* (Bl.) Bakh. f. | + | | | | | | | | |
| *Schefflera scandens* (Bl.) Vig. | + | + | + | | | | | | |
| *Schefflera* sp. | + | | | | + | + | + | | |
| *Usnea* sp. | + | | | | | + | | | |
| *Vaccinium laurifolium* var. *ellipticum* (Bl.) Sleum. | + | | | | | | | | |
| *V. laurifolium* (Bl.) Miq. var. *laurifolium* | + | | + | + | + | + | + | | |
| *V. lucidum* (Bl.) Miq. | + | | + | + | | | | | |

## Table A-10
### Tree-Epiphytic Climbers of Mount Pangrango

| Species | P-1 (1,600) | P-2 (1,700) | P-3 (1,900) | P-4 (2,100) | P-5 (2,300) | P-6 (2,400) | P-7 (2,600) | P-8 (2,800) | P-9 (3,000) |
|---|---|---|---|---|---|---|---|---|---|
| *Actinidia callosa* Lindl. var *callosa* | + | | | | | | | | |
| *Alyxia reinwardti* Bl. | + | | | | | | | | |
| *Cissus adnata* Roxb. | + | | | | | | | | |
| *Dendrotrophe umbellata* (Bl.) Miq. | + | | | | | | | | |
| *Dissochaeta leprosa* (Bl.) Bl. | + | | | | | | | | |
| *Elaeagnus conferta* Roxb. | + | | | | | | | | |
| *Embelia pergamacea* DC. | + | | | | | | | | |
| *E. ribes* Burm. f. | + | | | | | | | | |
| *Embelia* sp. | | | + | | | | | | |
| *Ficus lanata* Bl. | + | + | | | | | | | |
| *F. sagittata* Vahl | + | | + | | | | | | |
| *F. trichocarpa* Bl. | + | | | | | | | | |
| *Freycinetia insignis* Bl. | + | + | | | | | | | |
| *Kadsura scandens* (Bl.) Bl. | + | + | + | + | | | | | |
| *Lonicera javanica* (Bl.) DC. | | | | | | + | + | | |
| *Mussaenda frondosa* L. | + | | | | | | | | |
| *Piper baccatum* Bl. | + | | | | | | | | |

*(continued on next page)*

Table A-10    (continued)

| SPECIES | P-1 (1,600) | P-2 (1,700) | P-3 (1,900) | P-4 (2,100) | P-5 (2,300) | P-6 (2,400) | P-7 (2,600) | P-8 (2,800) | P-9 (3,000) |
|---|---|---|---|---|---|---|---|---|---|
| *P. cilibracteum* DC. | + | | | | | | | | |
| *Psychotria sarmentosa* Bl. | + | | + | + | | | | | |
| *Pyrus* sp. | | | + | + | | | | | |
| *Rhaphidophora pinnata* (L.f.) Schott | + | | | | | | | | |
| *Rubia cordifolia* L. | + | | | | | | | | |
| *Rubus lineatus* Reinw. ex Bl. | | | | + | | | | | |
| *Smilax macrocarpa* Bl. | + | | | | | | | | |
| *S. odoratissima* Bl. | | | | + | | | | | |
| *S. zeylanica* L. | | + | | | | | | | |
| *Tetrastigma dichotomum* (Bl.) Planch. | + | + | | + | | | | | |

Table A-11
Tree-Epiphytic Ferns of Mount Pangrango

| SPECIES | P-1 (1,600) | P-2 (1,700) | P-3 (1,900) | P-4 (2,100) | P-5 (2,300) | P-6 (2,400) | P-7 (2,600) | P-8 (2,800) | P-9 (3,000) |
|---|---|---|---|---|---|---|---|---|---|
| Asplenium caudatum Forst. | + | + | + | + | + | + |  | + |  |
| A. longissimum Bl. |  |  | + | + |  |  |  |  |  |
| A. nidus L. | + | + |  |  |  |  |  |  |  |
| A. thunbergii Kunze | + |  |  |  |  |  |  |  |  |
| Asplenium sp. | + |  |  |  |  |  |  |  |  |
| Coniogramme fraxinea (Don) Diels |  | + |  |  |  |  |  |  |  |
| Crypsinus macrochasmus (Bak.) Copel. | + |  |  |  |  |  |  |  |  |
| Ctenopteris millefolia (Bl.) Copel. |  |  |  |  |  |  | + |  |  |
| Davallia trichomanoides Bl. | + | + |  |  |  |  |  |  |  |
| Davallia sp. | + |  |  |  |  |  |  |  |  |
| Elaphoglossum callifolium (Bl.) Moore | + | + | + |  |  | + | + |  |  |
| E. petiolatum (Sw.) Urban | + |  |  |  |  |  |  |  |  |
| Elaphoglossum sp. | + |  |  | + | + |  |  | + |  |
| Humata sp. |  |  |  | + |  |  | + |  |  |
| Hymenophyllum junghuhnii v.d.B. | + | + |  |  |  |  |  |  |  |
| Hymenophyllum sp. | + |  | + | + | + | + | + | + |  |
| Lomariopsis spectabilis (Kunze) Mett. | + |  |  |  |  |  |  |  |  |

(continued on next page)

Table A-11   (*continued*)

| SPECIES | P-1 (1,600) | P-2 (1,700) | P-3 (1,900) | P-4 (2,100) | P-5 (2,300) | P-6 (2,400) | P-7 (2,600) | P-8 (2,800) | P-9 (3,000) |
|---|---|---|---|---|---|---|---|---|---|
| *Lycopodium phlegmaria* L. | + | | + | | | | | | |
| *L. piscium* (Hert.) Tagawa & K. Iwats. | | + | | | | | | | |
| *Nephrolepis acuminata* (Houtt.) Kuhn | + | + | | | | | | | |
| *Nephrolepis* sp. | | | + | | | | | | |
| *Oleandra musifolia* (Bl.) Presl. | + | | | | | | | | |
| *Polypodium feei* (Bory) Mett. | | | + | + | | + | | + | |
| *Polypodium* sp. | + | | + | + | | + | | + | |
| *Selliguea heterocarpa* Bl. | | + | | | | | | | |
| *Trichomanes* sp. | + | | | | | | | | |

## Table A-12
### Tree-Epiphytic Orchids of Mount Pangrango

| SPECIES | P-1 (1,600) | P-2 (1,700) | P-3 (1,900) | P-4 (2,100) | P-5 (2,300) | P-6 (2,400) | P-7 (2,600) | P-8 (2,800) | P-9 (3,000) |
|---|---|---|---|---|---|---|---|---|---|
| Agrostophyllum sp. | + | | | | | | | | |
| Appendicula ramosa Bl. | + | | + | | | | | | |
| Bulbophyllum uniflorum (Bl.) Hassk. | + | | | | | | | | |
| Bulbophyllum sp. | + | | | + | | | | | |
| Cyperorchis rosea (J. J. S.) Schltr. | + | | | | | | | | |
| Dendrobium conspicuum Bakh. f. | | | + | | | | | | |
| D. hasseltii (Bl.) Lindl. | | | | | | + | | + | + |
| D. montanum J. J. S. | | | - | | | | | | |
| Dendrobium sp. | | | + | | | | | | |
| Eria flavescens (Bl.) Lindl. | | | | + | | | | | |
| Eria sp. | | | + | + | | + | | | |
| Liparis pallida (Bl.) Lindl. | + | | | | | | | | |
| Liparis sp. | + | | | | | | | | |
| Malaxis blumei (Boerl. & J. J. S.) Bakh. f. | + | | | | | | | | |

# Bibliography

Abdul Razak, B. M. A., C. K. Low, and A. Abu Said. 1981. Determination of relative tannin contents of the barks of some Malaysian plants. *Malay. For.* 44 (1): 87–92.

Aksornkoae, S. 1975. Structure, regeneration, and productivity of mangroves in Thailand. Ph.D. diss., Michigan State University.

Anderson, J. A. R. 1961a. The destruction of *Shorea albida* forest by an unidentified insect. *Emp. For. Rev.* 40:19–29.

———. 1961b. The ecology and forest types of the peat swamp forests of Sarawak and Brunei in relation to their silviculture. Ph.D. diss., Edinburgh University.

———. 1963. The flora of the peat swamp forests of Sarawak and Brunei, including a catalogue of all recorded species of flowering plants, ferns, and fern allies. *Gardens Bulletin, Singapore* 20:131–228.

———. 1964a. The structure and development of the peat swamps of Sarawak and Brunei, including a catalogue of all recorded species of flowering plants, ferns, and fern allies. *J. Trop. Geogr.* 18:7–16.

———. 1964b. Observations on climatic damage in peat swamp forest in Sarawak. *Commonw. For. Rev.* 43:145–158.

———. 1976. Observations on the ecology of five peat swamp forests in Sumatra and Kalimantan. In *Peat and podzolic soils and their potential for agriculture in Indonesia,* pp. 45–55. Proceedings ATA 106 Midterm Seminar, Tugu, 13–14 October 1976. Bulletin 3, Soil Res. Inst. Bogor.

Anderson and Marsden (Forestry Consultants) Ltd. 1984. Brunei forest resources and strategic planning study. 2 vols. Draft report submitted to the Government of Brunei.

Andriesse, J. P. 1974. *The characteristics, agricultural potential and reclamation problems of tropical lowland peats in South-East Asia.* Amsterdam: Koninklijk Instituut voor de Tropen.

Aoki, M., K. Yabuki, and H. Koyama. 1975. Micrometeorology assessment of a tropical rainforest in West Malaysia. *J. Agr. Met. Japan* 31:115–124.

Appanah, S., and H. T. Chan. 1982. Methods of studying the reproductive biology of some Malaysian primary forest trees. *Malay. For.* 45 (1): 10–20.

Armitage, F. B., and J. Burley. 1980. *Pinus kesiya.* Tropical Forestry Papers No. 9. Commonwealth Forestry Institute, Oxford.

Ashton, P. S. 1964a. Ecological studies in the mixed dipterocarp forests of Brunei State. *Oxf. For. Mem.* 25.

———. 1964b. *Manual of the dipterocarp trees of Brunei State.* London: Oxford University Press.

———. 1976. Factors affecting the development and conservation of tree genetic resources in South-East Asia. In *Tropical Trees,* ed. J. Burley and B. T. Style, pp. 189–198. London: Academic Press.

Ashton, P. S., T. J. Givnish, and S. Appanah. 1988. Staggered flowering in the Diptero-
    carpaceae: A new insight into floral induction and the evolution of mast fruiting in
    the aseasonal tropics. *Amer. Nat.* 132(1): 44–66.
Backer, C. A., and R. C. van den Brink Bakhuizen, eds. 1963–1968. *Flora of Java.* 3 vols.
    Groningen: N.V.P. Noordhoff.
Baker, H. G., K. S. Bawa, G. W. Frankie, and P. A. Opler. 1983. Reproductive biology of
    plants in tropical forests. In *Ecosystems of the world 14A. Tropical rain forest ecosystems–
    structure and function,* ed. F. B. Golley, pp. 183–215. Amsterdam: Elsevier.
Baltzer, F. 1969. Les formations végétales associées au delta de la Dumbea. *Cah. Orstrom,
    Ser. Geol.* 1 (1): 59–84.
Becker, P. F. 1982. Shorea seedling ecology in a lowland Malaysian rain forest. Ph.D.
    diss., University of Michigan.
Becking, J. H., L. G. Den Berger, and H. W. Meindersma. 1922. Vloed-of mangrove-
    bosschen in Ned.-Indië. *Tectona* 15:561–611.
Bennett, E. 1970. Adaptation in wild and cultivated plant populations. In *Genetic resources
    in plants,* ed. O. H. Frankel and E. Bennett, pp. 115–129. Oxford and Edinburgh:
    Blackwell.
Blume, C. D. 1849. Eucalyptus deglupta. *Bl. Mus. Bot Ludg.-Bat.* 1:83–84.
Borchert, R. 1980. Phenology and ecophysiology of a tropical tree: *Erythrina poeppigiana*
    O. F. Cook. *Ecology* 61 (5): 1065–1074.
————. 1983. Phenology and control of flowering in tropical trees. *Biotropica* 15 (2):
    81–89.
Bouvarel, P. 1970. The conservation of gene resources of forest trees. In *Genetic resources in
    plants,* ed. O. H. Frankel and E. Bennett, pp. 523–529. Oxford and Edinburgh:
    Blackwell.
Bradshaw, A. D. 1975. Population structure and the effects of isolation and selection. In
    *Crop genetic resources for today and tomorrow,* ed. O. H. Frankel and J. G. Hawkes, pp.
    37–51. Cambridge: Cambridge University Press.
Brass, L. J. 1941. The 1938–1939 expedition to the Snow Mountains, Netherlands New
    Guinea. *J. Arnold Arbor.* 22:271–342.
Browne, F. G. 1955. *Forest trees of Sarawak and Brunei and their products.* Kuching, Sara-
    wak: Government Printer.
Brunig, E. F. 1964. A study of damage attributed to lightning in two areas of *Shorea
    albida* forest in Sarawak. *Emp. For. Rev.* 43:134–144.
————. 1974. *Ecological studies in the Kerangas forests of Sarawak and Brunei.* Kuching,
    Sarawak: Government Printer.
Bryndum, K. 1974. *Pinus merkusii* provenance hybridization. Report of the Thai-Danish
    Pine Project, 1969–1974, pp. 38–41.
Burgess, P. F. 1972. Studies on the regeneration of hill forests of the Malay Peninsula:
    The phenology of dipterocarps. *Malay. For.* 35:103–123.
————. 1975. Silviculture in the hill forests of the Malay Peninsula. F.R.I. Research
    Pamphlet. No. 66.
Burkill, I. H. 1966. A dictionary of the economic products of the Malay Peninsula. 2
    vols. Reprinted 2nd ed. Kuala Lumpur: Ministry of Agriculture and Cooperatives.
Burley, J., and E. N. G. Cooling. 1972. Status of the C.F.I. international provenance trial
    of *Pinus merkusii* jungh. and de Vriese, September 1970. In *Selection and breeding to
    improve some tropical conifers,* ed. J. Burley and D. G. Nikles, pp. 153–160. Com-
    monwealth Forestry Institute, Oxford.

Burley, J., and G. Namkoong. 1980. Conservation of forest genetic resources. Paper for 11th Commonw. For. Conf., Trinidad.

Callaham, R. Z. 1970. Geographic variation in forest trees. In *Genetic resources in plants,* ed. O. H. Frankel and E. Bennett, pp. 43–47. Oxford and Edinburgh: Blackwell.

Chan, H. T. 1980. Reproductive biology of some Malaysian dipterocarps. II. Fruiting biology and seedling studies. *Malay. For.* 43:438–451.

Chan, H. T., and S. Appanah. 1980. Reproductive biology of some Malaysian dipterocarps. I. Flowering biology. *Malay. For.* 43:132–143.

Chapman, V. J. 1940. The botany of the Jamaica shoreline. *Geog. Journal* 96:312–327.

———. 1975. Mangrove biogeography. In *Proceedings of the International Symposium on Biology and Management of Mangroves,* ed. G. E. Walsh, S. C. Snedaker, and H. J. Teas, pp. 3–22. Gainesville, Florida: University of Florida.

———. 1976. *Mangrove vegetation.* Vaduz: J. Cramer.

———, ed. 1977. *Wet coastal ecosystems.* Ecosystems of the world, ed. D. W. Goodall, vol. 1. Amsterdam: Elsevier.

Christensen, B. 1978. Biomass and primary production of *Rhizophora apiculata* Bl. in a mangrove in southern Thailand. *Aquatic Botany* 4:43–52.

Cintrón, G., A. E. Lugo, D. J. Pool, and G. Morris. 1978. Mangroves of arid environments in Puerto Rico and adjacent Islands. *Biotropica* 10 (2): 110–121.

Cockburn, P. F. 1975. Phenology of dipterocarps in Sabah. *Malay. For.* 38 (3): 160–170.

Connor, D. J. 1969. Growth of grey mangrove *(Avicennia marina)* in nutrient culture. *Biotropica* 1 (2): 36–40.

Cooke, F. C. 1930. Coconuts on peat. *Malay. Agr. Journal* 18:587–595.

Cooling, E. N. G. 1968. *Pinus merkusii.* Commonwealth Forestry Institute, Oxford.

———. 1972. Nomenclature of *Pinus merkusii* Jungh. and de Vriese and *Pinus merkusii* Cooling and Gaussen. In *Selection and breeding to improve some tropical conifers,* ed. J. Burley and P. G. Nikles, pp. 89–97. Commonwealth Forestry Institute, Oxford.

Cooling, E. N. G., and E. I. Gaussen. 1970. In Indochina: *Pinus merkusiana* sp. nov. et non *P. merkusii* Jungh. et de Vriese. *Trav. Lab. For. Toulouse,* vol. 1, p. 8.

Corner, E. J. H. 1978. *The freshwater swamp-forest of south Johore and Singapore.* Singapore: Botanic Gardens Parks & Recreation Department.

Coster, C. 1923. Lauberneuerung und andere periodische Lebensprozesse in dem trockenen monsun-gebjet Ost-Java's. *Ann. Jard. Bot. Buitenzorg* 33:117–187.

———. 1926. Periodische blueteerscheinungen in den tropen. *Ann. Jard. Bot. Buitenzorg* 35:125–162.

Coulter, J. K. 1950. Peat formations in Malaya. *Malay. Agr. J.* 33:66–81.

———. 1957. Development of the peat soils of Malaya. *Malay. Agr. J.* 40:188–199.

Cox, A. J. 1911. Philippine firewood. *Philip. J. Sci.* Sec. A, p. 6.

Croat, T. B. 1975. Phenological behavior of habit and habitat classes on Barro Colorado Island (Panama Canal Zone). *Biotropica* 7:270–277.

Daljeet Singh, K. 1976. The influence of seed predators on the development and yield of dipterocarp seeds. Paper presented at the 5th Malay. For. Conf., Sarawak.

Davidson, J. 1977. Exploration, collection, evaluation, conservation, and utilization of the gene resources of tropical *Eucalyptus deglupta* Bl. Third world consultation on forest tree breeding, Canberra, pp. 75–102.

Davis, J. H. 1940. The ecology and geologic role of mangroves in Florida. *Pap. Tortugas Lab.* 32 (Carnegie Inst. Publ. no. 517): 305–412.

Daubenmire, R. F. 1972. Phenology and other characteristics of tropical semi-deciduous forest in northwestern Costa Rica. *J. Ecol.* 60:147–170.

De Haan, J. H. 1931. Het een en ander over de Tijlatjapsche vloedbosschen. *Tectona* 24:39–76.

Ding Hou. 1958. Rhizophoraceae. In *Flora Malesiana* ser. 1, vol. 5, pt. 4:429–493.

——. 1960. A review of the genus *Rhizophora* with special reference to the Pacific species. *Blumea* 10:625–634.

Dolianiti, E. 1955. Frutos de Nypa no Palaeocene du Pernambuco, Brasil. *Div. Geol. Mineral Brazil Bol.* 158:1–36.

Dobby, E. H. G. 1961. *Tōnan ajia* [Southeast Asia], trans. Kobori Iwao. Tokyo: Kokon Shoin.

Driessen, P. M. 1978. Peat soils. In *Soils and rice,* pp. 768–779. Los Baños: I.R.R.I.

Edeling, A. C. J. 1870. *Botanische wandeling in dem outrek van Vidara Tjina. Nat. Tijdschr. Ned. Indie,* p. 31.

Edwards, P. J., and P. J. Grubb. 1977. Studies of mineral cycling in a montane rain forest in New Guinea. Pt. I: The distribution of organic matter in the vegetation and soil. *J. Ecol.* 65:943–969.

Egler, F. E. 1948. The dispersal and establishment of red mangroves, *Rhizophora,* in Florida. *Carib. Forest.* 9 (4): 299–319.

Ellenberg, H. 1956. *Grundlagen der vegetationsgliederung.* Vol. 1: *Aufgaben und Methoden der Vegetationskunde.* Stuttgart: Eugen Ulmer.

Endert, F. H. 1920. De woudbloomflora van Palembang. *Tectona* 13:113–160.

Food and Agriculture Organization (FAO). 1975. *The methodology of conservation of forest genetic resources. Report on a pilot study.* Rome: FAO.

——. 1981. *Report on the FAO/UNEP expert consultation on in situ conservation of forest genetic resources, held in Rome, Italy, 2–4 December 1980.* Rome: FAO.

——. 1984. *In situ conservation of wild plant genetic resources: A status review and action plan.* Rome: FAO.

Fleming, T. H., C. F. Williams, F. J. Bonaccorso, and L. H. Herbst. 1985. Phenology, seed dispersal, and colonization in *Muntingia calabura,* a neotropical pioneer tree. *Amer. J. Bot.* 72 (3): 383–391.

Ford-Lloyd, B., and M. Jackson. 1986. *Plant genetic resources.* Bristol: Edward Arnold.

Frankel, O. H. 1970. Genetic conservation in perspective. In *Genetic resources in plants,* ed. O. H. Frankel and E. Bennett, pp. 469–489. Oxford and Edinburgh: Blackwell.

——. 1975. Genetic resources survey as a basis for exploration. In *Crop genetic resources for today and tomorrow,* ed. O. H. Frankel and J. G. Hawkes, pp. 99–109. Cambridge: Cambridge University Press.

Frankel, O. H., and E. Bennett, eds. 1970. *Genetic resources in plants: Their exploration and conservation.* Oxford and Edinburgh: Blackwell.

Frankel, O. H., and J. G. Hawkes, eds. 1975. *Crop genetic resources for today and tomorrow.* Cambridge: Cambridge University Press.

Frankel, O. H., and M. E. Soulé. 1981. *Conservation and evolution.* Cambridge: Cambridge University Press.

Frankie, G. W., H. G. Baker, and P. H. Opler. 1974a. Tropical plant phenology: Applications for studies in community ecology. In *Phenology and seasonality modeling,* ed. H. Lieth, pp. 287–296. Berlin: Springer-Verlag.

——. 1974b. Comparative phenological studies of trees in tropical wet and dry forests in the lowlands of Costa Rica. *J. Ecol.* 62:881–919.

Franklin, I. R. 1980. Evolutionary change in small populations. In *Conservation biology*, ed. M. E. Soulé and B. A. Wilcox, pp. 135–150. Sunderland, Mass.: Sinauer Associates.

Franklin, J. F., and J. M. Trappe. 1967. Natural areas: Needs, concepts, and criteria. *J. For.* 66:456–461.

Furukawa, H. 1988. Stratigraphic and geomorphic studies of peat and giant podzols in Brunei. Pt. 1: Peat. *Pedologist* 32 (1): 26–42.

Gentry, A. H. 1974. Flowering phenology and diversity in tropical Bignoniaceae. *Biotropica* 6 (1): 64–68.

Gill, A. M. 1971. Studies on the growth of red mangrove (*Rhizophora mangle* L.). Pt. 3: Phenology of the shoot. *Biotropica* 3 (2): 109–124.

———. 1977. Studies on the growth of red mangrove (*Rhizophora mangle* L.). Pt. 4: The adult root system. *Biotropica* 9 (3): 145–155.

Gill, A. M., and P. B. Tomlinson. 1969. Studies on the growth of red mangrove (*Rhizophora mangle* L.). Pt. 1: Habitat and general morphology. *Biotropica* 1 (1): 1–9.

———. 1971a. Studies on the growth of red mangrove (*Rhizophora mangle* L.). Pt. 2: Growth and differentiation of aerial roots. *Biotropica* 3 (1): 63–77.

———. 1971b. Studies on the growth of red mangrove (*Rhizophora mangle* L.). Pt. 3: Phenology of the shoot. *Biotropica* 3 (2): 109–124.

Golley, F. B., J. T. McGinnis, R. G. Clements, G. I. Child, and M. J. Duever. 1975. *Mineral cycling in a tropical moist forest ecosystem.* Athens, Georgia: University of Georgia Press.

Golley, F. B., H. T. Odum, and R. F. Wilson. 1962. The structure and metabolism of a Puerto Rican red mangrove forest in May. *J. Ecol.* 43 (1): 9–19.

Guldager, P. 1975. Ex situ conservation stands in the tropics. In *The methodology of conservation of forest genetic resources*, pp. 85–92. Rome: Food and Agriculture Organization.

Guppy, H. B. 1906. *Observations of a naturalist in the Pacific between 1896 and 1899.* Vol. 2: *Plant dispersal.* London and New York: Macmillan.

Hamilton, W. 1979. Tectonics of the Indonesian region. U.S. Geol. Survey Prof. paper 1078.

Hedberg, I., and O. Hedberg. 1968. Conservation of the vegetation in Africa south of the Sahara. *Acta Phytogeogr. Suec.* 54.

Hewitt, B. R. 1967. The occurrence, origin, and vegetation of lowland peat in Malaya. *Proc. Lin. Soc. N.S.W.* 92 (1): 58–66.

Heyne, K. 1950. *De nuttige planten van Indonesië.* 2 vols. Bandung: van Hoeve.

Hicks, D. B., and L. A. Burns. 1975. Mangrove metabolic response to alterations of natural freshwater drainage to southwestern Florida estuaries. In *Proceedings of the International Symposium on Biology and Management of Mangroves,* ed. G. E. Walsh, S. C. Snedaker and H. J. Teas, pp. 238–255. Gainesville, Florida: University of Florida.

Holttum, R. E. 1931. On periodic leaf change and flowering of trees in Singapore. *Gardens Bulletin, Singapore* 5:173–211.

———. 1940. Periodic leaf change and flowering of trees in Singapore (II). *Gardens Bulletin, Singapore* 11:119–175.

Hopkins, B. 1968. Vegetation of the Olokemeji forest reserve, Nigeria. Pt. 5: The vegetation on the savanna site with special reference to its seasonal changes. *J. Ecol.* 56: 97–115.

Howes, F. N. 1962. Tanning materials. In *Wiesner's die rohstoffe des pflanzenreichs,* vol. 1, pp. 178–184. Leutershausen: Cramer.

Hozumi, K., K. Shinozaki, and Y. Tadaki. 1968. Studies on the frequency distribution of the weight of individual trees in a forest stand. Pt. 1: A new approach toward the

analysis of the distribution function and the -3/2 power distribution. *Jap. J. Ecol.*
18 (1): 10–20.

Imanishi, K. 1937. Nihon arupusu no suichokubunputai [Vertical distribution zones in the Japan Alps]. *Sangaku* 32:269–364.

Ingram, G. B. 1984. *In situ conservation of genetic resources of plants: The scientific and technical base.* Rome: Food and Agriculture Organization.

Janzen, D. H. 1970. Herbivores and the number of tree species in tropical forests. *Amer. Natur.* 104:501–528.

————. 1974. Tropical blackwater rivers, animals, and mast fruiting by the Dipterocarpaceae. *Biotropica* 6:69–103.

Johnstone, I. M. 1981. Consumption of leaves by herbivores in mixed mangrove stands. *Biotropica* 13 (4): 252–259.

Karsten, G. 1891. Ueber die mangrove-vegetation im Malayischen archipel. *Bibliogr. Bot.* 22:1–71.

Keiding, H. 1972. Collection of pine seed in Southeast Asia with emphasis on provenance sampling. In *Selection and breeding to improve some tropical conifers,* ed. J. Burley and D. G. Nikles, pp. 17–28. Oxford: Commonwealth Forestry Institute.

————. 1980. Constraints and prospects in international cooperation on tree improvement and seed procurement. In *Proceedings of the conference on Southeast Asian tree improvement and seed procurement cooperative programme,* pp. 20–30. Bangkok: Royal Forest Department.

Kemp, R. H. 1972. Seed sources and seed procurement of low-altitude tropical pines in Central America. In *Selection and breeding to improve some tropical conifers,* ed. J. Burley and D. G. Nikles, pp. 9–16. Oxford: Commonwealth Forestry Institute.

Kemp, R. H., L. Roche, and R. L. Winan. 1976. Current activities and problems in the exploration and conservation of tropical forest gene resources. In *Tropical trees,* ed. J. Burley and B. T. Styles, pp. 223–233. London: Academic Press.

Kira, T. 1949. *Nihon no shinrintai* [Forest zones of Japan]. Tokyo: Ringyōgijutsu kyōkai.

————. 1967. Mangurōbu no seitai [Mangrove ecology]. *Nettairingyō* 5:1–16.

————. 1976. *Rikujō seitaikei* [Terrestrial ecosystems]. Tokyo: Kyōritsu Shuppan.

————. 1978. Minamata shōyōjurin no dōtai [Dynamics of Minamata laurel forests]. *Shizen:* 26–39.

————. 1983. *Nettairin no seitai* [Tropical forest ecology]. Kyoto: Jinbunshoin.

Kira, T., and T. Shidei. 1967. Primary production and turnover of organic matter in different forest ecosystems of the western Pacific. *Jap. J. Ecol.* 17:70–87.

Koelmeyer, K. O. 1959. The periodicity of leaf change and flowering in the principal forest communities of Ceylon. *Ceylon Forester* 4 :157–189.

Komiyama, A., K. Ogino, S. Aksornkoae, and S. Sabhasri. 1987. Root biomass of a mangrove forest in southern Thailand. Pt. 1: Estimation by the trench method and the zonal structure of root biomass. *J. Trop. Ecol.* 3:97–108.

Komiyama, A., H. Moriya, S. Prawiroatmodho, T. Toma, and K. Ogino. 1988. Primary productivity of mangrove forest. In *Biological system of mangroves,* ed. K. Ogino and M. Chihara, pp. 96–97. Ehime: Ehime University.

Koriba, K. 1947. Marai tokuni Shingapōru niokeru jubokuseichō no shūki ni tsuite (1) [On the periodicity of tree growth in Malaya, particularly Singapore, part 1]. *Seiri-seitai* 1–2:93–109.

————. 1958. On the periodicity of tree-growth in the tropics, with reference to the mode of branching, the leaf-fall, and the formation of the resting bud. *Gardens Bulletin, Singapore* 17:11–81.

Koski, V. 1974. Effective population size in a really continuous forest. Proc. of the joint meeting of IUFRO working parties on population genetics and breeding theory, Stockholm.

Lai, K. K. 1976. Performance of planted *Shorea albida.* An interim summary. Forest research report no. S.R. 14. Forest Department, Sarawak, Malaysia.

————. 1978. Investigation 56 research plot 84. Line planting trials of *Shorea albida.* An interim report. Forest research report no. S.R. 21. Forest Department, Sarawak, Malaysia.

Larsen, E., and D. A. N. Cromer. 1970. Exploration, evaluation, utilization, and conservation of Eucalypt gene resources. In *Genetic resources in plants,* ed. O. H. Frankel and E. Bennett, pp. 381–388. Oxford and Edinburgh: Blackwell.

Lawrence, D. B. 1949. Self-erecting habit of seedling red mangroves (*Rhizophora mangle* L.). *Amer. J. Bot.* 36 (5): 426–427.

Lee H. S. 1972. The role of silviculture in the management of the peat swamp reserves in Sarawak. Paper presented at the Fourth Pan-Malaysian Forestry Conference, Kuala Lumpur.

————. 1976. Trees poisoned during G1 silvicultural treatment in mixed peatswamp forest in the Sibu forest section. Forest research report no. S.R. 12. Forest Department, Sarawak, Malaysia.

————. 1977. Manipulation and regeneration of the mixed swamp forest in Sarawak. *Malayan Nature Journal* 31 (1): 1–9.

————. 1979. Natural regeneration and reforestation in the peat swamp forests of Sarawak. Tropical Agriculture Research Series No. 12, pp. 51–60. Tropical Agriculture Research Center, Tsukuba, Japan.

Leo, C., and H. S. Lee. 1971. The alan *(Shorea albida)* resources, their properties and utilization. *Malay. For.* 34:20–35.

Lieth, H., ed. 1974. *Phenology and seasonality modeling.* New York: Springer-Verlag.

Liew, T. C., and F. O. Wong. 1973. Density, recruitment, mortality, and growth of dipterocarp seedlings in virgin and logged-over forests in Sabah. *Malay. For.* 36:3–15.

Liming, F. G. 1957. Homemade dendrometers. *J. For.* 55:575–577.

Linne, C. von. 1753. *Species Plantarum.* Vol. 1, *Lundae.*

Lugo, A. E. 1980. Mangrove ecosystems: Successional or steady state? *Biotropica* 12, supplement (tropical succession): 65–72.

Lugo, A. E., and G. Cintrón. 1975. The mangrove forests of Puerto Rico and their management. In *Proceedings of the International Symposium on Biology and Management of Mangroves,* ed. G. E. Walsh, S. C. Snedaker, and H. J. Teas, pp. 825–846. Gainesville, Florida: University of Florida.

Lugo, A. E., and S. C. Snedaker. 1974. The ecology of mangroves. *Ann. Rev. of Ecol. and Systematics* 5:39–65.

————. 1975. Properties of a mangrove forest in southern Florida. In *Proceedings of the International Symposium on Biology and Management of Mangroves,* ed. G. E. Walsh, S. C. Snedaker, and H. J. Teas, pp. 170–212. Gainesville, Florida: University of Florida.

Lugo, A. E., G. Evink, M. M. Brinson, A. Broce, and S. C. Snedaker. 1975. Diurnal rates of photosynthesis, respiration, and transpiration in mangrove forests of south Florida.

In *Tropical ecological systems,* ed. F. B. Goney and E. Medina, pp. 335–350. New York: Springer-Verlag.

Luytjes, A. 1923. De vloedbosschen in Atjeh. *Tectona* 16:575–601.

McClure, H. E. 1966. Flowering, fruiting, and animals in the canopy of a tropical rain forest. *Malay. For.* 29:182–203

McKenzie, D. P., and J. G. Sclater. 1973. The evolution of the Indian Ocean. *Sci. Amer.* 228:62–74.

Macnae, W. 1968. A general account of the fauna and flora of mangrove swamps and forests in the Indo-West-Pacific region. *Advan. Mar. Biol.* 6:73–270.

Malaisse, F. R. 1974. Phenology of the Zambezian woodland area with emphasis on the Miombo ecosystem. In *Phenology and Seasonality Modeling,* ed. H. Lieth, pp. 269–286. New York: Springer-Verlag.

Marshall, D. R., and A. H. D. Brown. 1975. Optimum sampling strategies in genetic conservation. In *Crop genetic resources for today and tomorrow,* ed. O. H. Frankel and J. G. Hawkes, pp. 53–80. Cambridge: Cambridge University Press.

Mead, J. P. 1912. *The mangrove forests of the west coast of Federated Malay States.* Kuala Lumpur: Government Printer.

Medway, Lord. 1972. Phenology of a tropical rain forest in Malaya. *Biol. J. Linnean Soc.* 4:117–146.

Meijer, W. 1959. Plant sociological analysis of montane rainforest near Tjibodas, West Java. *Acta bot. Neerl.* 8:277–291.

Mohr, E. C. J. 1922. *De grand van Java en Sumatra.* Amsterdam: J. H. de Bussy.

Montford, H. H. 1970. The terrestrial environment during upper Cretaceous and Tertiary times. *Proc. Geol. Ass.* 81:181–204.

Mueller-Dombois, D., K. W. Bridges, and H. L. Carson, eds. 1981. *Island ecosystems: Biological organization in selected Hawaiian communities.* US/IBP synthesis series 15. Woods Hole, Mass.: Hutchinson Ross.

Muller, J. 1965. Palynological study of Holocene peat in Sarawak. In *Symposium on ecological research in humid tropics vegetation,* pp. 147–156. Kuching, Sarawak: Government of Sarawak and UNESCO.

———. 1970. Palynological evidence on early differentiation of angiosperms. *Biol. Rev.* 45:417–450.

———. 1972. Palynological evidence for change in geomorphology, climate, and vegetation in the Mio-Pliocene of Malesia. In "The quaternary era in Malesia," ed. P. S. Ashton and M. Ashton, Misc. Series 13, Dept. of Geography, University of Hull (mimeo.).

Ng, F. S. P. 1977a. Gregarious flowering of dipterocarps in Kepong 1976. *Malay. For.* 40:126–137.

———. 1977b. The problems of forest seeds production with reference to dipterocarps. In *Seed technology in the tropics,* pp. 181–186. Serdang Malaysia: Universiti Pertaninan Malaysia.

Nicholson, D. I. 1958. One year's growth of *Shorea smithiana* in North Borneo. *Malay. For.* 21:193–196.

Nikles, D. G. 1980a. A S.E. Asian regional (international) programme and its relation to a global structure for genetic improvement of tropical forest trees. In *Proceedings of the conference on Southeast Asian tree improvement and seed procurement cooperative programme,* pp. 31–47. Bangkok: Royal Forest Department.

————. 1980b. Role of international cooperation in establishing and improving the genetic foundations of forest plantations in the tropics. In *Proceedings of the conference on Southeast Asian tree improvement and seed procurement cooperative programme,* pp. 48–53. Bangkok: Royal Forest Department.

Norton, B. G., ed. 1986. *The preservation of species.* Princeton: Princeton University Press.

Oey Djoen Seng. 1964. *Specific gravity of Indonesian woods and its significance for practical use.* Bogor: Forest Research Institute.

Ogawa, F. 1974. *Nettai no seitai I–shinrin* [Tropical ecology I: Forests]. Tokyo: Kyōritsu shuppan.

Ogawa, H., K. Yoda, T. Kira, K. Ogino, T. Shidei, D. Ratanawongse, and C. Apasutaya. 1965. Comparative ecological study on three main types of forest vegetation in Thailand. Pt. 1: Structure and floristic composition. *Nature and life in Southeast Asia* 4:13–48.

Ogino, K. 1974. Tai no shinrinshokusei to ringyō ni kansuru shinrinseitaigakuteki kōsatsu [A consideration of forest vegetation and forestry in Thailand from the standpoint of forest ecology]. Ph.D. diss., Kyoto University.

Ogino, K., D. Ratanawongse, T. Tsutsumi, and T. Shidei. 1967. Taikokushinrin no dai-ichijiseisanryoku [Primary productivity of Thai forests]. *Tōnanajia kenkyū {Southeast Asian Studies}* 5:121–54.

Ogura, K. 1940. "Mangurōvu" oyobi shitchisan shokubutsu no ijōkon no shokata ni tsuite [On the various forms of modified roots of "mangrove" and swamp vegetation]. *Shokubutsugaku zasshi* 54:389–404.

Okimori, Y. 1987. Nettaijurin ni okeru seichō no kisetsusei ni kansuru kenkyū [A study of the seasonality of growth in tropical forest]. Ph.D. diss., Kyoto University.

Okutomi, K. 1977. Nettai anettai no shinrin to teibokurin [Forest and scrub forest of the tropics and subtropics]. In *Gunraku no bunpu to kankyō* [Distribution and environment of communities], ed. K. Ishizuka, pp. 28–71. Tokyo: Asakura shoten.

Omura, S., and M. Ando. 1970. Kojiirin no rakuyōryō: Shōyō jurin no shokubutsuseisan ni kansuru kenkyū [Litter volume in *Castanopsis* forest: A study of vegetation production in laurel forest]. JIBP-PT-Minamata special study area report, pp. 50–55.

Opler, P. A., G. W. Frankie, and H. G. Baker. 1976. Rainfall as a factor in the release, timing, and synchronization of anthesis by tropical trees and shrubs. *J. Biogeog.* 3: 231–236.

————. 1980. Comparative phenological studies of treelet and shrub species in tropical wet and dry forests in the lowlands of Costa Rica. *J. Ecol.* 68:167–188.

Paijmans, K. 1976. *New Guinea vegetation.* Amsterdam: Elsevier.

Palmberg, C. 1987. *Conservation of genetic resources of woody species.* Rome: Food and Agriculture Organization.

Polak, B. 1950. Occurrence and fertility of tropical peat soils in Indonesia. *Contr. Cen. Agr. Res. Sta.* 104 (Bogor).

————. 1951. Construction and origin of floating islands in the Rawa Pening (Central Java). *Contr. Cen. Agr. Res. Sta.* 121 (Bogor): 1–11.

————. 1975. Character and occurrence of peat deposits in the Malaysian tropics. In *Modern Quaternary research in Southeast Asia,* vol. 1, ed. G. J. Bartstra and W. A. Casparie, pp. 71–81. Rotterdam: A. A. Balkema.

Polak, E. 1933. Ueber Torf und moor in Niederländisch Indien. *Verh. Kon. Akad. V. Wetensch.* 30:1–85.

Pool, D. J., S. C. Snedaker, and A. E. Lugo. 1977. Structure of mangrove forests in Florida, Puerto Rico, Mexico, and Costa Rica. *Biotropica* 9 (3): 195–212.

Putz, F. E. 1979. Aseasonality in Malaysian tree phenology. *Malay. For.* 42:1–24.

Rabinobitz, D. 1978. Early growth of mangrove seedlings in Panama, and an hypothesis concerning the relationship of dispersal and zonation. *J. Biogeog.* 5:113–133.

Reich, P. B., and R. Borchert. 1982. Phenology and ecophysiology of the tropical tree *Tabebuia neochrysantha* (Bignoniaceae). *Ecol.* 63 (2): 294–299.

Richards, P. W. 1952. *The tropical rain forest.* Cambridge: Cambridge University Press.

———. 1971. Some problems of nature conservation in the tropics. *Bull. Jard. Bot. Nat. Belg.* 41:173–187.

Roche, L. R. 1975a. Guidelines for the methodology of conservation of forest genetic resources. In *The methodology of conservation of forest genetic resources,* pp. 107–113. Rome: Food and Agriculture Organization.

———. 1975b. Tropical hardwoods. In *The methodology of conservation of forest genetic resources,* pp. 65–78. Rome: Food and Agriculture Organization.

Roche, L., and M. J. Dourojeanni. 1984. *A guide to in situ conservation of genetic resources of tropical woody species.* Rome: Food and Agriculture Organization.

Rumphius, G. E. 1743. *Herbarium Amboinensis* 3:122.

Saito, H. 1972. Shinrin no ritāfōruryō no suitei ni kansuru kenkyū [A study on the estimation of forest litter-fall volume]. Ph.D. diss., Kyoto University.

Sandrasegaran, K. 1966. A note on the growth of *Melaleuca leucadendron* L. (Gelam). *Malay. For.* 29 (1): 21–25.

Sasaki, S. 1976. The physiology, storage, and germination of timber seeds. In *Seed technology in the tropics,* pp. 111–115. Kuala Lumpur: Universiti Pertanian Malaysia.

Sasaki, S., C. H. Tan, and A. R. Zulfatah. 1979. Some observations on unusual flowering and fruiting of dipterocarps. *Malay. For.* 42:38–45.

Schimper, A. F. W. 1891. *Die Indo-malayische Strandflora.* Bot. mitt Trop. Jena 3.

Scholander, P. F., L. van Dam, and S. I. Scholander. 1955. Gas exchange in the roots of mangroves. *Amer. J. Bot.* 42:92–98.

Schuster, W. H. 1952. *Fish culture in brackish water ponds in Java.* Spec. Pub. Indo-Pac. Fish Counc., 1.

Seifriz, W. 1923. The altitudinal distribution of plants on Mt Gedeh, Java. *Bull. Torrey Bot. Cl.* 50:283–305.

Sewandono, M. 1938. Het veengebied van Bengkalis. *Tectona* 31:99–135.

Shanklin, J. P. 1951. Scientific use of natural areas. *J. For.* 49:793–794.

Shidei, T., and T. Kira, eds. 1977. *Primary productivity of Japanese forests: Productivity of terrestrial communities.* JIBP Synthesis, vol. 16. Tokyo: University of Tokyo Press.

Shimizu, Y. 1983. Phenological studies of the subtropical broad-leaved evergreen forests at Chichijima Island in the Bonin (Ogasawara) Islands. *Jap. J. Ecol.* 33:135–147.

Smits, W. Th. M. 1983. Vegetative propagation of *Shorea* cf. *obtusa* and *Agathis dammara* by means of leaf-cuttings and stem-cuttings. *Malay. For.* 46 (2): 175–185.

Snedaker, S. C. 1982. Mangrove species zonation: Why? In *Tasks for vegetation science,* vol. 2, ed. D. N. Sen and K. S. Rajpurohit, pp. 111–125. The Hague: Junk.

Srivastava, P. B. L. 1980. Research proposals for mangrove vegetation in Malaysia. In *Workshop on mangrove and estuarine vegetation, 10 December 1977,* ed. P. B. L. Srivastava and R. A. Kader, pp. 64–75. Serdang, Malaysia: Faculty of Forestry, Universiti Pertanian Malaysia.

Stamp, L. D. 1925. *The vegetation of Burma.* Calcutta: Thacker, Spink.

Stern, K., and L. Roche. 1974. *Genetics of forest ecosystems.* New York: Springer-Verlag.

Sukardjo, S., and K. Kartawinata. 1978. Mangrove forest in Banyuasin estuary, South Sumatra. Seminar on mangrove and estuarine vegetation in S.E. Asia. Selangor, Malaysia.

Synge, H., ed. 1980. *The biological aspects of rare plant conservation.* Chichester: John Wiley & Sons.

Tamai, S., T. Nakasuga, R. Tabuchi, and K. Ogino. 1986. Standing biomass of mangrove forests in southern Thailand. *J. Jpn. For. Soc.* 68 (9): 384–388.

Tamari, C. 1976. *Phenology and seed storage trials of dipterocarps.* FRI Research Pamphlet 69. Kuala Lumpur: Forestry Department.

Tamari, C., and I. L. Domingo. 1979. Phenology of Philippine dipterocarps. *Tropical Agriculture Research Series,* no. 12:131–139.

Tamari, C., and D. V. Jacalne. 1984. Fruit dispersal of dipterocarps. *Bull. For. and For. Prod. Res. Inst.,* no. 325:127–140.

Taylor, B. W. 1959. The classification of lowland swamp communities in northeastern Papua. *Ecology* 40 (4): 703–711.

Thanom Premrasmi. 1980. Proposal for the establishment of South East Asian Tree Improvement and Seed Procurement Cooperative Programme. In *Proceedings of the conference on Southeast Asian tree improvement and seed procurement cooperative programme,* pp. 13–18. Bangkok: Royal Forest Department.

Theophrastus. (305 B.C.) *Historia Plantarum* 4 (7): 4–7.

Toda, R. 1965. Preservation of gene pool in forest tree populations. Pro. IUFRO Working Party, Zagreb.

Tomlinson, P. B. 1986. *The botany of mangroves.* Cambridge: Cambridge University Press.

Tomlinson, P. B., R. B. Primack, and J. S. Bunt. 1979. Preliminary observations on floral biology in mangrove Rhizophoraceae. *Biotropica* 11 (4): 256–277.

Troll, W., and O. Dragendorff. 1931. Uber die luftwurzeln von *Sonneratia* L. und ihre biologische bedeutung. *Planta* 13:311–473.

Turnbull, J. W. 1972. *Pinus kesiya* Royle ex Gordon (Syn. *P. khasya* Royle; *P. insularis* Endliche.) in the Philippines: Distribution, characteristics and seed sources. In *Selection and breeding to improve some tropical conifers,* ed. J. Burley and D. G. Niklcs, pp. 43–53. Commonwealth Forestry Institute, Oxford.

van Bodegon, A. H. 1929. De vloedbosschen in het Gewest Riouw en Onderhoorigheden. *Tectona* 22:1302–1332.

van Steenis, C. G. G. J. 1957. Outline of vegetation types in Indonesia and some adjacent regions. *Proc. Pacif. Sci. Congr.* 8 (4): 61–97.

—————. 1958. Rhizophoraceae (Introductory matter). *Flora Malesiana* ser. 1, vol. 5, part 4:429–447.

—————. 1962. The distribution of mangrove plant genera and its significance for paleogeography. *Kon. Neder. Akad. van Wetench.* 65:164–169.

—————. 1965. Concise plant-geography of Java. In *Flora of Java,* vol. 2, ed. C. A. Backer and R. C. van den Brink Bakhuizen, pp. 1–72. Groningen: N. V. P. Noordhoff.

—————. 1971. Plant conservation in Malaysia. *Bull. Jard. Bot. Nat. Belg.* 41:189–202.

van Steenis, C. G. G. J., and M. J. van Steenis–Kruseman. 1953. Brief sketch of the Tjibodas Mountain Garden Flora. *Malay. Bull.* 10:313–351.

van Steenis, C. G. G. J., A. Hamzah, and M. Toha. 1972. *The mountain flora of Java.* Leiden: E. J. Brill.

Walsh, G. E. 1974. Mangroves: A review. In *Ecology of halophytes,* ed. R. J. Reimold and W. H. Queen, pp. 51–174. New York: Academic Press.

———. 1977. Exploitation of mangal. In *Wet coastal ecosystems,* ed. V. J. Chapman, pp. 347–362. Vol. 1 of *Ecosystems of the world,* ed. D. W. Goodall. Amsterdam: Elsevier.

Walsh, G. E., K. A. Ainsworth, and R. Rigby. 1979. Resistance of red mangrove (*Rhizophora mangle* L.) seedlings to lead, cadmium, and mercury. *Biotropica* 11 (1): 22–27.

Walter, H. 1955. Klimagramme als mittes zur beurteilung der Klimarerhaltnisse für ökologische, vegetationskundliche und landwirdschaftliche zwecke. *Ber. d. deutsch. bot. Gaz.* 68:331–344.

Walter, H., and M. Steiner. 1936. Die ökologie der ost-afrikanischen mangroven. *Z. Bot.* 30:65–193.

Wang, B. S. P. 1975. Tree seed and pollen storage for genetic conservation: Possibilities and limitations. In *The methodology of conservation of forest genetic resources,* pp. 93–103. Rome: Food and Agriculture Organization.

Watanabe, M. 1959. *Tōnanajia* [Southeast Asia]. Tokyo: Asakura shoten.

Watson, J. G. 1928. *Mangrove forests of the Malay Peninsula.* Malayan Forest Records No. 6. Kuala Lumpur: Forest Department, Federated Malay States.

Webber, M. L. 1954. The mangrove ancestry of a freshwater swamp forest suggested by its diatom flora. *Malay. For.* 17:25–26.

Whitmore, T. C. 1975. *Tropical rain forests of the far east.* Oxford: Oxford University Press.

———. 1977. *A first look at Agathis.* Tropical forestry papers, No. 11. University of Oxford.

Whitmore, T. C., and M. R. Bowen. 1983. Growth analysis of some *Agathis* species. *Malay. For.* 46 (2): 186–196.

Whittaker, R. H. 1956. Vegetation of the Great Smoky Mountains. *Ecol. Monogr.* 26:1–80.

———. 1970. *Communities and Ecosystems.* New York: Macmillan.

Wilford, G. E. 1960. Radiocarbon age determinations of Quaternary sediments in Brunei and northeast Sarawak. British Borneo Geological Survey Annual Report, 1959.

Williams, J. T., C. H. Lamoureux, and N. Wulijarni-Soetjipto. 1975. *South East Asian plant genetic resources.* Bogor: IBPGR/SEAMEO/LIPI.

Wium-Anderson, S., and B. Christensen. 1978. Seasonal growth of mangrove trees in southern Thailand. Pt. 2: Phenology of *Bruguiera cylindrica, Ceriops tagal, Lumnitzera littorea,* and *Avicennia marina. Aquatic Botany* 5:383–390.

Wolff von Wülfing, H. E. 1935. Stamtafels voor enkele eiken-en Kastanjesoorten van West Java en voor Kihoedjan. *Tectona* 28:733–843.

Wong, M. 1983. Understory phenology of the virgin and regenerating habitats in Pasoh Forest Reserve, Negeri Sembilan, West Malaysia. *Malay. For.* 46 (2): 197–223.

Woodwell, G. M., and E. V. Pecan, eds. 1973. *Carbon and the biosphere.* Brookhaven Symposia in Biology 24. Springfield.

Wyatt-Smith, J. 1961. A note on the freshwater swamp, lowland, and hill forest types of Malaya. *Malay. For.* 24:110–121.

———. 1963. *Manual of Malayan silviculture for inland forests.* 2 vols. Malayan Forest Records No. 23. Kuala Lumpur: Forest Department, Federated Malay States.

———. 1964. A preliminary vegetation map of Malaya with description of the vegetation types. *J. Trop. Geogr.* 18:200–213.

Wycherley, P. R. 1973. The phenology of plants in the humid tropics. *Micronesica* 9 (1): 75–96.

Yamada, I. 1982. Nettairin no genjō karamita rinboku ikushū no arikata [State of silviculture as judged from the present state of tropical forests]. In *Rinboku no ikushū tokubetsugō* [Special volume on silviculture], pp. 11–13. Tokyo: Rinbokuikushūkyōkai.

Yamada, I., and S. Soekardjo. 1979. Minami Sumatora teishitchi no shinrinshokusei [Forest vegetation of the South Sumatra wetlands]. *Tōnanajia kenkyū {Southeast Asian Studies}* 17 (3): 121–154.

Yap, C. L. 1966. A mechanised method of log extraction in peat swamp forest. *Malay. For.* 29 (1): 34–36.

Yap, S. K. 1980. Jelutong: Phenology, fruit, and seed biology. *Malay. For.* 43 (3): 309–315.

———. 1982. The phenology of some fruit tree species in a lowland dipterocarp forest. *Malay. For.* 45 (1): 21–35.

Yoda, K. 1971. *Shinrin no seitaigaku* [Forest ecology]. Tokyo: Tsukuji shokan.

Zobel, B. 1970. Mexican pines. In *Genetic resources in plants,* ed. O. H. Frankel and E. Bennett, pp. 367–373. Oxford and Edinburgh: Blackwell.

# Index

Page numbers for illustrations and tables are in italics.

MONOGRAPHS OF THE CENTER FOR SOUTHEAST ASIAN STUDIES
KYOTO UNIVERSITY


**English-language Series:**

1. Takashi Sato, *Field Crops in Thailand,* 1966
2. Tadayo Watabe, *Glutinous Rice in Northern Thailand,* 1967
3. Kiyoshi Takimoto (ed.), *Geology and Mineral Resources in Thailand and Malaya,* 1969
4. Keizaburo Kawaguchi and Kazutake Kyuma, *Lowland Rice Soils in Thailand,* 1969
5. Keizaburo Kawaguchi and Kazutake Kyuma, *Lowland Rice Soils in Malaya,* 1969
6. Kiyoshige Maeda, *Alor Janggus: A Chinese Community in Malaya,* 1967
7. Shinichi Ichimura (ed.), *The Economic Development of East and Southeast Asia,* 1975
8. Masashi Nishihara, *The Japanese and Sukarno's Indonesia,* 1976
9. Shinichi Ichimura (ed.), *Southeast Asia: Nature, Society and Development,* 1977
10. Keizaburo Kawaguchi and Kazutake Kyuma, *Paddy Soils in Tropical Asia,* 1977
11. Kunio Yoshihara, *Japanese Investment in Southeast Asia,* 1978
12. Yoneo Ishii (ed.), *Thailand: A Rice-Growing Society,* 1978
13. Lee-Jay Cho and Kazumasa Kobayashi (eds.), *Fertility Transitions of the East Asian Populations,* 1979
14. Kuchiba, Tsubouchi and Maeda, *Three Malay Villages: A Sociology of Paddy Growers in West Malaysia,* 1979
15. Cho, Suharto, McNicoll and Mamas, *Population Growth of Indonesia,* 1980
16. Yoneo, Ishii, *Sangha, State and Society: Thai Buddhism in History,* 1986
17. Yoshikazu Takaya, *Agricultural Development of a Tropical Delta,* 1987
18. Kenji Tsuchiya, *Democracy and Leadership: The Rise of the Taman Siswa Movement in Indonesia,* 1987
19. Hayao Fukui, *Food and Population in a Northeast Thai Village,* 1993


**Japanese-language Series:**

Available from the Center for Southeast Asian Studies, Kyoto, Japan.

1. Joji Tanase, *Primitive Form of the Idea of the Other World,* 1966
2. Toru Yano, *Modern Political History of Thailand and Burma,* 1968
3. Takeshi Motooka, *Agricultural Development of Southeast Asia,* 1968
4. Yoshihiro and Reiko Tsubouchi, *Divorce,* 1970
5. Shigeru Iijima, *Social and Cultural Change of Karens,* 1971
6. H. Storz (trans. by H. Nogami), *Burma: Land, History and Economy,* 1974
7. Shinichi Ichimura (ed.), *Southeast Asia: Nature, Society and Economy,* 1974

8. Yoneo Ishii, *Thailand: A Rice-Growing Society,* 1975
9. Yoneo Ishii, *Political Sociology of Theravada Buddhism,* 1975
10. Shinichi Ichimura (ed.), *The Economic Development of East and Southeast Asia,* 1975
11. Takeshi Motooka, *Rice in Indonesia,* 1975
12. Kuchiba, Tsubouchi and Maeda, *The Structure and Change of Malayan Villages,* 1976
13. Masashi Nishihara (ed.), *Political Corruption in Southeast Asia,* 1976
14. A. Eckstein (trans. by S. Ichimura et al.), *Economic Trends in Communist China,* 1979
15. Tadayo Watabe (ed.), *The World of Southeast Asia: Verification of Its Images,* 1980
16. Koichi Mizuno, *Social Organization of Thai Villages,* 1980
17. Kenji Tsuchiya, *A Study of Indonesian Nationalism: Evolution and Development of Taman Siswa,* 1982
18. Yoshikazu Takaya, *Agricultural Evolution in the Tropical Delta: The Case of the Menam Chao Phraya Delta,* 1982
19. Kazumasa Kobayashi, *Population in Southeast Asia,* 1984
20. Yoneo Ishii (ed.), *The Structure and Change of Southeast Asia,* 1986
21. Yumio Sakurai, *The Formation of a Vietnamese Village Community with Special Reference to the Historical Development of the Communal Padi-Field or the Cong-Dien,* 1987
22. Hayao Fukui, *Don Daeng: Agroecology of a Northeast Thai Village,* 1988
23. Masuo Kuchiba (ed.), *Traditional Structure and Its Change in Don Daeng Village,* 1990
24. Isamu Yamada, *Tropical Rain Forest World in Southeast Asia,* 1991